Current Topics in Behavioral Neurosciences

Volume 33

About this Series

Current Topics in Behavioral Neurosciences provides critical and comprehensive discussions of the most significant areas of behavioral neuroscience research, written by leading international authorities. Each volume offers an informative and contemporary account of its subject, making it an unrivalled reference source. Titles in this series are available in both print and electronic formats.

With the development of new methodologies for brain imaging, genetic and genomic analyses, molecular engineering of mutant animals, novel routes for drug delivery, and sophisticated cross-species behavioral assessments, it is now possible to study behavior relevant to psychiatric and neurological diseases and disorders on the physiological level. The *Behavioral Neurosciences* series focuses on "translational medicine" and cutting-edge technologies. Preclinical and clinical trials for the development of new diagnostics and therapeutics as well as prevention efforts are covered whenever possible.

More information about this series at http://www.springer.com/series/7854

Andrew J. Lawrence • Luis de Lecea

Editors

Behavioral Neuroscience of Orexin/Hypocretin

 Springer

Editors
Andrew J. Lawrence
Florey Institute of Neuroscience &
    Mental Health, Melbourne
    Brain Centre
University of Melbourne
Parkville, Victoria
Australia

Luis de Lecea
Dep Psychiatry & Behav Sciences
Stanford University
Stanford, California
USA

ISSN 1866-3370   ISSN 1866-3389   (electronic)
Current Topics in Behavioral Neurosciences
ISBN 978-3-319-57534-6   ISBN 978-3-319-57535-3   (eBook)
DOI 10.1007/978-3-319-57535-3

Library of Congress Control Number: 2017942750

Printed on acid-free paper

This Springer imprint is published by Springer Nature
The registered company is Springer International Publishing AG
The registered company address is: Gewerbestrasse 11, 6330 Cham, Switzerland

Preface

Twenty years have passed since the discovery of the Hypocretins/Orexins [1, 2], as we have witnessed an amazing set of discoveries demonstrating key roles for these transmitters in arousal and beyond. This volume entitled *Behavioral Neuroscience of Orexin/Hypocretin* in the series *Current Topics in Behavioral Neurosciences* focuses on the most recent advances on this neurotransmitter system and brings together scientists from around the world who provide a timely discussion of how this peptide regulates different aspects of behavior. We have attempted to give readers a broad taste of the current knowledge base surrounding the evolutionary conservation of orexin/hypocretin systems as demonstrated in zebrafish (see [3]), orexin/hypocretin receptor structure (see [4]), orexin/hypocretin signaling (see [5]), and the role of orexin/hypocretin in numerous physiological and behavioral processes. For example, the critical role of orexin/hypocretin in arousal is canvassed (see [6]) as well as the evidence for the involvement of orexin/hypocretin systems in Alzheimer's disease (see [7]). From a systems physiology perspective we cover the role of orexin/hypocretin in energy balance (see [8]), plus how orexin/hypocretin modulates cardiovascular and respiratory function (see [9]). From a clinical perspective this issue details the development of OX2 receptor antagonists as sleep aids (see [10]).

Over the past decade or so there has been intense interest regarding how orexin/hypocretin regulates both natural and aberrant reward-seeking. Included in this issue are chapters on orexin/hypocretin and addiction (see [11]), plastic changes associated with drug abuse (see [12]), and orexin/hypocretin and alcohol use/abuse (see [13]). Related to this facet of orexin/hypocretin neurobiology we also address how this system impacts stress reactivity (see [14]).

The discovery of the hypocretins/orexins has led to many important advances in the neurosciences: identifying a molecular marker to untangle the cellular composition of the lateral hypothalamus, providing a mechanism for sleep/wake regulation, a diagnostic tool for narcolepsy, a first-in-class chemical entity approved for the treatment of primary insomnia, and the first implementation of optogenetics in vivo. Overall this issue captures many of these discoveries. We acknowledge that this is a fast-moving field and therefore cannot encompass every relevant topic; however our intention was to appeal to a broad cross section of basic scientists and clinicians.

Parkville, VIC, Australia   Andrew J. Lawrence
Stanford, CA, USA   Luis de Lecea

References

1. Sakurai T, Amemiya A, Ishii M, Matsuzaki I, Chemelli RM, Tanaka H, Williams SC, Richardson JA, Kozlowski GP, Wilson S, Arch JR, Buckingham RE, Haynes AC, Carr SA, Annan RS, McNulty DE, Liu WS, Terrett JA, Elshourbagy NA, Bergsma DJ, Yanagisawa M (1998) Orexins and orexin receptors: a family of hypothalamic neuropeptides and G protein-coupled receptors that regulate feeding behavior. Cell 92:573–585
2. de Lecea L, Kilduff TS, Peyron C, Gao X, Foye PE, Danielson PE, Fukuhara C, Battenberg EL, Gautvik VT, Bartlett FS 2nd, Frankel WN, van den Pol AN, Bloom FE, Gautvik KM, Sutcliffe JG (1998) The hypocretins: hypothalamus-specific peptides with neuroexcitatory activity. Proc Natl Acad Sci U S A 95:322–327
3. Elbaz I, Levitas-Djerbi T, Appelbaum L (2017) The hypocretin/orexin neuronal networks in zebrafish. Curr Topics Behav Neurosci 33:75–92. doi:10.1007/7854_2016_59
4. Yin J, Rosenbaum DM (2017) The human orexin/hypocretin receptor crystal structures. Curr Topics Behav Neurosci 33:1–16. doi:10.1007/7854_2016_52
5. Kukkonen JP (2017) Orexin/hypocretin signaling. Curr Topics Behav Neurosci 33:17–50. doi:10.1007/7854_2016_49
6. Li S-B, Giardino WJ, de Lecea L (2017) Hypocretins and arousal. Curr Topics Behav Neurosci 33:93–104. doi:10.1007/7854_2016_58
7. Liguori C (2017) (2017) Orexin and Alzheimer's disease. Curr Topics Behav Neurosci 33:305–322. doi:10.1007/7854_2016_50
8. Goforth PB, Myers MG (2017) Roles for orexin/hypocretin in the control of energy balance and metabolism. Curr Topics Behav Neurosci 33:137–156. doi:10.1007/7854_2016_51
9. Carrive P, Kuwaki T (2017) Orexin and central modulation of cardiovascular and respiratory function. Curr Topics Behav Neurosci 33:157–196. doi:10.1007/7854_2016_46
10. Jacobson LH, Chen S, Mir S, Hoyer D (2017) Orexin OX_2 Receptor antagonists as sleep aids. Curr Topics Behav Neurosci 33:105–136. doi:10.1007/7854_2016_47
11. James MH, Mahler SV, Moorman DE, Aston-Jones G (2017) A decade of orexin/hypocretin and addiction: where are we now? Curr Topics Behav Neurosci 33:247–282. doi:10.1007/7854_2016_57
12. Baimel C, Borgland SL (2017) Hypocretin/orexin and plastic adaptations associated with drug abuse. Curr Topics Behav Neurosci 33:283–304. doi:10.1007/7854_2016_44
13. Walker LC, Lawrence AJ (2017) The role of orexins/hypocretins in alcohol use and abuse. Curr Topics Behav Neurosci 33:221–246. doi:10.1007/7854_2016_55
14. James MH, Campbell EJ, Dayas CV (2017) Role of the orexin/hypocretin system in stress-related psychiatric disorders. Curr Topics Behav Neurosci 33:197–220. doi:10.1007/7854_2016_56

Contents

The Human Orexin/Hypocretin Receptor Crystal Structures

Jie Yin and Daniel M. Rosenbaum

Abstract The human orexin/hypocretin receptors (hOX$_1$R and hOX$_2$R) are G protein-coupled receptors (GPCRs) that mediate the diverse functions of the orexin/hypocretin neuropeptides. Orexins/hypocretins produced by neurons in the lateral hypothalamus stimulate their cognate GPCRs in multiple regions of the central nervous system to control sleep and arousal, circadian rhythms, metabolism, reward pathways, and other behaviors. Dysfunction of orexin/hypocretin signaling is associated with human disease, and the receptors are active targets in a number of therapeutic areas. To better understand the molecular mechanism of the orexin/hypocretin neuropeptides, high-resolution three-dimensional structures of hOX$_1$R and hOX$_2$R are critical. We have solved high-resolution crystal structures of both human orexin/hypocretin receptors bound to high-affinity antagonists. These atomic structures have elucidated how different small molecule antagonists bind with high potency and selectivity, and have also provided clues as to how the native ligands may associate with their receptors. The orexin/hypocretin receptor coordinates, now available to the broader academic and drug discovery community, will facilitate rational design of new therapeutics that modulate orexin/hypocretin signaling in humans.

Keywords Antagonist • Crystal structure • GPCR • High-resolution • Hypocretin • Orexin

Contents

J. Yin and D.M. Rosenbaum (✉)
Department of Biophysics, University of Texas Southwestern Medical Center, Dallas, TX, USA
e-mail: dan.rosenbaum@utsouthwestern.edu

© Springer International Publishing AG 2016
Curr Topics Behav Neurosci (2017) 33: 1–16
DOI 10.1007/7854_2016_52
Published Online: 26 December 2016

1 Goals Behind Solving Structures of hOX$_1$R and hOX$_2$R

The motivation for structural studies of the orexin/hypocretin receptors is twofold. First, we would like to understand at the atomic level how the native ligands (orexin-A and orexin-B) and synthetic antagonists bind to the receptors, and how ligand binding stabilizes distinct receptor conformations. This goal is fundamentally a problem of understanding GPCR allostery for the specific case of the orexin/hypocretin receptors. Second, by studying receptor–ligand interactions, we hope to create knowledge and tools that translate into the design of more potent and selective modulators of orexin/hypocretin signaling.

hOX$_1$R and hOX$_2$R exhibit different physiological functions and pharmacology [1]. The two receptor subtypes are expressed differentially in various CNS regions [2], and pharmacological and genetic studies have uncovered differences in their behavioral functions. The hOX$_2$R is the more evolutionarily ancient of the two subtypes [3], and plays a more significant role in controlling circadian rhythms, sleep, and arousal [4, 5]. Thus modulation of hOX$_2$R has become an attractive therapeutic strategy for sleep and wake disorders such as insomnia (e.g., the FDA-approved drug suvorexant) and narcolepsy (for potential small molecule orexin mimetics). The hOX$_1$R functions in modulating reward [6, 7], nociception [8], and stress [9], and inhibition of this subtype has developed into an active therapeutic area for disorders such as addiction [10].

2 Challenges for Solving High-Resolution Crystal Structures of GPCRs

The key to determining high-resolution structures of the hOX$_1$R and hOX$_2$R was the ability to obtain diffraction-quality crystals. A decade ago, ligand-activated GPCRs were thought to be largely intractable targets for structure determination. However crystallization of GPCRs has recently become possible due to a number of breakthrough technologies that were developed for GPCRs and other integral membrane proteins. First, methods of expression and purification from recombinant systems such as *Spodoptera frugiperda* (Sf9 insect cells) [11], *Pichia pastoris* (yeast) [12, 13], and mammalian cells (such as HEK293) [14] have allowed labs to purify milligram quantities of many GPCRs. Second, protein engineering methods including crystallizable domain chimeras [15], thermostabilizing mutations [16], and antibodies [12, 17] and nanobodies [18, 19] have made purified GPCRs more stable and amenable to forming three-dimensional crystals. Third, detergents such as the neopentyl glycols [20] were developed which stabilize GPCRs during solubilization, purification, and reconstitution. Finally, lipid-mediated crystallization techniques such as lipidic cubic phase (LCP) [21] and bicelles [22] have facilitated GPCR crystal formation by promoting lateral crystal contacts between the receptors' transmembrane (TM) regions.

Despite these advances, GPCRs are still highly challenging targets for X-ray crystallography. To highlight this point, the only previous successful crystallography efforts for GPCRs in the β branch of the rhodopsin family were for thermostabilized mutants of the rat neurotensin receptor (rNTSR1) bound to neurotensin agonist peptides [23–25]. In order to crystallize hOX_1R and hOX_2R, we solubilized the protein out of Sf9 cell membranes in lauryl maltose neopentyl glycol (LMNG), and purified milligram quantities of the receptors to homogeneity. Further, we created chimeras with domains such as T4L that had previously facilitated GPCR crystallization [15]. After using these methods, we were still not able to obtain diffraction-quality crystals. To finally overcome this barrier, we had to identify and develop a new fusion domain – PGS (*Pyrococcus abysii* glycogen synthase) – that yielded high-resolution diffracting crystals for both hOX_1R [26] and hOX_2R [27] when fused at the third intracellular loop (ICL3). The ICL3 fusion strategy is successful for GPCR crystallization because it removes an inherently flexible region of the receptor (ICL3), which may hinder crystal contact formation, and adds a stable folded structure in its place that can successfully mediate lattice contacts [15]. While other crystallizable domains have been developed as fusion protein partners [28], the new PGS domain proved indispensible for our efforts to crystallize both orexin receptor subtypes.

3   Comparison of hOX_1R and hOX_2R to Other GPCRs

The crystal structures that we obtained of the orexin/hypocretin receptors each have high-affinity antagonists bound (see Sect. 5). The extracellular surfaces of hOX_1R and hOX_2R broadly resemble those of other peptide-binding GPCRs. Figure 1 shows identical views of four different GPCRs activated by peptide hormones that have been structurally characterized: hOX_2R [27], rat neurotensin receptor rNTSR1 [25], opioid receptor μOR [29], and the chemokine receptor CXCR4 [30]. Along with other related examples [31, 32], these structures support the idea that all peptide-activated GPCRs in the rhodopsin family (Class A) contain a short β-hairpin (two antiparallel β-strands) in the second extracellular loop (ECL2) situated above the orthosteric pocket. In the "partially active" conformation shown for rNTSR1, the agonist peptide Neurotensin$_{8-13}$ packs against the β-hairpin motif. This motif may be a general platform for GPCR-peptide interaction due to the large flat surface area it presents, although other peptide-bound structures are needed to

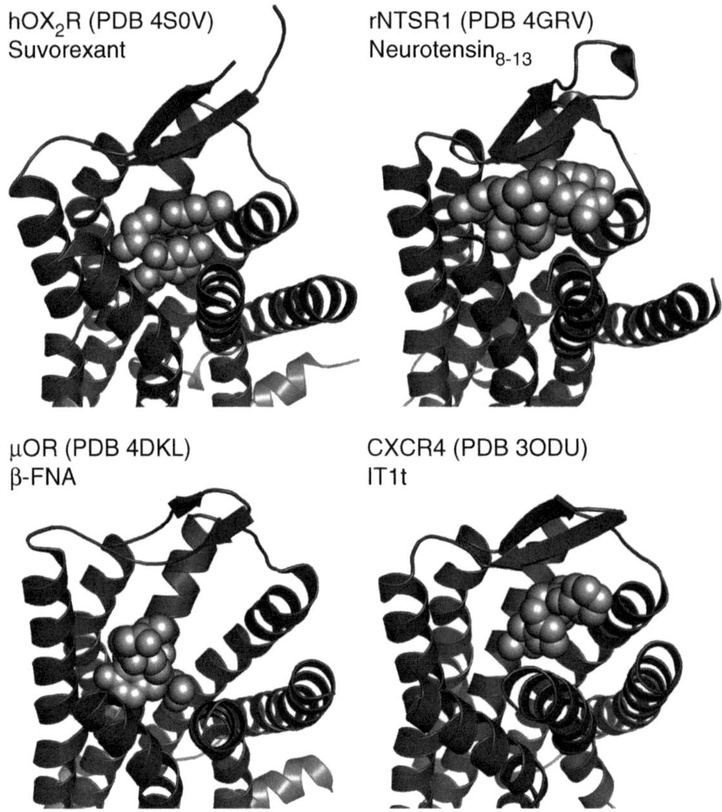

Fig. 1 Crystal structures of different peptide-activated GPCRs. Receptors are depicted as dark gray cartoons, and antagonist ligands are depicted as light gray spheres

confirm this prediction. The sites of several of the most deleterious reported mutations for orexin/hypocretin affinity and potency are on this β-hairpin [33]. In contrast, in the secretin family of GPCRs (Class B), high affinity peptide recognition requires a large folded N-terminal extracellular domain [34].

As illustrated in Fig. 1, small molecule antagonists for hOX_2R, μOR, and CXCR4 are each positioned in the orthosteric pocket that is created by the TM α-helices embedded within the membrane. In fact, the position of the drug suvorexant (used for co-crystallization of both hOX_1R and hOX_2R) is very similar to the binding site for β-blocker inverse agonists such as carazolol in the $β_2$ adrenergic receptor ($β_2AR$), the most extensively studied model system for ligand-activated GPCRs (Fig. 2a). Structures of $β_2AR$ with antagonists [15], agonists [18, 35], and G proteins [19] led to a model in which agonist-induced inward movements of the TM α-helices at the orthosteric binding pocket initiate conformational changes through the transmembrane 7TM bundle. These changes ultimately lead to outward movement of TMs 5 and 6 at the intracellular surface, which is necessary for the engagement of G proteins. Based on this model, also recapitulated in other members

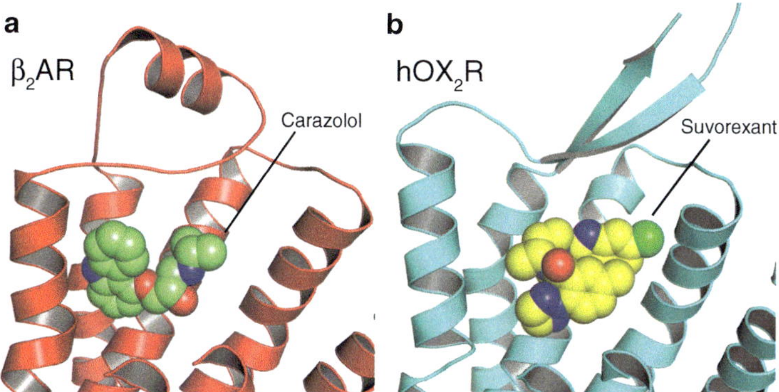

Fig. 2 Analogous positions of antagonist binding in the orthosteric pockets of different GPCRs. (**a**) Carazolol inverse agonist (spheres with green carbons) bound to the β_2 adrenergic receptor (red cartoon). (**b**) Suvorexant antagonist (spheres with yellow carbons) bound to hOX$_2$R (cyan cartoon)

of the GPCR superfamily [36], binding of antagonists such as suvorexant (Fig. 2b) are hypothesized to stabilize the inactive conformation by preventing inward movement of the TM α-helices at the orthosteric site, as well as competing for binding surfaces with the native orexin/hypocretin ligands. Comparison of the published structures of inactive [23] and partially active [25] conformations of rNTSR1 showed that a peptide agonist can promote an outward shift of TM6 at the intracellular surface, through binding of a surface that overlaps with suvorexant's binding site in the orexin receptors.

4 Overall Structures of hOX$_1$R and hOX$_2$R

The global structures of hOX$_1$R and hOX$_2$R are shown in Fig. 3. To date, we have reported the structure of hOX$_1$R bound to the dual orexin receptor antagonist (DORA) suvorexant at 2.75 Å resolution (Fig. 3a); the structure of hOX$_1$R bound to the type 1-selective antagonist SB-674042 at 2.8 Å resolution (Fig. 3b); and the structure of hOX$_2$R bound to suvorexant at 2.5 Å resolution (Fig. 3c) [27, 37]. The root mean squared deviation (rmsd) between superimposed hOX$_1$R and hOX$_2$R is 0.4 Å over 282 Cα's, indicating that these two receptors (with 64% sequence identity) are very similar in three-dimensional structure (Fig. 3d). Beyond this similarity, the rmsd between the hOX$_2$R structure (Fig. 3c) and the inactive-state structure of the β_2AR (with 23% sequence identity) is only 2.2 Å, highlighting the strong structural conservation within the GPCR superfamily. The major difference between the crystal structures lies in the extracellular region containing the ECL2 and N-terminus. The hOX$_1$R has a short α-helix preceding TM1, which packs against the ECL2 (Fig. 2a, b). We did not observe such a motif in our hOX$_2$R structure (Fig. 3c); however, this α-helix may exist but have too much flexibility to

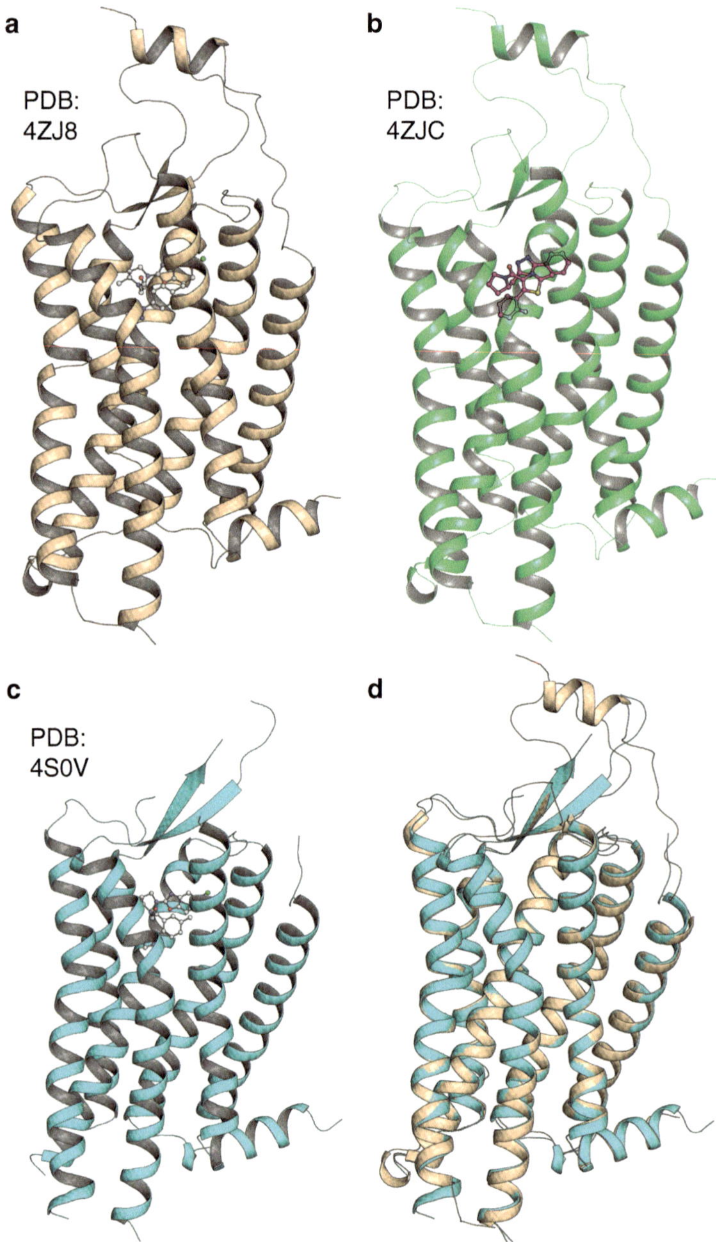

Fig. 3 Structures of orexin receptors bound to small molecules. (**a**) hOX$_1$R (wheat cartoon) bound to suvorexant (balls and sticks with gray carbons). (**b**) hOX$_1$R (green cartoon) bound to SB-674042 (balls and sticks with magenta carbons). (**c**) hOX$_2$R (cyan cartoon) bound to suvorexant (balls and sticks with gray carbons). (**d**) Superposition of hOX$_1$R and hOX$_2$R

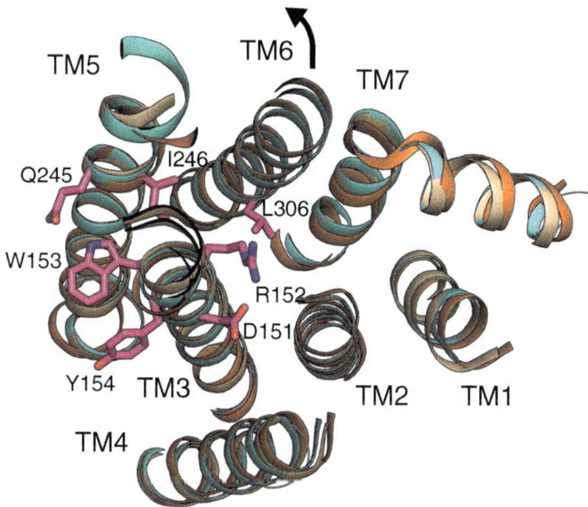

Fig. 4 Structures of inactive-state GPCRs at the intracellular surface of the membrane. Superposition of suvorexant-bound hOX$_1$R (wheat cartoon), suvorexant-bound hOX$_2$R (cyan cartoon), and tiotropium-bound M3 muscarinic acetylcholine receptor (orange cartoon, PDB 4DAJ), viewed from intracellular side. Intracellular loops are removed for clarity. The DRWY residues on TM3 and interacting residues on TMs 5 and 6 are shown as magenta sticks. *Arrow* indicates outward movement of TM6 during activation

visualize in the crystal's electron density. We believe that this N-terminal α-helix is directly involved in orexin/hypocretin recruitment and receptor activation (see Sect. 6).

Like all GPCRs, hOX$_1$R and hOX$_2$R translate agonist binding into functional responses through receptor-mediated activation of intracellular heterotrimeric G proteins, principally G$_{q/11}$ and G$_{i/o}$ for the orexin/hypocretin receptors [38]. Based on previous structural and biophysical studies of other GPCRs [18, 35, 36], the binding of agonists should stimulate outward movement of the TM α-helices at the cytoplasmic surface to facilitate binding of G proteins [19]. In the inactive state of the β$_2$AR, TM5 and TM6 pack against TM3 bearing the conserved DRY motif at the intracellular surface, blocking epitopes involved in G protein binding. The DRY motif is conserved throughout Class A GPCRs, and these residues are important for maintaining a stable inactive-state conformation with low basal activity [39, 40]. For hOX$_1$R and hOX$_2$R, the DRY motif is changed to DRWY, and these residues are tightly packed against residues from TM5 and TM6 (Fig. 4), analogous to the antagonist-bound M3 acetylcholine receptor (another G$_q$-coupled GPCR). Therefore we can conclude that the antagonist-bound crystal structures of hOX$_1$R and hOX$_2$R represent inactive conformations [26, 27].

5   Binding of Small Molecule Antagonists

Orexin receptor antagonists are prospective therapeutics for a number of different human diseases, as detailed elsewhere in this volume. So far, the only such molecule to be approved by the FDA is suvorexant (Belsomra) [41] for treatment of insomnia. An important goal of characterizing the structures of the receptors is to understand the precise mechanisms by which antagonists bind and prevent activation. The solvent-exposed orthosteric binding sites where small molecule antagonists bind to hOX_1R and hOX_2R are well ordered in the crystal structures, along with the bound small molecules. Figure 5 shows the ligand binding poses and detailed interactions with the receptors for all three crystal structures we have reported: hOX_1R with suvorexant (Fig. 5a), hOX_1R with SB-674042 (Fig. 5b), and hOX_2R with suvorexant (Fig. 5c) [26, 27]. The binding sites include contributions from the extracellular ends of all TMs except TM1, as well as from the ECL2. Several of the amino acids that make the greatest contact with the ligands (in terms of buried surface area) have been previously characterized in mutagenesis studies as contributing greatly to antagonist affinity, adding further functional support to our structural data [42–45].

The two ligands that we have so far co-crystallized with the orexin receptors both adopt a compact horseshoe-like bound conformation, in which two aromatic groups, separated by a spacer group, engage in intramolecular aromatic stacking interactions (Fig. 5). For suvorexant analogs, a related 3D conformation of the isolated ligand in solution was previously reported, and suggested to be relevant to the receptor-binding conformation [46]. Our structures support the idea that these molecules and related antagonists prepay some of the entropic cost of ligand binding by constraining their 3D conformations through intramolecular packing. Indeed, a large number of the small molecule orexin/hypocretin receptor antagonists discovered by different laboratories have the same basic form, in which two aromatic moieties are separated and presented by a small ring scaffold [47]. We predict that many of these molecules will bind in a similar mode as we have elucidated in our crystal structures.

In conjunction with solving the hOX_1R structures, we carried out several different computational analyses to better understand how subtype-selective ligands such as SB-674042 discriminate between hOX_1R and hOX_2R. These studies demonstrated how the very highly conserved orthosteric binding pockets, which have only two subtle differences in amino acid composition (Fig. 4d), create small differences in pocket volume and shape that can be exploited to achieve selectivity toward either subtype [26]. The hOX_1R-selective SB-674042 occupies slightly more space in the orthosteric pocket and clashes with the two larger divergent residues (Thr111 and Thr135) in the resulting slightly smaller hOX_2R pocket. In contrast, docking and simulation of the hOX_2R-selective antagonist 2-SORA-DMP indicated that better shape complementarity and van der Waals contacts with hOX_2R lead to greater affinity compared to hOX_1R with its larger orthosteric pocket. While our structural observations and calculations are consistent with the

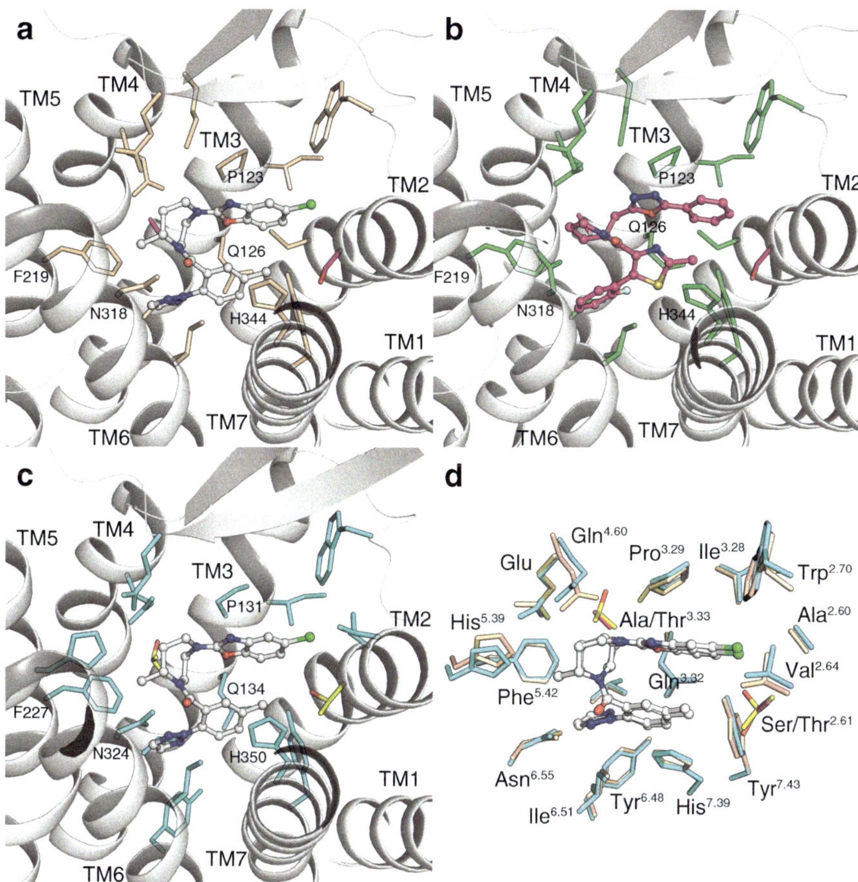

Fig. 5 Binding pockets and antagonist interactions with the orexin/hypocretin receptors. (**a**) Contact residues within 4 Å between suvorexant (balls and sticks with gray carbons) and hOX$_1$R (gray cartoon with wheat sidechains). (**b**) Contact residues within 4 Å between SB-674042 (balls and sticks with magenta carbons) and hOX$_1$R (gray cartoon with green sidechains). (**c**) Contact residues within 4 Å between suvorexant (balls and sticks with gray carbons) and hOX$_1$R (gray cartoon with cyan sidechains). (**d**) Superposition of binding pocket residues of suvorexant-bound hOX$_1$R (wheat sticks) and suvorexant-bound hOX$_2$R (cyan sticks). Labels use Ballesteros–Weinstein numbering in superscript. The two divergent binding site residues are displayed as magenta sticks for hOX$_1$R and yellow sticks for hOX$_2$R

subtype selectivity of these antagonists, one caveat is that the crystallographic coordinates represent saturated complexes with high ligand occupancy and do not inform kinetic mechanisms that influence binding selectivity. Intriguingly, the antagonist-bound orexin receptor structures revealed a "lid" over the binding pocket formed by multiple salt bridges, leaving only a constricted solvent channel to the orthosteric site. This feature implies that the receptor's extracellular surface must breath in order to allow access for antagonists, which may influence both association and dissociation. The precise contributions of binding pocket residues and the extracellular structure to ligand binding kinetics can now be probed with pharmacological studies of mutant receptors guided by the high-resolution structures. Another important factor that is not captured by our structures is ligand and binding pocket desolvation that occurs during complex formation [48]. In future studies, interactions of water molecules with the receptor, ligand, and bound complex can be simulated using our crystallographic coordinates as a framework, to achieve a more complete understanding of the large differences in subtype affinity displayed by selective orexin receptor antagonists.

6 Clues for Orexin/Hypocretin Interaction with hOX$_1$R and hOX$_2$R

We currently lack a clear understanding of how the orexin/hypocretin neuropeptides bind and activate their cognate GPCRs. One intriguing clue from the hOX$_1$R structure was the ordered N-terminal region prior to TM1, containing a short amphipathic α-helix that is positioned over the orthosteric binding pocket (Fig. 6a). This region is conserved in all known vertebrate orexin/hypocretin receptor sequences, from fish and amphibians to humans [49] (Fig. 6b). Given that the NMR structures of orexin-A and orexin-B also revealed amphipathic α-helices [50, 51], we hypothesized that the structured N-terminal region is involved in binding and recruitment of the neuropeptides through interactions of α-helices. Using a combination of binding and receptor activation assays, we showed that this element is essential for potent orexin-A activation for both hOX$_1$R and hOX$_2$R. We also found that mutation of a polar residue in the orthosteric binding pocket (hOX$_1$R N318 or hOX$_2$R N324) severely diminished orexin-A potency [26]. Previous site-directed mutagenesis experiments demonstrated that residues in the ECL2 β-hairpin (e.g., hOX$_1$R D203A or hOX$_2$R D211A) are critical for orexin-A potency [33]. Putting these findings together, we propose that orexin-A binds to hOX$_1$R or hOX$_2$R through a polytopic interface involving all three of these receptor sites (Fig. 6a). In this context, the published result that 17 or more amino acids in the orexin/hypocretin neuropeptides are required to reach low-nanomolar potency [52] can be easily rationalized. A detailed picture of how this interface forms and influences receptor conformation will await structures of the neuropeptide-bound GPCRs. It is worth noting that juxtamembrane N-terminal regions are involved in binding peptide agonists for a number of other rhodopsin family GPCRs, including formyl peptide receptor [53], cholecystokinin (CCK) receptor [54], and the tachykinin receptors [55]. The N-terminal region also plays a key role in chemokine receptors, providing an extended epitope that interacts with the folded globular domain of the chemokine hormone [56, 57].

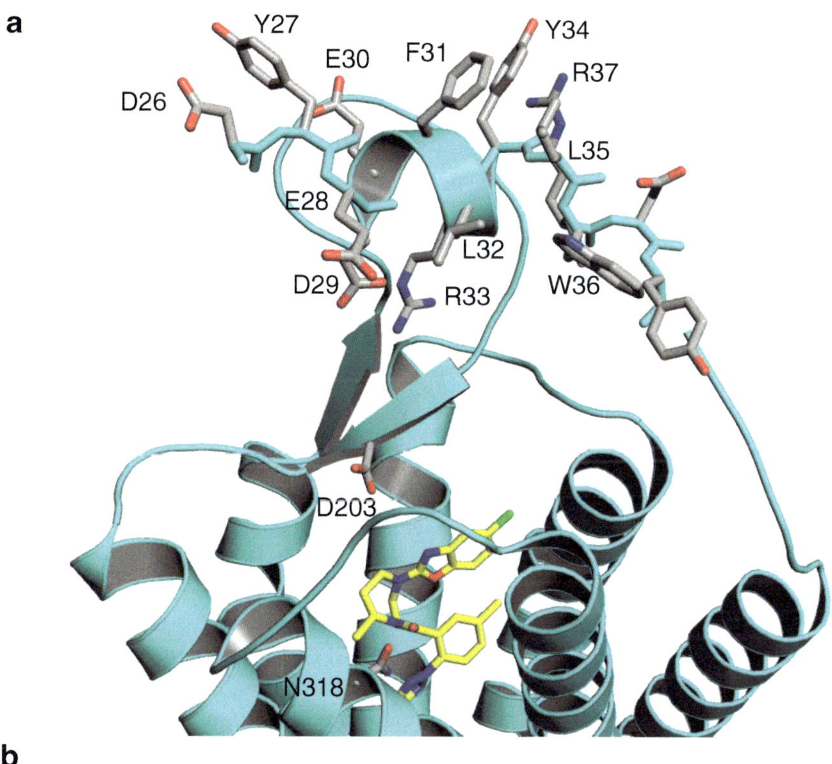

Fig. 6 Structure of the N-terminal extracellular region of hOX$_1$R. (**a**) The amphipathic α-helix at the N-terminus preceding TM1 in hOX$_1$R structure (cyan cartoon, sidechains with gray carbons). (**b**) Sequence alignment of orexin receptor N-termini, including *Mus musculus* (mouse), *Homo sapiens* (human), *Bos Taurus* (bovine), *Danio rerio* (fish), *Xenopus laevis* (frog), and *Canis lupus* (dog) sequences (adapted from Yin et al. [26])

7   Conclusions and Future Prospects

The crystal structures of the human orexin/hypocretin receptors have provided an atomic-level framework for understanding binding and subtype selectivity of small molecule antagonists, including the clinically used suvorexant and the hOX$_1$R-selective SB-674042 [26, 27]. In addition, the structures have revealed a previously unknown role for the receptor N-terminal region in recruitment of the orexin/ hypocretin neuropeptides [26]. We anticipate that our publicly deposited co-ordinates (PDB accession codes 4S0V, 4ZJ8, 4ZJC) and future co-crystal structures using our reported constructs and protocols will aid the design and optimization of orexin/hypocretin receptor antagonists with improved affinity and subtype selectivity.

To fully elucidate the mechanism for orexin hormone activation of the orexin receptors, we must obtain structures bound to agonists, as well as complexes with G proteins or G protein-mimetic antibodies. Crystallographic and biophysical studies of β_2AR other ligand-activated GPCRs have shown that extracellular agonist binding and intracellular conformational changes leading to signaling are weakly coupled [35, 58]. Therefore structures of the receptors bound to the orexin neuro-peptides alone (akin to Egloff et al. [23]) may not reveal the propagated conforma-tional changes through the membrane that ultimately result in signal propagation. In studies of active β_2AR [18], M3 muscarinic acetylcholine receptor [36], and μ opioid receptor [59], this problem was overcome by selecting nanobodies (small single-chain antibody domains derived from llamas) that stabilize the active con-formation by binding at the G protein coupling site [60]. Discovery of active state-stabilizing nanobodies for the orexin receptors may similarly enable structure determination of an active neuropeptide-bound GPCR, illuminating how the pep-tide agonist allosterically promotes the active conformation. Further, the structures of the orexin receptors in different conformations will allow design of biophysical experiments to measure the dynamic changes between states, for example, by fluorescence [61] and NMR spectroscopy [62]. Finally, the biochemical precedents for homogeneous purification and crystallization of the orexin receptors may facilitate attempts to co-crystallize these GPCRs with G proteins, such as the G_q heterotrimer. While this goal is highly challenging (there is only one GPCR-G protein complex structure solved to date, for β_2AR) [19], only a complex with a G protein will ultimately explain how orexin neuropeptides stimulate G protein signaling. Importantly, these developments would pave the way for structure-guided design of small molecule activators of orexin signaling, which have so far been extremely challenging to isolate. The coming years promise to be an excit-ing time for biophysical characterization and manipulation of orexin/hypocretin signaling.

References

1. Li J, Hu Z, de Lecea L (2014) The hypocretins/orexins: integrators of multiple physiological functions. Br J Pharmacol 171:332–350. doi:10.1111/bph.12415
2. Marcus JN, Aschkenasi CJ, Lee CE et al (2001) Differential expression of orexin receptors 1 and 2 in the rat brain. J Comp Neurol 435:6–25

3. Wong KKY, Ng SYL, Lee LTO et al (2011) Orexins and their receptors from fish to mammals: a comparative approach. Gen Comp Endocrinol 171:124–130. doi:10.1016/j.ygcen.2011.01.001

4. Lin L, Faraco J, Li R et al (1999) The sleep disorder canine narcolepsy is caused by a mutation in the hypocretin (orexin) receptor 2 gene. Cell 98:365–376

5. Willie JT, Chemelli RM, Sinton CM et al (2003) Distinct narcolepsy syndromes in orexin receptor-2 and orexin null mice: molecular genetic dissection of non-REM and REM sleep regulatory processes. Neuron 38:715–730

6. Boutrel B, Kenny PJ, Specio SE et al (2005) Role for hypocretin in mediating stress-induced reinstatement of cocaine-seeking behavior. Proc Natl Acad Sci U S A 102:19168–19173. doi:10.1073/pnas.0507480102

7. Harris GC, Wimmer M, Aston-Jones G (2005) A role for lateral hypothalamic orexin neurons in reward seeking. Nature 437:556–559. doi:10.1038/nature04071

8. Bingham S, Davey PT, Babbs AJ et al (2001) Orexin-A, an hypothalamic peptide with analgesic properties. Pain 92:81–90

9. Johnson PL, Truitt W, Fitz SD et al (2010) A key role for orexin in panic anxiety. Nat Med 16:111–115. doi:10.1038/nm.2075

10. Aston-Jones G, Smith RJ, Moorman DE, Richardson KA (2009) Role of lateral hypothalamic orexin neurons in reward processing and addiction. Neuropharmacology 56(Suppl 1):112–121. doi:10.1016/j.neuropharm.2008.06.060

11. Kobilka BK (1995) Amino and carboxyl terminal modifications to facilitate the production and purification of a G protein-coupled receptor. Anal Biochem 231:269–271

12. Hino T, Arakawa T, Iwanari H et al (2012) G-protein-coupled receptor inactivation by an allosteric inverse-agonist antibody. Nature 482:237–240. doi:10.1038/nature10750

13. Shimamura T, Shiroishi M, Weyand S et al (2011) Structure of the human histamine H1 receptor complex with doxepin. Nature 475:65–70. doi:10.1038/nature10236

14. Kang Y, Zhou XE, Gao X et al (2015) Crystal structure of rhodopsin bound to arrestin by femtosecond X-ray laser. Nature 523:561–567. doi:10.1038/nature14656

15. Rosenbaum DM, Cherezov V, Hanson MA et al (2007) GPCR engineering yields high-resolution structural insights into beta2-adrenergic receptor function. Science 318:1266–1273. doi:10.1126/science.1150609

16. Vaidehi N, Grisshammer R, Tate CG (2016) How can mutations thermostabilize G-protein-coupled receptors? Trends Pharmacol Sci 37:37–46. doi:10.1016/j.tips.2015.09.005

17. Rasmussen SGF, Choi H-J, Rosenbaum DM et al (2007) Crystal structure of the human beta2 adrenergic G-protein-coupled receptor. Nature 450:383–387. doi:10.1038/nature06325

18. Rasmussen SGF, Choi H-J, Fung JJ et al (2011) Structure of a nanobody-stabilized active state of the β2 adrenoceptor. Nature 469:175–180. doi:10.1038/nature09648

19. Rasmussen SGF, DeVree BT, Zou Y et al (2011) Crystal structure of the β2 adrenergic receptor-Gs protein complex. Nature 477:549–555. doi:10.1038/nature10361

20. Chae PS, Rasmussen SGF, Rana RR et al (2010) Maltose–neopentyl glycol (MNG) amphiphiles for solubilization, stabilization and crystallization of membrane proteins. Nat Methods 7:1003–1008. doi:10.1038/nmeth.1526

21. Caffrey M, Cherezov V (2009) Crystallizing membrane proteins using lipidic mesophases. Nat Protoc 4:706–731. doi:10.1038/nprot.2009.31

22. Faham S, Boulting GL, Massey EA et al (2005) Crystallization of bacteriorhodopsin from bicelle formulations at room temperature. Protein Sci 14:836–840. doi:10.1110/ps.041167605

23. Egloff P, Hillenbrand M, Klenk C et al (2014) Structure of signaling-competent neurotensin receptor 1 obtained by directed evolution in *Escherichia coli*. Proc Natl Acad Sci U S A 111:E655–E662. doi:10.1073/pnas.1317903111

24. Krumm BE, White JF, Shah P, Grisshammer R (2015) Structural prerequisites for G-protein activation by the neurotensin receptor. Nat Commun 6:7895. doi:10.1038/ncomms8895

25. White JF, Noinaj N, Shibata Y et al (2012) Structure of the agonist-bound neurotensin receptor. Nature 490:508–513. doi:10.1038/nature11558

26. Yin J, Babaoglu K, Brautigam CA et al (2016) Structure and ligand-binding mechanism of the human OX1 and OX2 orexin receptors. Nat Struct Mol Biol 23:293–299. doi:10.1038/nsmb.3183
27. Yin J, Mobarec JC, Kolb P, Rosenbaum DM (2015) Crystal structure of the human OX2 orexin receptor bound to the insomnia drug suvorexant. Nature 519:247–250. doi:10.1038/nature14035
28. Chun E, Thompson AA, Liu W et al (2012) Fusion partner toolchest for the stabilization and crystallization of G protein-coupled receptors. Structure 20:967–976. doi:10.1016/j.str.2012.04.010
29. Manglik A, Kruse AC, Kobilka TS et al (2012) Crystal structure of the μ-opioid receptor bound to a morphinan antagonist. Nature 485:321–326. doi:10.1038/nature10954
30. Wu B, Chien EYT, Mol CD et al (2010) Structures of the CXCR4 chemokine GPCR with small-molecule and cyclic peptide antagonists. Science 330:1066–1071. doi:10.1126/science.1194396
31. Granier S, Manglik A, Kruse AC et al (2012) Structure of the δ-opioid receptor bound to naltrindole. Nature 485:400–404. doi:10.1038/nature11111
32. Tan Q, Zhu Y, Li J et al (2013) Structure of the CCR5 chemokine receptor-HIV entry inhibitor maraviroc complex. Science 341:1387–1390. doi:10.1126/science.1241475
33. Malherbe P, Roche O, Marcuz A et al (2010) Mapping the binding pocket of dual antagonist almorexant to human orexin 1 and orexin 2 receptors: comparison with the selective OX1 antagonist SB-674042 and the selective OX2 antagonist N-ethyl-2-[(6-methoxy-pyridin-3-yl)-(toluene-2-sulfonyl)-amino]-N-pyridin-3-ylmethyl-acetamide (EMPA). Mol Pharmacol 78:81–93. doi:10.1124/mol.110.064584
34. Pioszak AA, Xu HE (2008) Molecular recognition of parathyroid hormone by its G protein-coupled receptor. Proc Natl Acad Sci U S A 105:5034–5039. doi:10.1073/pnas.0801027105
35. Rosenbaum DM, Zhang C, Lyons JA et al (2011) Structure and function of an irreversible agonist-β(2) adrenoceptor complex. Nature 469:236–240. doi:10.1038/nature09665
36. Kruse AC, Ring AM, Manglik A et al (2013) Activation and allosteric modulation of a muscarinic acetylcholine receptor. Nature 504:101–106. doi:10.1038/nature12735
37. Yin J, Li L, Shaw N et al (2009) Structural basis and catalytic mechanism for the dual functional endo-beta-N-acetylglucosaminidase A. PLoS One 4:e4658. doi:10.1371/journal.pone.0004658
38. Zhu Y, Miwa Y, Yamanaka A et al (2003) Orexin receptor type-1 couples exclusively to pertussis toxin-insensitive G-proteins, while orexin receptor type-2 couples to both pertussis toxin-sensitive and -insensitive G-proteins. J Pharmacol Sci 92:259–266
39. Ballesteros JA (2001) Activation of the beta 2-adrenergic receptor involves disruption of an ionic lock between the cytoplasmic ends of transmembrane segments 3 and 6. J Biol Chem 276:29171–29177. doi:10.1074/jbc.M103747200
40. Rasmussen SG, Jensen AD, Liapakis G et al (1999) Mutation of a highly conserved aspartic acid in the beta2 adrenergic receptor: constitutive activation, structural instability, and conformational rearrangement of transmembrane segment 6. Mol Pharmacol 56:175–184
41. Cox CD, Breslin MJ, Whitman DB et al (2010) Discovery of the dual orexin receptor antagonist [(7R)-4-(5-chloro-1,3-benzoxazol-2-yl)-7-methyl-1,4-diazepan-1-yl][5-methyl-2-(2H-1,2,3-triazol-2-yl)phenyl]methanone (MK-4305) for the treatment of insomnia. J Med Chem 53:5320–5332. doi:10.1021/jm100541c
42. Heifetz A, Morris GB, Biggin PC et al (2012) Study of human orexin-1 and -2 G-protein-coupled receptors with novel and published antagonists by modeling, molecular dynamics simulations, and site-directed mutagenesis. Biochemistry 51:3178–3197. doi:10.1021/bi300136h
43. Langmead CJ, Jerman JC, Brough SJ et al (2004) Characterisation of the binding of [3H]-SB-674042, a novel nonpeptide antagonist, to the human orexin-1 receptor. Br J Pharmacol 141:340–346. doi:10.1038/sj.bjp.0705610

44. Putula J, Kukkonen JP (2012) Mapping of the binding sites for the OX1 orexin receptor antagonist, SB-334867, using orexin/hypocretin receptor chimaeras. Neurosci Lett 506:111–115. doi:10.1016/j.neulet.2011.10.061
45. Tran D-T, Bonaventure P, Hack M et al (2011) Chimeric, mutant orexin receptors show key interactions between orexin receptors, peptides and antagonists. Eur J Pharmacol 667:120–128. doi:10.1016/j.ejphar.2011.05.074
46. Cox CD, McGaughey GB, Bogusky MJ et al (2009) Conformational analysis of N,N-disubstituted-1,4-diazepane orexin receptor antagonists and implications for receptor binding. Bioorg Med Chem Lett 19:2997–3001. doi:10.1016/j.bmcl.2009.04.026
47. Lebold TP, Bonaventure P, Shireman BT (2013) Selective orexin receptor antagonists. Bioorg Med Chem Lett 23:4761–4769. doi:10.1016/j.bmcl.2013.06.057
48. Biela A, Nasief NN, Betz M et al (2013) Dissecting the hydrophobic effect on the molecular level: the role of water, enthalpy, and entropy in ligand binding to thermolysin. Angew Chem Int Ed Engl 52:1822–1828. doi:10.1002/anie.201208561
49. Isberg V, Vroling B, van der Kant R et al (2014) GPCRDB: an information system for G protein-coupled receptors. Nucleic Acids Res 42:D422–D425. doi:10.1093/nar/gkt1255
50. Kim H-Y, Hong E, Kim J-I, Lee W (2004) Solution structure of human orexin-A: regulator of appetite and wakefulness. J Biochem Mol Biol 37:565–573
51. Lee JH, Bang E, Chae KJ et al (1999) Solution structure of a new hypothalamic neuropeptide, human hypocretin-2/orexin-B. Eur J Biochem 266:831–839
52. German NA, Decker AM, Gilmour BP et al (2013) Truncated orexin peptides: structure-activity relationship studies. ACS Med Chem Lett 4:1224–1227. doi:10.1021/ml400333a
53. Perez HD, Vilander L, Andrews WH, Holmes R (1994) Human formyl peptide receptor ligand binding domain(s). Studies using an improved mutagenesis/expression vector reveal a novel mechanism for the regulation of receptor occupancy. J Biol Chem 269:22485–22487
54. Kennedy K, Gigoux V, Escrieut C et al (1997) Identification of two amino acids of the human cholecystokinin-A receptor that interact with the N-terminal moiety of cholecystokinin. J Biol Chem 272:2920–2926
55. Valentin-Hansen L, Park M, Huber T et al (2014) Mapping substance P binding sites on the neurokinin-1 receptor using genetic incorporation of a photoreactive amino acid. J Biol Chem 289:18045–18054. doi:10.1074/jbc.M113.527085
56. Burg JS, Ingram JR, Venkatakrishnan AJ et al (2015) Structural basis for chemokine recognition and activation of a viral G protein-coupled receptor. Science 347:1113–1117. doi:10.1126/science.aaa5026
57. Qin L, Kufareva I, Holden LG et al (2015) Crystal structure of the chemokine receptor CXCR4 in complex with a viral chemokine. Science 347:1117–1122. doi:10.1126/science.1261064
58. Manglik A, Kim TH, Masureel M et al (2015) Structural insights into the dynamic process of β2-adrenergic receptor signaling. Cell 161:1101–1111. doi:10.1016/j.cell.2015.04.043
59. Huang W, Manglik A, Venkatakrishnan AJ et al (2015) Structural insights into μ-opioid receptor activation. Nature 524:315–321. doi:10.1038/nature14886
60. Steyaert J, Kobilka BK (2011) Nanobody stabilization of G protein-coupled receptor conformational states. Curr Opin Struct Biol 21:567–572. doi:10.1016/j.sbi.2011.06.011
61. Yao X, Parnot C, Deupi X et al (2006) Coupling ligand structure to specific conformational switches in the beta2-adrenoceptor. Nat Chem Biol 2:417–422. doi:10.1038/nchembio801
62. Nygaard R, Zou Y, Dror RO et al (2013) The dynamic process of β(2)-adrenergic receptor activation. Cell 152:532–542. doi:10.1016/j.cell.2013.01.008

Orexin/Hypocretin Signaling

Jyrki P. Kukkonen

Abstract Orexin/hypocretin peptide (orexin-A and orexin-B) signaling is believed to take place via the two G-protein-coupled receptors (GPCRs), named OX_1 and OX_2 orexin receptors, as described in the previous chapters. Signaling of orexin peptides has been investigated in diverse endogenously orexin receptor-expressing cells – mainly neurons but also other types of cells – and in recombinant cells expressing the receptors in a heterologous manner. Findings in the different systems are partially convergent but also indicate cellular background-specific signaling. The general picture suggests an inherently high degree of diversity in orexin receptor signaling.

In the current chapter, I present orexin signaling on the cellular and molecular levels. Discussion of the connection to (potential) physiological orexin responses is only brief since these are in focus of other chapters in this book. The same goes for the post-synaptic signaling mechanisms, which are dealt with in Burdakov: Postsynaptic actions of orexin. The current chapter is organized according to the tissue type, starting from the central nervous system. Finally, receptor signaling pathways are discussed across tissues, cell types, and even species.

Keywords Adenylyl cyclase • Calcium • Diacylglycerol lipase • Endocannabinoid • G_i • G_q • G_s • Heterotrimeric G-protein • Hypocretin • K^+ channel • Na^+/Ca^{2+}-exchanger • Non-selective cation channel • Orexin • OX_1 receptor • OX_2 receptor • Phospholipase A_2 • Phospholipase C • Phospholipase D • Protein kinase C

J.P. Kukkonen (✉)
Biochemistry and Cell Biology, Department of Veterinary Biosciences, POB 66, FIN-00014 University of Helsinki, Finland
e-mail: jyrki.kukkonen@helsinki.fi

© Springer International Publishing AG 2016
Curr Topics Behav Neurosci (2017) 33: 17–50
DOI 10.1007/7854_2016_49
Published Online: 30 January 2017

Contents

Abbreviations

2-AG	2-Arachidonoyl glycerol
σ_1	Sigma 1 receptor
κOR	κ Opioid receptor
AC	Adenylyl cyclase
ACTH	Adrenocorticotropic hormone
BRET and FRET	Bioluminescence and fluorescence/Förster resonance energy transfer, respectively
CHO	Chinese hamster ovary (cells)
CNS	Central nervous system
CRF and CRF_1	Corticotropin-releasing factor and corticotropin-releasing factor receptor 1, respectively
DAG	Diacylglycerol
ER	Endoplasmic reticulum
GEF	Guanine nucleotide exchange factor
GPCR	G-protein-coupled receptor
G$\alpha\beta\gamma$, Gα and G$\beta\gamma$	Heterotrimeric G-protein heterotrimeric complex, α-subunit, and the $\beta\gamma$-subunit complex, respectively
IP_3	Inositol-1,4,5-trisphosphate
JNK	c-Jun N-terminal kinase
MAPK	Mitogen-activated protein kinase
MEF	Mouse embryonic fibroblast

mTORC1	Mammalian target of rapamycin complex 1
NCX	Na^+/Ca^{2+}-exchanger
NSCC	Non-selective cation channel
$p70^{S6K}$	p70 ribosomal S6 kinase
$p90^{RSK}$	p90 ribosomal S6 kinase
PI3K	Phosphoinositide-3-kinase
PIP_2	Phosphatidylinositol-4,5-bisphosphate
PKA, PKB, and PKC	Protein kinase A, B, and C, respectively
PLA_2, PLC, and PLD	Phospholipase A_2, C, and D, respectively
PP5	Protein phosphatase 5
TRP	Transient receptor potential (channel)
VTA	Ventral tegmental area

1 Introduction to the Signaling of G-Protein-Coupled Receptors

As their name implies, GPCRs classically signal via G-proteins, specifically heterotrimeric G-proteins composed of α-, β-, and γ-subunits. In the resting state, the G-protein subunits remain in complex (also called Gαβγ) and GDP is bound to the α-subunit. The activated receptor interacts with the G-protein catalyzing the release of GDP and binding of GTP (which is dominant inside the cell), upon which the G-protein structure changes (reviewed in [1]). In most cases this is thought to lead to dissociation of the α-subunit (Gα) from the β- and γ-subunits (Gβγ), but there also are suggestions of structural reorganization without dissociation (reviewed in [2]). Nevertheless, both Gα and Gβγ should be able to interact with downstream effectors. Upon hydrolysis of GTP to GDP, the G-protein can return to the resting state.

Heterotrimeric G-proteins are divided into four families based on Gα: G_i, G_s, G_q, and $G_{12/13}$. Gα's of each family mediate characteristic responses (Fig. 1). These are, in simplified terms, for $Gα_i$ and $Gα_s$ inhibition and stimulation, respectively, of adenylyl cyclase (AC). ACs produce cAMP, which is a classical second messenger with protein kinase A (PKA) as the most ubiquitous target. $Gα_q$-family members stimulate phospholipase Cβ (PLCβ). PLC hydrolyzes inositol phospholipids producing diacylglycerol (DAG) and inositol phosphates. In case of hydrolysis of phosphatidylinositol-4,5-bisphosphate (PIP_2), the latter is inositol-1,4,5-trisphosphate (IP_3). Both DAG and IP_3 are second messengers, with the most well-known roles in the activation of protein kinase C (PKC) and Ca^{2+} release from the endoplasmic reticulum (ER), respectively. PLC also has a role in regulating membrane levels of inositol phospholipids, which, especially PIP_2, have been shown to interact and regulate the activity of membrane proteins such as ion channels (reviewed in [3–5]). $Gα_{12/13}$ have been shown to directly activate protein phosphatase 5 and some guanine nucleotide exchange factors (GEFs) for Rho

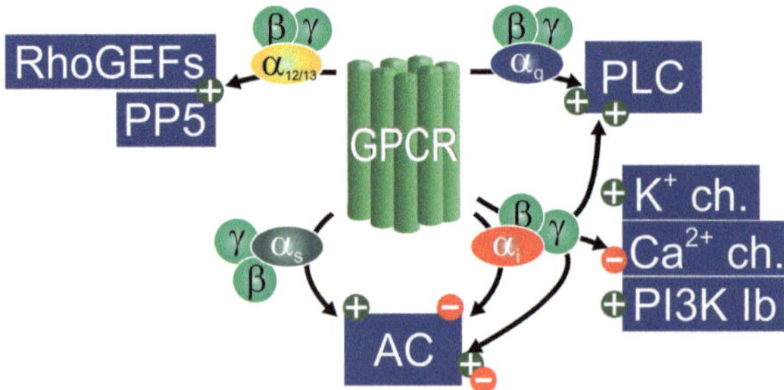

Fig. 1 Heterotrimeric G-protein signaling pathways. Only the most central pathways are shown; the pathways also end at the G-protein target. $G\alpha_s$ is a stimulant of all membrane-bound AC isoforms, while some G_i family $G\alpha$'s inhibit some AC isoforms. Some AC isoforms are also stimulated or inhibited by $G\beta\gamma$, and additional regulation comes from PKC and Ca^2, e.g. downstream of the PLC cascade. AC produces cAMP, which acts as a second messenger via PKA, some ion channels and EPAC, a GEF for the monomeric G-protein Rap. $G\beta\gamma$ signaling is usually supposed to originate from G_i-family proteins. $G\beta\gamma$ can, in addition to AC, inhibit some voltage-gated Ca^{2+} channels and stimulate class 1b PI3K, some voltage-gated K^+ channels, some PLC isoforms and some GEFs. $G\alpha_q$ stimulates all $PLC\beta$ species, which lead to phosphoinositide hydrolysis (a message in itself) and generation of the second messengers DAG and IP_3. DAG activates, for instance, PKC while IP_3 induces Ca^{2+} release from the endoplasmic reticulum. $G\alpha_{12}$ and $G\alpha_{13}$ have been shown to directly interact with protein phosphatase 5 (PP5) and certain GEFs for Rho (e.g., lsc/p115 RhoGEF and LARG), but RhoGEFs may also be regulated by $G\alpha_q$ and $G\beta\gamma$

(Fig. 1). Most well-known targets for $G\beta\gamma$ signaling are some PLC isoforms and, in the brain, K^+ and Ca^{2+} channels (Fig. 1). $G\beta$ and $G\gamma$ appear rather promiscuous in coupling to different $G\alpha$'s. However, $G\beta\gamma$ signaling is often assumed to originate from the G_i-family, in part based on the use of the G_i-family-inactivating toxin, Pertussis toxin. $G\beta\gamma$ has lower potency for many effectors than $G\alpha$'s; as G_i-family proteins are assumed to be present at higher levels than the other heterotrimeric G-proteins, they should also be better able to supply the higher levels of $G\beta\gamma$ required for activity. For reviews on heterotrimeric G-protein signaling, see [6–9].

GPCRs have also been shown to interact with other proteins. A ubiquitous signaling partner is β-arrestin (reviewed in [10]). Also a number of other proteins interact with specific receptors; in some cases, these have been shown relevant for the signaling, trafficking, etc. while their role in other cases is unclear (reviewed in [11, 12]).

Different receptors are able to preferentially couple to specific G-proteins. For some receptors this selectivity can be very high while other receptors – like orexin receptors – may be strongly promiscuous. The determinants of selectivity are not known; there are, for instance, no obvious general receptor features behind this. Also, for some promiscuous receptors the preferential G-protein coupling has been shown to be different in different tissues, possibly determined by tissue-specific receptor-interacting proteins (reviewed in [12]). Promiscuous signaling, already at the level of receptor coupling to primary signal transducers (G-proteins or other), is one way of explaining the differential effects of receptor activation in different tissues, but also downstream signal transducers and such can affect this. Oligomerization of GPCRs may affect the signal transduction or other receptor properties (Sect. 5). Yet a complicating factor is so-called *biased agonism* [13]. This implies that each chemically different endo- or exogenous receptor activator produces a complex with the receptor with a potentially unique conformation and ability to interact with signal transducers. Thus, one ligand may preferentially stimulate activation of G_q and β-arrestin, while another might direct the signaling towards G_s. This complicates the analysis of native (physiological) signaling, if there is more than one endogenous ligand and also constitutes a challenge in drug discovery. This is a highly relevant issue for orexin receptors as well, as these have been shown to connect to three families of heterotrimeric G-protein – G_q, G_i, and G_s – and other proteins like β-arrestin as well ([14, 15]; reviewed in [16]), and the coupling also seems to be different in different tissues, as discussed below.

2 Orexin Signaling in Native Neurons

2.1 *Impact on Electrical Activity*

In central nervous system (CNS) neurons, usually investigated in rat or mouse slice preparations, orexins are excitatory. Both pre- and post-synaptic mechanisms have been suggested to be operational. Dozens of highly relevant papers have been published, and we apologize for not being able to cite these here; the readers are referred to other chapters of this book (especially Burdakov: Postsynaptic actions of orexin) and our own recent reviews [16, 17] for full citations and detailed analyses.

Orexin receptor activation is associated with post-synaptic depolarization in a number of neurons (reviewed in [16, 17]). Three different mechanisms have been suggested for this depolarization. In some cases, depolarization is associated with an increase in input resistance, indicative of K^+ channel inhibition (see, e.g., [18–20]; reviewed in [16, 17]). This has been verified by I–V curve analysis. The identity of the channels has not been directly determined, but as these seem to be open at rest, the two-pore-domain K^+ (K_{2P}) channels or possibly inward rectifier (K_{ir}) channels are the most likely channel types regulated by orexin receptors. Both interact with inositol phospholipids (e.g. PIP_2), and hydrolysis of such upon PLC activity would inhibit the channels (reviewed in [21, 22]). For K_{2P} channels, phosphorylation by PKA and PKC and direct interaction with $G\alpha_q$ are inhibitory factors (reviewed in [6, 21]). K_{ir} channels can also be inhibited by phosphorylation and direct interaction with $G\alpha_q$ (reviewed in [6, 22]). There is evidence that orexin receptor activation indeed can inhibit both K_{ir} and K_{2P} channels ([23–25]; reviewed in [16]). The G_q – PLC(–PKC) cascade is an especially likely signal pathway for orexin receptors (see Sects. 2.2, 3, and 4), but the mechanisms have not been fully investigated for orexin responses in CNS neurons.

Another common finding upon orexin receptor-mediated depolarization is activation of non-selective cation fluxes. Usually, this is ascribed to activation of non-selective cation channels (NSCCs) (see, e.g., [19, 26–29]), but sometimes the Na^+/Ca^{2+} exchanger (NCX) has been suggested as an alternative explanation (see, e.g., [30–32]). However, we have analyzed this for another recent review [17], and most often the methods employed do not allow distinction of NCX from NSCC. It should also be pointed out that it is unclear whether NCX is likely to produce currents high enough to allow significant depolarization of most neurons, but, on the other hand, even a minor Na^+ influx may be enough to increase the excitability, especially if K^+ channels are inhibited at the same time (below). Nevertheless, there are some findings which clearly exclude one or the other and thus strongly point in the direction of either NCX or NSCCs (reviewed in [17]). A "normal" mode for NCX implies exchange of intracellular Ca^{2+} for extracellular Na^+, which produces a depolarizing current due to the 1:3 exchange ratio. NCX can also act in "reverse" mode, exchanging intracellular Na^+ for extracellular Ca^{2+} and hyperpolarizing the cell. The direction of the exchange is thus dependent on the extra- and intracellular concentrations of Na^+ and Ca^{2+} and the membrane potential. The normal mode of action would likely require an elevated intracellular Ca^{2+} concentration. This has not been investigated in orexin studies, but one likely source for this could be the G_q – PLC cascade-dependent inositol-1,4,5-trisphosphate (IP_3) production and Ca^{2+} release from the ER (see Sects. 3 and 4). In addition, NCX activity is regulated by phosphorylation by, e.g., PKC and by PIP_2 (reviewed in [33]), offering additional links to orexin receptor signaling. See further discussion on the collaboration of NCX and NSCCs in recombinant cells under Sect. 4.3.

Most cloned NSCCs are predicted to structurally resemble Shaker-type K^+ channels with six transmembrane helices per subunit and four subunits per channel. The most likely channels among NSCCs to be regulated by orexin receptors (and other GPCRs) belong to the transient receptor potential (TRP) channel family (reviewed in [34, 35]). However, this is speculation since the channel identities have not been assessed for orexin receptors in the CNS, except for one study, where Na^+ influx is suggested to be TRPC channel-mediated [36]. Also the receptor mechanisms involved in the regulation of NSCCs are unclear. We have speculated that the activation of the channels could take place via lipids, regulated via the "classical" GPCR-mediated regulation of phospholipases such as PLC but also phospholipase A_2 or D (PLA_2 and PLD, respectively) (reviewed in [37]), as is known for TRP channels (reviewed in [34]). Such signals could relate to the reduction or increase in PIP_2 or release of messengers such as DAG or arachidonic acid, which are known to affect TRP channel activity (reviewed in [34]). There are indications of engagement of TRP channels in orexin signaling in recombinant cell models (Sect. 4), supporting the idea of at least a principal ability of orexin receptors to couple to TRP channels; however, firm evidence from native neurons is lacking.

In some cases, orexins have been found to affect synaptic events, i.e. miniature synaptic potentials or currents. Especially interesting are the findings of increased frequency of events in the presence of action potential block, as this suggests a direct presynaptic effect of orexins ([38–49]; reviewed in [16, 17]). However, one also needs to keep in mind that such an effect could also be obtained upon engagement of new (silent) post-synaptic receptors/synapses. Further studies are required to resolve this.

Also other types of currents have been shown to be affected by orexin signaling, e.g. producing enhancement or inhibition of other depolarizing input (reviewed in

[17]). This is in agreement with the spectrum of signaling orexin receptors show (see, e.g., Sect. 4.1), which is likely to affect many ion channels. Orexin signaling has also been shown to be involved in plasticity, i.e. long-term potentiation in hippocampal circuitry [50–52], but the cellular targets and molecular mechanisms are thus far unclear.

In many sites, several mechanisms of depolarization have been seen to be utilized by orexin receptors in parallel, e.g. K^+ channel inhibition and NSCC/NCX activation co-occur ([19, 53–63]; reviewed in [17]). This is logical if we are to assume that basic orexin receptor signaling and in some degree also the principal channel/transporter expression profiles are rather similar in different neurons. However, as indicated above, details of the channel identities and signaling are not known. In addition, pre- and post-synaptic mechanisms may co-occur in the same site, as on the orexinergic neurons themselves [44, 49, 64, 65], neocortex [39, 66], nucleus tractus solitarius [48, 61, 62], dorsal motor nucleus of vagus [19, 42, 67], and POMC-ergic neurons of arcuate nucleus [38, 68]; however, it should be noted that it is in most cases unclear whether the pre- and post-synaptic actions of orexins take place in the same synapses.

The studies of molecular details of signaling in primary neurons are often hampered by methodological obstacles. Acutely isolated neurons are not an easy preparation (limitations of patch-clamp technique such as space-clamp problems and the different problems in whole-cell and perforated patch recordings, poor pharmacological tools, difficulties in applying molecular biological tools). In some neurons, the depolarizing inputs from orexins have been assessed using inhibitors of, e.g., PLC and protein kinases, but the targets are not clear. It also should be noted that PLC inhibitors often show both false negative and positive responses, and not all protein kinase or ion channel inhibitors are as selective as optimistically assumed by the users. Some novel pharmacological tools are now available for such receptor and channel studies, and these should be applied in orexin research. The tools are discussed under Sect. 6.4.

2.2 Other Responses in Neurons

Ca^{2+} is a second messenger often associated with orexin responses (Sects. 2.1 and 4). Few neuronal preparations have been investigated for Ca^{2+} elevations, but when done, this has been observed (e.g., a recent report of [69]; reviewed [17]). The source of Ca^{2+} elevations could be intracellular release or influx. The influx has been either shown or suggested in several cases for orexin, but there are no direct data demonstrating orexin-mediated Ca^{2+} release in native neurons – some indirect indications though (reviewed [17]). The pathway for Ca^{2+} influx is usually not clear, but it may be a result of activation of voltage-gated Ca^{2+} channels secondary to depolarization, direct influx via NSCCs or even NCX-mediated Ca^{2+} influx secondary to NSCC-mediated Na^+ influx. Among voltage-gated Ca^{2+} channels, L-type channels have been implicated in orexin responses in several sites by Ca^{2+}

or electrophysiological measurements. While Ca^{2+}-measurements do not separate depolarization-mediated and directly enhanced flux as such, patch-clamp measurements have often identified a PLC – PKC-dependent enhancement of depolarization-induced L-type channel current [70–72]. Ca^{2+} and PKC have been shown to be central for the activation of AMP-activated protein kinase in rat arcuate nucleus neurons in culture, partially via the L-type channels [72]. Ca^{2+} elevation has also been indirectly assessed using the intracellular Ca^{2+} chelator BAPTA, but one must remember that the heavy intracellular Ca^{2+}-buffering, required to effectively block Ca^{2+}-dependent orexin signaling, may also affect other (constitutive) processes. Reduction of extracellular Ca^{2+} concentration should block Ca^{2+} influx, but this also seems to reduce orexin binding as we have recently shown [73], and thus this may not be a suitable method to assess the role of Ca^{2+} influx in orexin responses. Ca^{2+} elevation is indirectly supported by the concept of orexin signaling involving NCX, since the normal (depolarizing) mode of action of this transporter requires intracellularly elevated Ca^{2+}.

Orexin receptor coupling to G-proteins has been investigated in a few studies. GTPγS binding in slice preparations suggests orexin receptor regulation of G_i-family G-proteins [74, 75], while direct measurements in the hypothalamus show coupling to G_i, G_s, and G_q family proteins [76]. Also potential responses downstream of the latter two, cAMP elevation and IP_3 generation, respectively, have been observed in the hypothalamus [76].

The PLC pathway is involved in orexin-mediated endocannabinoid generation [77, 78]. PLC produces DAG which can be hydrolyzed in the sn-1 position by DAG lipase, producing sn-2-monoacylglycerol. In cases where the sn-2 fatty acid is arachidonic acid, the resulting monoacylglycerol is 2-arachidonoyl glycerol (2-AG), which is a potent endocannabinoid, i.e. agonist at CB_1 and CB_2 cannabinoid receptors. Orexin signaling has been shown to rely on CB_1 receptor activation downstream of orexin receptors and DAG lipase in rat dorsal raphé nucleus and periaqueductal gray matter, suggesting the cascade orexin receptor $\rightarrow G_q \rightarrow$ DAG (PLC) $\rightarrow$ 2-AG (DAG lipase) $\rightarrow CB_1$ receptor. Endocannabinoid action elicits retrograde homo- or heterosynaptic presynaptic inhibition, which, depending on the neurons affected, produces either inhibition or disinhibition of the post-synaptic output (reviewed in [79]).

Also other CNS cells than neurons, such as astrocytes and microglia, have been suggested to express orexin responses, but the findings are thus far rather preliminary.

2.3 Conclusions

Orexin receptor signaling is well mapped at the level of the potential impact on neuronal electrical activity, but the molecular identities of the target channels/transporters as well as the signal cascades between the activated receptors and the channels/transporters are only weakly known, and the few observations to date of the second messenger generation or kinase activation in neurons are not well connected to the "final" outputs. This is typical for GPCR signaling, and is partially caused by poor tools for such investigations, but there clearly also is a lack of consequent approaches. As discussed under Sect. 6.4, novel tools should assist in the studies.

3 Orexin Signaling Features in Other Native Cell Types and Cell Lines

Different peripheral cell types and tumor cell lines express orexin receptors and show functional responses to orexins. The physiological significance of this is rather unknown, since the utilization of orexins as transmitters in the periphery has not been demonstrated, and in most sites where the receptors are expressed, prepro-orexin protein expression has not been demonstrated – though often not even investigated. Studies utilizing these cells and cell lines are, under any circumstance, useful in revealing orexin receptor signaling mechanisms and may also pave the way for discoveries of the physiological role of orexin signaling in these tissues.

3.1 Endocrine Cells

Several types of endocrine cells display responses to orexins. The most thoroughly investigated ones are adrenal cortical (i.e., zona fasciculata and reticularis) cells, which have been studied in native preparations of both man and rat ([76, 80–82]; reviewed in [17]). In different studies, both OX_1 and OX_2 receptors are found. Depending on the species and developmental stage, orexin receptor activation leads to activation of G-proteins of the families G_i, G_s, and G_q, and activation of both AC and PLC is seen. Further downstream, a strong stimulation of glucocorticoid release is observed. The release is similar in magnitude to that induced by adrenocorticotropic hormone (ACTH) and apparently utilizes the same classical cascade $AC \rightarrow cAMP \rightarrow PKA$. However, it is not known, whether the orexin response is simply a result of $G\alpha_s$-mediated AC stimulation or whether also G_i and G_q proteins contribute to this; different AC isoforms can be stimulated by also $G\beta\gamma$, Ca^{2+}, and PKC (reviewed in [83]). Steroidogenesis has been further investigated in the human tumor cell line H295R, which expresses both OX_1 and OX_2 receptors. Orexin stimulation leads to activation of multiple kinase pathways, i.e. PKC and the mitogen-activated protein kinases (MAPKs) ERK1/2 and p38, and also Ca^{2+} elevation [84]. Orexin stimulation induces an increased steroidogenesis but also a generally more mature phenotype [84, 85], and thus it is difficult to judge whether the signaling cascades identified relate to the former or the latter (discussed in [17]). In adrenal medulla, as investigated in both native rat chromaffin cells and human tumor cells, orexins stimulate catecholamine release [86, 87]. As investigated in the latter, orexin signaling activates PLC, and the catecholamine release appears to be mediated by the PKC cascade [87], which is a classical enhancer of catecholamine release (via voltage-gated Ca^{2+} channels?). Prepro-orexin mRNA and protein have been found in human adrenal gland [82, 88].

Orexin responses in the different endocrine cell types of the pituitary gland have been investigated in many studies but few of these offer insights in the cellular and molecular mechanisms. Two studies suggest involvement of PKC in the response [89, 90].

Orexin responses in the endocrine pancreas have been investigated in explants, isolated cells, and tumor cell lines. In InR1-G9 Chinese hamster glucagon-producing tumor cells, orexin signaling induces activating phosphorylation of phosphoinositide-dependent kinase 1 (PDK1) and protein kinase B (PKB a.k.a. Akt), indicating activation of phosphoinostide-3-kinase (PI3K) [91]. GPCRs can activate class I PI3K either via $G\beta\gamma$ (Fig. 1) or via the Ras cascade (reviewed in [92]).

Orexin receptors are also expressed in the male reproductive tract of human and rat. In undefined testicular cells, their activation stimulates PLC [93]. Increased testosterone production as well as altered gene transcription is also seen [94, 95], but whether these responses are connected to the PLC signaling is not known. Expression of prepro-orexin mRNA and protein has been detected in the male reproductive tract [93, 94, 96–98], thus suggesting a possibility of paracrine orexin signaling.

3.2 Other Cell Types

The gastrointestinal tract was early on found to be a target of orexin responses and possibly also a site for orexin production (prepro-orexin mRNA detected) [99]. Both OX_1 and OX_2 receptor mRNA is expressed in neurons and ileal smooth muscle cells, and orexins seem to display direct actions on these cell types. The responses are complex and do not allow a clear-cut analysis; for instance, many different players are involved in the orexin-induced ileal contraction [100, 101].

Acutely patient-derived human colon carcinoma cells (and their metastases) express OX_1 receptors while the native colon epithelial cells do not [102, 103]. AR42J rat pancreatic acinar carcinoma cells express OX_2 receptors, and are also induced to undergo programmed cell death upon orexin challenge [104]. Thus induction of cell death is a possible response for both orexin receptors. See Sect. 4.1 for the cell death mechanisms.

Mesenchymal stem cells give rise to a number of different cell types, including brown and white adipocytes. Differentiation of mouse brown adipose tissue has been shown to require placentally produced orexins likely acting via OX_1 receptors [105, 106]. Orexin signaling was suggested to proceed in part via p38 MAPK and in part by induction of bone morphogenic protein 7 and its receptor, BMPR1A, which generates a potential autocrine loop [105]. No other details of the signal cascades are known. White adipocytes and white adipocyte precursor cell lines also express orexin receptors. Whether orexins induce differentiation of adipocytes is questionable. Upon orexin challenge, human subcutaneous adipocytes, rat (undefined) adipocytes, and differentiated 3T3-L1 adipocytes show apparent maturation as indicated by expression of peroxisome proliferator-activated receptor γ2 (PPARγ2), hormone-sensitive lipase and lipoprotein lipase, and increased glucose uptake and triglyceride synthesis, while the human omental adipocytes rather show the opposite [107, 108]. 3T3-L1 preadipocytes are not differentiated by orexin-A, which is suggested to be due to the lack of PI3K – PKB activation [109]. Intracellular signals detected are activation of the MAPK pathways ERK1/2 and p38, while the PI3K – PKB cascade seems to be activated in differentiated 3T3-L1 cells [108]. cAMP is one of the major stimulants of differentiation of both brown and white adipocytes, and it is surprising orexin signaling via cAMP has not been assessed in any of these studies. Adipocyte studies also show an unexpected finding: while orexin-A stimulates the proliferation of 3T3-L1 preadipocytes, orexin-B does the opposite [109, 110].

Utilizing human, mouse, and rat models (vascular endothelial cells, and in and ex vivo), it was shown that orexin-A, at the high nanomolar range, induces heme oxygenase 1 expression and apparent activity as well as angiogenesis [111].

4 Orexin Signaling Features in Recombinant Orexin Receptor-Expressing Cells

Native orexin receptor-expressing cells are difficult to isolate, do not survive long in culture, and are present in rather small amounts – this is most obvious for the native neurons. Thus, researchers have resorted to recombinant cell lines, as in much other research as well. These cells have the additional advantage of expressing only a single orexin receptor subtype, which allows analysis of the receptor subtype differences. However, the cell background is also very different from the native cells (especially CNS neurons) and the results may thus not be directly physiologically applicable. Receptors from species other than human have barely been investigated in recombinant cells.

4.1 CHO Cells

Chinese hamster ovary K1 cells (CHO-K1; usually cited just as CHO) and HEK293 human embryonic kidney cells are the two most used cell lines in orexin receptor studies to date. The exact cell type is unclear for both cells, but they appear epithelial. Both are widely used cells for heterologous expression, especially of receptors. Orexin signaling in CHO cells involves many second messenger pathways. The PLC pathway releases DAG and inositol phosphates including IP_3, which releases Ca^{2+} from the ER and leads to secondary (capacitative) Ca^{2+} influx (see, e.g., [112, 113]). The PLC pathway also gives rise to the endocannabinoid 2-AG (Sect. 2.2); recombinant CHO-K1 cells (and similarly recombinant HEK293 and neuro-2a cells; Sects. 4.2 and 4.3) are indeed the only cells where 2-AG production has been directly shown [114, 115], while the findings in native neurons (Sect. 2.2) are solely based on the use of pharmacological inhibitors. Breakdown of 2-AG gives rise to free arachidonic acid, but the latter is also produced upon action of PLA_2 (likely specifically $cPLA_2$) [114, 115]. PLD (primarily PLD1) is also activated upon orexin receptor stimulation [114, 116]. The G-protein family behind the entire Ca^{2+} response (see below), and PLA_2, PLC, and PLD activation was recently shown to be G_q, based on the use of the depsipeptide inhibitor of $G_{q/11/14}$, UBO-QIC a.k.a. FR900359 [117]. G_i signaling is seen in the inhibition and G_s in the stimulation of AC, respectively ([114, 118], a); however, it is possible that also the G_q family proteins play a part in the AC inhibition [117]. These responses are seen for both OX_1 and OX_2 receptors. The studies may not be directly used to assess the efficiency of coupling, but OX_2 receptors show weaker signaling to PLD and G_s. OX_1 signaling has been assessed for a wider set of responses than OX_2 responses. These include activation of PKC, MAPKs ERK1 and -2, p38 (α/β isoform), and c-jun N-terminal kinase (JNK) [119, 120]. PKC is likely activated downstream of PLC – DAG, though there also are conflicting findings [116], and even PI3K and Src are indirectly associated with the ERK cascade, though not directly measured [119]. Ras activation seems to be necessary for ERK activity [119].

One characteristic finding in OX_1 signaling is Ca^{2+} influx. There is an apparent receptor-operated Ca^{2+} influx that occurs at low orexin concentration and that does not require IP_3 [113, 121]. This influx is also seen as a non-selective cation conductance in patch-clamp [122]. Some inhibitor studies have been conducted but these are not conclusive for the channel type, and the patch-clamp studies are technically demanding as the conductance is not easily seen. The influx is partially inhibited by dominant-negative TRPC1 and TRPC3 channels, but these have not been tested in combination. In the lack of an effective and selective inhibitor, Ca^{2+} influx has in many studies been inhibited by reduction of extracellular Ca^{2+} concentration or depolarization, which reduces the driving force for Ca^{2+} entry. These procedures but especially the former, produce a significant inhibition of many downstream orexin responses [73, 113, 115, 116, 119, 123, 124], suggesting that intracellular Ca^{2+} elevation is somehow required for them. However, we have recently shown that extracellular Ca^{2+} is also required for orexin-A binding [73], and thus some earlier conclusions are not valid. Nevertheless, depolarization also produces significant inhibition of some orexin responses, and thus the issue requires further studies. It should be noted that activation of a non-selective cation conductance is also seen in native neurons (Sect. 2.1) and in other recombinant cells (Sects. 4.2 and 4.3), and thus we assume that this reflects the same general coupling ability (though not necessarily a single mechanism; Sect. 6.1) for orexin receptors.

We have recently seen that the $cPLA_2$ cascade, also activated by orexin receptors at high potency, is required for this influx, offering a clue to its regulation [115]. Unfortunately, we have not been able to continue these studies yet.

Another interesting feature of orexin signaling in CHO cells is the range of orexin concentrations it involves. The most high-potency responses are AC inhibition, Ca^{2+}/cation influx and $cPLA_2$ activation in the subnanomolar or low nanomolar range. These are followed by apparently two different PLC activities, PLD and DAG lipase in the low – mid-nanomolar range, and then finally G_s in the high nanomolar range (reviewed for OX_1 in [17]; see [114] for OX_2). As mentioned earlier, we do not know the physiological orexin concentrations, but one might assume that peptides usually are present at rather low levels. Thus, a response requiring rather high orexin concentration, such as G_s activation in CHO cells, would not be physiologically likely, unless mediated by rather direct synaptic contact. However, the apparent G_s-coupling takes place at high potency in other cells, such as hypothalamic neurons (Sect. 2.2) and adrenal cortex (Sect. 3.1). Thus we may assume that different cell backgrounds may direct orexin signaling to different outputs (Sect. 6.2). One such background factor is expression of other receptors capable of dimerizing/oligomerizing with orexin receptors (Sect. 5).

Programmed cell death is also seen in CHO-K1 cells (only investigated for OX_1). The cell death shows the hallmarks of apoptosis by requirements of gene transcription and protein synthesis and nuclear condensation [120]. However, upon inhibition of the caspases, the profile of the cell death changes, no longer requiring gene transcription. p38 MAPK is driving the cell death, but the activation mechanism for this kinase is not known in this context. In contrast, in the suspension-adapted CHO cell clone, CHO-S, orexin receptor-mediated signaling in this response has been suggested to involve activation of the tyrosine kinase Src, which phosphorylates the OX_1 receptor at Tyr83 within *protein immunoreceptor tyrosine-based switch motif* (ITSM), which then can attract protein phosphatase SHP-2 [125]. The cascades between the receptor and Src and from SHP-2 to cell death are thus far unclear. OX_1 receptor was previously suggested to also contain another similarly signaling motif, ITIM, but this finding may have alternative explanations [16]. The findings of the orexin-induced programmed cell death in CHO-K1 and CHO-S cells are also different in the perspective that the former is inhibited by serum while the latter is not. Nevertheless, the responses should be examined across the cell types, to judge, for instance, whether the conclusions are distorted by compromised selectivity of some of the inhibitors used. Interestingly, it is not known whether neurons are susceptible to this cell death. If not, then the question arises, why not, and whether this response might be contextually activated even in neurons.

4.2 HEK293 Cells

Both human orexin receptor subtypes are able to elevate intracellular Ca^{2+} and activate PLC in HEK293 cells [97, 126, 127], but the role of Ca^{2+} influx has not been assessed in such detail as for CHO-K1 cells. An oscillatory Ca^{2+} response is obtained at low orexin-A concentrations, and this response is nearly fully blocked by the non-selective serine hydrolase and protease inhibitor, MAFP [128], linking this to the findings with both MAFP and a more selective $cPLA_2$ inhibitor in CHO-K1 cells (Sect. 4.1) [115, 124]. It is not known whether only the PLC – DAG lipase cascade is operational in these cells or also the PLA_2 cascade [115, 128]. Several dominant-negative TRPC channel constructs, i.e. of TRPC1, and -3 and -4, reduce the Ca^{2+} response magnitude at low orexin-A concentrations (1 nM).

Both orexin receptor subtypes activate ERK1 and -2, and OX_2 receptor stimulation has additionally been shown to activate p38 MAPK (not investigated for OX_1) [15, 129]. Also cAMP elevation has been measured in OX_2 cells [129]. The MAPK cascades appear to be regulated by several signal cascades, similar to CHO-K1 cells [119]; even if one disregards the results obtained with dominant-negative G-proteins, which are uncertain, the use of other inhibitors demonstrates cooperation of the cascades [129]. OX_1 receptors also couple to β-arrestin in HEK293 cells, and this contributes to the ERK1/2 activation [14, 15]. Upon expression of GIRK1 and -2 (G-protein-regulated K_{IR}) channels together with orexin receptors, orexin receptor activation inhibits the channels; however, the potency of orexin-A is in the high nanomolar range [24].

One aspect distinguishes the studies with CHO-K1 and HEK293 cells: while all the studies in Sect. 4.1 for CHO-K1 cells have been performed with the same clones of OX_1 and OX_2 cells, different clones (some stable and some even transient) have been used for HEK293 cells. It is thus possible that even background clonal differences affect the results.

4.3 Neuron-Like Recombinant Models

A few other cell lines and transient expression models have been utilized in orexin studies. We generated PC12 rat pheochromocytoma and neuro-2a mouse neuroblastoma cells as neuronal models for orexin investigations. Unfortunately, these cells have not been used much since the original characterization of Ca^{2+} elevation and PLC activation in both cell types [130]. Ca^{2+} elevation in OX_1-expressing neuro-2a cells seems to involve IP_3-independent Ca^{2+} influx similar to CHO-K1 cells [121]. We recently also observed 2-AG generation in OX_1-expressing neuro-2a cells [115].

Another neuronal cell model, utilized in a single study, are the BIM hybridoma cells [131]. This study is often cited as reference for the ability of OX_2 receptors but not OX_1 receptors to couple to G_i proteins. Unfortunately the authors miss the classical finding with receptors coupling to AC in multiple ways; usually nothing can be seen under basal conditions. Thus, a most likely explanation is that the positive and negative signaling from the receptors take each other out under basal conditions but can be seen when each pathway is selectively inhibited; this is evident with, e.g., CHO-K1 cells [114].

IMR-32 neuroblastoma cells, differentiated with 5-Br-2′-deoxyuridine and transiently transduced with baculoviral vectors harboring the human OX_1 receptor under cytomegalovirus promoter, have been utilized in two studies of orexin-A-induced cation fluxes by Åkerman & co-workers [132, 133]. It was found that orexin-A triggers a Ca^{2+} influx that is partially sensitive to extracellular Na^+: in most cells Ca^{2+} influx was either strongly inhibited by removal of extracellular Na^+ or not affected by it. In the former case, the studies indicate that Ca^{2+} influx is secondary to Na^+ entry (i.e., Ca^{2+} taken in by the reverse mode of NCX) while in the latter case it is caused by a direct channel-mediated Ca^{2+} influx. Expression of dominant-negative TRP channel subunits and RNA interference suggest that the apparent Na^+ influx in the former case is in most part mediated by TRPC3 channels. In contrast, the channels mediating the direct Ca^{2+} influx remain unknown, but their pharmacology is clearly different from the other channels. In OX_1-expressing CHO-K1 or HEK293 cells, where partial sensitivity to dominant-negative TRPC channels has been noted for the orexin-A-induced Ca^{2+} influx (Sects. 4.1 and 4.2), the role of Na^+ has not been examined [122].

Hypothalamic N41 & N42 cells, i.e. SV40 large T antigen-immortalized mouse embryonic hypothalamic neurons, stably transduced with OX_1 and OX_2 receptors (species unclear), respectively, have been reported in a recent study [134]. The cells were used to demonstrate orexin-mediated activation of mammalian target of rapamycin complex 1 (mTORC1) as indicated by, for instance, rapamycin-sensitive activation of p70 ribosomal S6 kinase ($p70^{S6K}$). The same was seen in HEK293 cells expressing OX_1 and OX_2 receptors. In N41 and -42 cells, orexin-A and -B stimulus activated the p90 ribosomal S6 kinase ($p90^{RSK}$) and ERK1/2 but not PKB, but neither $p90^{RSK}$ nor ERKs were involved in mTORC1 activation. In contrast, Ca^{2+} elevation appeared sufficient to activate mTORC1 as this could also be induced by thapsigargin or ionomycin and was, for orexin, abolished by intracellular Ca^{2+} chelation with BAPTA. Ca^{2+} was determined to elicit its effect via the lysosomal pathway of vacuolar H^+-translocating ATPase (v-ATPase) – Ragulator – Rag. Ragulator is a GEF for the monomeric G-protein Rag. How Ca^{2+} works in this cascade is unclear.

SV40 large T antigen has also been used to immortalize rat olfactory epithelium neurons to create the Odora cell line, which has been used for orexin studies upon transient expression of human orexin receptors [135]. Weak IP_3 and Ca^{2+} elevations were observed, while the cAMP elevation was more robust, though this was only tested at 100 nM orexin-A/B. ERK1 and -2 phosphorylation was also increased at this concentration.

4.4 Other Recombinant Cells

Some other cell types have also been used in orexin studies, but these have mainly offered supporting information. Mouse embryonic fibroblasts (MEF) expressing OX_1 receptors have been used in two studies. Laburthe & co-workers have used the wild-type and $G\alpha_{q/11}$ knock-out cells to demonstrate that orexin-promoted cell death requires $G_{q/11}$ signaling [136]. MEF cells were also used for RNA interference studies in [134] (Sect. 4.3).

4.5 Species Variants Other Than Human

Canine OX_2 receptors induce elevation of intracellular Ca^{2+} [137], and murine OX_1 and the two OX_2 splice variants IP_3 release in HEK293 cells [138, 139].

5 Orexin Receptor Oligomerization

GPCRs are assumed to exist in complexes (reviewed in [140, 141]). Though this is often called dimerization, the actual number of units is usually unknown, but often either dimers or tetramers are assumed. The most often utilized resonance energy transfer-based techniques, FRET and BRET, usually assess dimerization, and neither show nor exclude higher order complexes. The findings imply that the receptors can make both homomeric and heteromeric complexes; the latter may have radically different properties as compared to homomeric ones as concerns trafficking, pharmacology and signaling. It should be noted that complex formation is not easily investigated and, even if a principal ability to form complexes is present, its physiological significance is seldom clear.

Orexin receptors have been shown to be able to form homomeric complexes as well as heteromers between the subtypes when expressed in recombinant systems [139, 142, 143]. This is a rather trivial finding for GPCRs, and the significance of this is unknown as also is the proportion of complexes in relation to receptor monomers (see below). Another complex formation can occur between either orexin receptor subtype and CB_1 cannabinoid receptors [142, 144]. The interaction between OX_1 and CB_1 has been investigated in detail in HEK293 and CHO-K1 cells, and the studies suggest clear effects on trafficking and signaling. The problem, however, is that since orexin signaling also generates 2-AG [115], there is also a functional interaction, and current studies have not separated these levels of interaction. We assume that the 2-AG-mediated interaction is more important [145], especially in the physiological context, since OX_1 and CB_1 receptors have not been demonstrated to be expressed in the same cells. However, it is possible that this could occur, and may be discovered once orexin receptor antibodies evolve and allow reliable analysis.

OX_1 receptors and κ-opioid receptors (κOR) have also been shown to display both direct molecular and functional interactions. In contrast to the $OX_1 - CB_1$ pairing, the functional interaction was suggested to occur through purely intracellular signaling pathways, i.e. OX_1 receptor interaction enhanced JNK-mediated phosphorylation of κOR in CHO-K1 cells, inhibiting κOR signaling to AC and enhancing its signaling via β-arrestin and p38 MAPK, apparently via G_q [146]. In another study, OX_1 and κOR were shown to heteromerize. The heteromerization reduced the coupling of OX_1 and κOR to G_q and G_i, respectively, while it instigated coupling of the receptor heteromers to G_s and thus to cAMP elevation in HEK293 cells [147]. OX_1 and κ-opioid receptor mRNA is expressed in the same cells in the hippocampus, thus suggesting a possible native interaction [147].

Another recent study demonstrated interaction of either orexin receptor subtype with GPR103 utilizing BRET and FRET techniques in HEK293 cells [148]. ERK1/2 phosphorylation in receptor co-expressing cells was stimulated by orexins as well as the GRP103 agonist, QRFP, and the effect of each peptide was blocked by orexin receptor antagonists. This indicates a very high degree of complex formation, while the FRET results are somewhat in disagreement with this. Expression of GPR103 correlates with the expression of the orexin receptor subtypes in the hippocampus, and the authors speculate that the receptor interaction could be relevant in Alzheimer's disease.

The most comprehensive study involves a receptor complex between OX_1, corticotropin-releasing factor receptor 1 (CRF_1) and the non-GPCR sigma 1 receptor (σ_1) [149]. The $OX_1 - CRF_1$ complex was shown in HEK293 by FRET and a functional response; the response to an agonist of each receptor was antagonized by an antagonist against either the same receptor or the other partner (cross-antagonism). Activation of each receptor was, in itself, antagonistic to the action of the other receptor with respect to β-arrestin recruitment and phosphorylation of PKB and ERK1/2 (negative crosstalk). The dimerization/oligomerization was blocked by cell-permeable peptides composed of either the OX_1 receptor transmembrane segment 1 or 5, which also abrogated the cross-antagonism and negative crosstalk. In rat ventral tegmental area (VTA) neurons, the dimerization/oligomerization was shown by cross-antagonism and negative crosstalk with respect to ERK1/2 phosphorylation. Dopamine release was stimulated by orexin-A, and this was inhibited by antagonists of both OX_1 and CRF_1 receptors and by CRF itself. However, whether the orexin-A response was enhanced by the disruption of the receptor complex by the cell-permeable peptides (above) is questionable. The complex could also harbor the σ_1 receptor, as was shown by the authors in HEK293 cells, and activation of σ_1 receptors with cocaine disrupted this. Application of cocaine on VTA neurons freed OX_1 and CRF_1 receptors from the cross-antagonism and negative crosstalk, suggesting a native $OX_1 - CRF_1 - \sigma_1$ interaction. Finally, the selectivity of the cocaine action for σ_1 receptor was verified and also the impact on VTA dopamine release, which both CRF and orexin-A were able to stimulate in the presence of cocaine (on top of the cocaine effect).

It is interesting to note that OX_1 receptor complex formation with either κOR, CRF_1 or GPR103 appears very efficient. For other complexes reported, i.e. those among orexin receptor subtypes or with CB_1 receptors, this is not known. The FRET and BRET methods mainly utilized do not easily give the efficiency of the complex formation, since resonance energy transfer efficiency is dependent on the orientation and distance of the donor and acceptor, which is also shown in orexin receptor studies [142]. While some studies indicate that receptor activation may affect the energy transfer [139, 142, 143, 147], this cannot be interpreted to suggest difference in dimerization/oligomerization, since any conformational change may also either increase or decrease the energy transfer efficiency. However, studies on GPCRs other than orexin receptors by, e.g., fluorescence correlation spectroscopy suggest that dimerization/oligomerization can be a dynamic process (reviewed in [150]).

6   Summary and Conclusions

6.1   Convergence of the Findings?

The first conclusion that arises is that orexin receptors have certain signaling properties that come to expression independent of the background, suggesting that these represent some strong intrinsic receptor properties. One such coupling is to PLC, which projects, in one hand, to DAG and therewith either to activation of PKC (or other targets) or breakdown of DAG to 2-AG, an endocannabinoid messenger. The other product of PLC activity is IP_3 and IP_3-dependent Ca^{2+} release, the role of which, however, is unclear for orexins in CNS neurons. The third role of PLC relates to the regulation of membrane phosphoinositide levels, which are known to affect membrane protein properties, especially ion channels but also, for instance, NCX. In this perspective it is interesting to note that phosphatidic acid, produced by, e.g., PLD activity, is able to stimulate phosphatidylinositol-4-phosphate 5-kinase, elevating PIP_2 levels (reviewed in [151]); we have observed activation of PLD1 in CHO-K1 cells expressing either OX_1 or OX_2 receptors (Sect. 4.1). PLD has also been suggested to be involved in orexin signaling in the brain [152], but the implications are thus far unclear. It is logical to assume that orexin receptors would couple to G_q proteins, which trigger the activation of PLCβ. Indeed, studies in native neurons and adrenal cortical cells indicate that orexin receptors are able to couple to G_q proteins (Sects. 2.2 and 3.1), and in CHO-K1 cells these seem to mediate most of the responses too (Sect. 4.1). In CHO-K1 cells expressing human OX_1 receptors, there seem to be two different PLC activities activated upon stimulation of these receptors [112], and we have also obtained indications that protein kinase Cδ is activated independent of PLC [116]. Thus, though it is logical to assume that the – for GPCRs classical – $G\alpha_q$ – PLCβ pathway is of high significance for orexin receptor signaling, there may be additional features of this signaling. We also have to admit that the involvement of this signaling has seldom been assessed in the brain, not even at the level of PKC (below).

Another general feature is the ability to activate non-selective cation fluxes. As reviewed for neurons (Sect. 2.1), there is an ongoing discussion of whether these depolarizing inward currents are mediated by NSCCs or the reverse mode of NCX, as most of the studies do not allow distinction. Nevertheless, it feels to the author of this review that NSCCs are the more likely candidate for the response in most cases. If so, what are the types of channels and how are they regulated by orexin receptors? The clearest identification of the channel originates from recombinant IMR-32 neuroblastoma cells, in which TRPC3 is suggested to lie behind the Na^+-dependent OX_1 response (Sect. 4.3) but also findings in CHO-K1 (Sect. 4.1) and HEK293 (Sect. 4.2) cells and native neurons (Sect. 2.1) indicate TRPC subfamily channels. The common theme for regulation of many TRP channels are lipid mediators, and thus, for instance the PLC pathway could offer activating signals by releasing DAG or reducing phosphatidylinositol species (reviewed in [34]) – remarkably, the PLC pathway also regulates NCX (Sect. 3.1). However, alike the channel identities, the regulatory mechanisms for the orexin-activated channel are unknown. The only direct pieces of evidence as to the regulatory mechanisms for the channels come from recombinant CHO-K1 cells, in which the $cPLA_2$ cascade is suggested to be essential for the activation (Sect. 4.1), and from rat thalamic neurons, in which Ca^{2+} influx, mediated by the T-type voltage-gated Ca^{2+} channels, has been suggested to drive the channel activation [36]. Clearly we have different channel types and mechanisms in different cells (see also Sect. 4.3). Studies in CHO-K1 cells show activation of multiple lipid signaling pathways by the receptors, and thus we might not find it surprising if somewhat different channel types were regulated by orexin receptors depending on the cell background. This also makes sense in the perspective of the multiple TRP channel genes (>20) and their regulatory mechanisms (reviewed in [34, 35]).

6.2   Divergence of the Findings?

Apparently divergent signaling is found in different tissues, especially in peripheral tissues as compared to the findings in CNS neurons. The obvious findings in peripheral cells include strong elevation of cAMP in the adrenal gland (Sect. 3.1), and activation of the PI3K – PKC cascade in InR1-G9 glucagonoma cells (Sect. 3.1) and 3T3-L1 adipocytes (Sect. 3.2) and indications of the same in some other cells. As discussed above, the physiological significance of orexin receptors in tissues outside the CNS is unclear. Are we then to assume that these responses may be irrelevant? G_s-coupling and cAMP elevation is, however, also seen in hypothalamic membranes (Sect. 2.2). In other sites this has not been investigated. It should also be noted that while the neuronal responses usually have only been assessed at high orexin concentrations (≥ 100 nM), the peripheral responses usually have been analyzed by concentration–response relationships and shown to display high potency. Thus many of these responses are significantly activated at low nanomolar orexins while most of the CNS neuron responses have been proven only at 100–1,000-fold higher orexin concentrations. Transmitter levels fall rapidly upon increasing distance (especially relevant for extrasynaptic signaling), so would this not rather question the physiological relevance of the orexin responses measured in CNS neurons? I guess we do not need go that far, but as long as we do not know the physiological orexin levels in different *loci* of the body (Sect. 6.2), we need to be more careful in our analyses and conclusions. For instance, we may need to test the concentration–response relationships for the responses we are measuring, and we should also look for potential other responses at different orexin concentrations; GPCR signaling is not only subject to an increasing amplification upon increased agonist concentration, but the signaling profile may change more radically according to the ligand concentration and time (see, e.g., [153–158]). Nevertheless, the studies in neurons and in recombinant cells are largely in agreement, and they are also in agreement with the more systemic orexin responses, which supports the notion that the responses measured both in neurons and other cell types are physiologically relevant.

Nevertheless it is clear that orexin signaling is also different in different cell types. For instance, G_s-coupling is apparently of high potency in hypothalamus and adrenal cortex (Sects. 2.2 and 3.1), while the potency is very low in CHO-K1 cells (Sect. 4.1) – quite the opposite to what would be expected from the assumed *overexpression artifacts* of recombinant cells. G_s is a stimulant of all AC isoforms. The expression of different AC isoforms in different tissues is unlikely to cause the apparent discrepancy, but we may assume that indeed the G-protein-coupling of orexin receptors is likely to be different in different tissues. Such has been observed for other GPCRs, but the mechanisms are unclear. One possibility is interaction with other proteins, maybe receptor oligomerization (Sect. 5).

The role of OX_1 and OX_2 receptors in neuronal responses was analyzed in our recent review [17]. In the analysis, we cannot see an obvious difference, but the material is too small to allow a firm conclusion. It should be noted that it is not easy to separate between the responses since many sites express both receptor subtypes and neither the subtype-selective agonists nor antagonists are quite as selective as hoped for. The recombinant cells have not been consequently analyzed but some studies with CHO-K1 cells suggest possible different coupling of OX_1 and OX_2 receptors (Sect. 4.1); however, this material also is too limited to allow conclusions.

6.3 *A Matter of Receptor Subtypes and Ligands*

The vast majority of the studies on the neuronal effects of orexins, from intact animals to isolated neurons, have been done with rodents (most often rat) while the native cell lines are of variable origin and the recombinant studies usually assess human receptors. Can we assume that these findings stand on a common ground?

This is always a relevant question and there are previous cases with other targets that would suggest either way. Orexin receptor sequences are very similar across species orthologs. This would suggest that findings at the pure receptor signaling and pharmacological level are likely to be rather similar. Another relevant general comparison may the different drug development campaigns that usually utilize animal models as first background. History offers examples of successes but also of failures due to species differences, but many successes – as concerns the physiological part – are encouraging even for the orexin field. For orexin receptors, there are species-specific identified (and predicted) splice variants, which may affect both the receptor expression and signaling [138, 159]. Based on the evidence accessible, we can assume that the findings can be extrapolated across species, although some converse findings are also likely to surface sooner or later. One significant physiological difference is already known: while incapacitation of the OX_2 receptor induces a strong narcoleptic phenotype in dogs, OX_2 knockout in mice gives a clearly milder phenotype [160–162]. However, this is more likely to relate to the receptor subtype expression profiles and not to the signaling properties of the receptors *per se*.

There are two native ligands for orexin receptors, orexin-A and orexin-B. These are utilized in most of the studies, but sometimes the synthetic peptide variant, Ala^{11}-D-Leu^{15}-orexin-B, has also been used. The latter peptide was originally reported as a highly OX_2-selective receptor activator [163], and has been used as such in a number of studies. The potency difference between orexin-A and orexin-B – the original study reports that orexin-A is tenfold more potent on OX_1 receptors while both peptides are equipotent on OX_2 for Ca^{2+} (and binding) in HEK293 cells [97] – has also been utilized as an indicator of involvement of a specific orexin receptor subtype in a given response. However, there is a built-in problem in this. For many GPCRs, it has been shown that chemically different ligands may induce/ stabilize distinct active receptor conformations, which may interact with the effectors with different affinity profiles, and thus lead to preferentially different response profiles (Sect. 1). This is termed *biased agonism* [13]. For GPCRs, this can effect interaction with G-proteins as well as with other signal transducers. Furthermore, receptor structure itself and signaling can be affected by interaction with other proteins. Thus the use of the so-called subtype-selective agonists with any receptors known to have several potentially physiologically important signal pathways is discouraged. Orexin receptors are such receptors; indeed, we have demonstrated this for the human variants. On one hand, orexin receptor signaling to apparently the same response, Ca^{2+} elevation, shows different relative potencies in two recombinant cell lines [127]. On the other, when the potencies of orexin-A and orexin-B are compared in CHO-OX_1 cells for different responses, the difference is not constant, and orexin-B induces some of the responses nearly as potently as orexin-A while the difference is 20-fold for other responses [164]. Thus, the use of any agonist-based pharmacology to separate orexin receptor subtypes is discouraged. An additional reason for this is that Ala^{11}-D-Leu^{15}-orexin-B shows much poorer selectivity even in recombinant cells than originally described [127].

The orexin-A sequence is identical in different mammalian species, and also the post-translational processing is assumed to be identical. However, orexin-B is not identical. Pig and dog express orexin-B, which differs by one amino acid from the human variant (Pro2 vs. Ser2, respectively), and mouse and rat one that differs by this and also another one (Asn18 vs. Ser18, respectively). Many studies in rat or mouse do not report whether the rodent variant of orexin-B was used. Based on mutagenesis and modeling studies, amino acid 2 is unlikely to have a major impact on orexin-B binding while amino acid 18 is suggested to enter the putative binding cleft of the receptor [165], and could thus contribute to binding. However, it is not conserved among orexin-A and orexin-B unlike many amino acids in the same region, and thus it may be of less importance. In fact, rodent orexin-B has been tested on human orexin receptors: Ca^{2+} response to the rodent peptide was not significantly different as compared to human orexin-B for either receptor subtype [166]. Although rodent receptors have not been exposed to human orexin-B, the rodent receptors are either completely (OX_2) or almost completely (OX_1) identical to the human orthologs for the transmembrane regions likely composing the binding cleft, and this together with the published data suggest that there is no major difference between the human and the rodent orexin-B peptides. However, there is no guarantee that this does not impact on the potency of the peptide to activate the receptor signaling, and thus constitutes a further reason not to utilize agonist-based pharmacology for orexin receptor subtype determination.

A fairer chance to obtain reliable information on the receptor subtype involvement in a response comes from the use of antagonists. Plenty of antagonists have been reported (please see Hoyer: OX_2 receptor antagonists as sleep aids of this book), and even though only very few are commercially available, luckily some subtype-selective ones are. Even with the antagonists one has to use common sense. The selectivities of the ligands even between orexin receptor subtypes are far from absolute, and thus the antagonist concentration needs to be adjusted to the agonist concentration to, on the one hand, get any effect at all, and on the other, to retain selectivity. One also needs to consider the access and stability issues, especially when working with whole animals; for instance, the much used OX_1-selective antagonist, SB-334867, has been reported to suffer from inherently low stability [167]. For a more complicated physiological response, it is unclear whether the response can be pin-pointed to a single orexin receptor subtype, as the measured output may be a result of more than one orexinergic pathway either in series or in parallel.

6.4 Where Are We Heading?

We can see that much has been done to resolve orexin signaling mechanisms, but also much remains to be done. It is understandable from the perspective of physiological interest and the methodological shortcomings that the major focus has been on the cell physiological details of orexin signaling. However, with the knowledge we now have – the signaling of other GPCRs in the CNS and the details of the signal cascade of orexin receptors as known from different cell types – and the tools we can access and which are actively developed – e.g., different TRP channel inhibitors – we should go for higher molecular detail. However, many signal pathways lack efficient and selective inhibitors and, as indicated above, work with especially native cells involves many obstacles related to the properties of these cells. Therefore it is not surprising that many questions regarding orexin signaling in the (at least currently) most interesting CNS neurons are open. There also is, however, a lack of consequent approaches. One such problem consists of pharmacological and molecular biological tools and poor use of these. For instance, many kinase inhibitors are not very selective at all, and thus a finding should rather be repeated with another one against the same assumed target but with a different chemical structure and off-target profile. The most non-selective inhibitors should altogether give way for more novel and selective ones. Little may be concluded from the PLC inhibitor U73122, which may affect other targets or be completely incapable of inhibiting PLC, or D609, the target of which is unknown. Use of the most non-selective inhibitors needs to be substantiated by a sensible tracking of the assumed pathway. Also general common sense needs to be used: for instance, when does a result, which indicates a nearly full abolishment of a response upon inhibition of almost any pathway make sense and when not? The other side of the coin is that the findings in cell lines, native or recombinant, are slowly translated to investigations of native cells (neurons). There are indications of, for instance, cAMP, PI3K, cPLA$_2$, and PLD signaling of orexins (Sects. 3.1, 3.2, 4.1), but little has been done to assess these in neurons, although there are decent inhibitors available. More selective inhibitors of the TRP channels are emerging, and a recently characterized inhibitor of G$_q$, UBO-QIC [168, 169] is already commercially available, and even tested for recombinant orexin receptors [117]. My hope thus is that the molecular mechanisms of orexin signaling in neurons are gradually revealed. In addition to its importance for understanding of orexin physiology, this would be highly valuable for understanding the actions of the entire superfamily of GPCRs in neurons.

The first orexin receptor antagonist, suvorexant, is on the market in the USA for insomnia (see Hoyer: OX$_2$ receptor antagonists as sleep aids). From a research point of view this finally gives us a glimpse into the role of orexins in human physiology, and thus the validity of the results mainly extrapolated from other species. Suvorexant is a rather non-subtype-selective antagonist, but pharmaceutical companies are also developing OX$_2$-selective ones for the same indication (Hoyer: OX$_2$ receptor antagonists as sleep aids) and, if reaching the market, these should also give equally valuable information. As noted above, there are significant species differences in orexin physiology, and thus we may expect other similar findings. Methods to reliably assess orexin levels (brain tissue, cerebrospinal fluid, plasma) should also be developed [170]. This would allow us to answer the question of whether the peripheral orexin receptors ever "see" orexins, and to adjust the experimentally used orexin concentrations to a sensible physiological level for the tissue under investigation. The ligand concentration affects the signaling, as shown in recombinant CHO-K1 cells, and thus we might see a different response profile even in neurons.

Determination of the crystal structures of antagonist-bound OX_1 and OX_2 receptors (Rosenbaum: OX receptor crystal structures; [171, 172]) gives a fresh start for the orexin receptor ligand discovery. It may not be too much to hope that also an orexin peptide-bound receptor could be crystallized in the hands of the utmost experts. This would give us much needed information on the active state of the receptor and help tremendously in agonist ligand discovery. Needless to say, such ligands would be much needed as drugs but also as tools in orexin signaling research.

References

1. Kukkonen JP (2004) Regulation of receptor-coupling to (multiple) G-proteins. A challenge for basic research and drug discovery. Receptors Channels 10(5-6):167–183. doi:10.1080/10606820490926151
2. Lambert NA (2008) Dissociation of heterotrimeric g proteins in cells. Sci Signal 1(25):re5. doi:10.1126/scisignal.125re5
3. Hansen SB (2015) Lipid agonism: The PIP2 paradigm of ligand-gated ion channels. Biochim Biophys Acta 1851(5):620–628. doi:10.1016/j.bbalip.2015.01.011
4. Hille B, Dickson EJ, Kruse M, Vivas O, Suh BC (2015) Phosphoinositides regulate ion channels. Biochim Biophys Acta 1851(6):844–856. doi:10.1016/j.bbalip.2014.09.010
5. Liu C, Montell C (2015) Forcing open TRP channels: mechanical gating as a unifying activation mechanism. Biochem Biophys Res Commun 460(1):22–25. doi:10.1016/j.bbrc.2015.02.067
6. Inanobe A, Kurachi Y (2014) Membrane channels as integrators of G-protein-mediated signaling. Biochim Biophys Acta 1838(2):521–531. doi:10.1016/j.bbamem.2013.08.018
7. Marinissen MJ, Gutkind JS (2001) G-protein-coupled receptors and signaling networks: emerging paradigms. Trends Pharmacol Sci 22(7):368–376. doi:10.1016/S0165-6147(00)01678-3
8. Milligan G, Kostenis E (2006) Heterotrimeric G-proteins: a short history. Br J Pharmacol 147 (Suppl 1):S46–S55. doi:10.1038/sj.bjp.0706405
9. Suzuki N, Hajicek N, Kozasa T (2009) Regulation and physiological functions of G12/13-mediated signaling pathways. Neurosignals 17(1):55–70. doi:10.1159/000186690
10. Shenoy SK, Lefkowitz RJ (2011) beta-Arrestin-mediated receptor trafficking and signal transduction. Trends Pharmacol Sci 32(9):521–533. doi:10.1016/j.tips.2011.05.002
11. Magalhaes AC, Dunn H, Ferguson SS (2012) Regulation of GPCR activity, trafficking and localization by GPCR-interacting proteins. Br J Pharmacol 165(6):1717–1736. doi:10.1111/j.1476-5381.2011.01552.x
12. Ritter SL, Hall RA (2009) Fine-tuning of GPCR activity by receptor-interacting proteins. Nat Rev Mol Cell Biol 10(12):819–830. doi:10.1038/nrm2803
13. Kenakin T (2014) What is pharmacological 'affinity'? Relevance to biased agonism and antagonism. Trends Pharmacol Sci 35(9):434–441. doi:10.1016/j.tips.2014.06.003
14. Dalrymple MB, Jaeger WC, Eidne KA, Pfleger KD (2011) Temporal profiling of orexin receptor-arrestin-ubiquitin complexes reveals differences between receptor subtypes. J Biol Chem 286(19):16726–16733. doi:10.1074/jbc.M111.223537
15. Milasta S, Evans NA, Ormiston L, Wilson S, Lefkowitz RJ, Milligan G (2005) The sustainability of interactions between the orexin-1 receptor and beta-arrestin-2 is defined by a single C-terminal cluster of hydroxy amino acids and modulates the kinetics of ERK MAPK regulation. Biochem J 387(Pt 3):573–584. doi:10.1042/BJ20041745
16. Kukkonen JP, Leonard CS (2014) Orexin/hypocretin receptor signalling cascades. Br J Pharmacol 171(2):294–313. doi:10.1111/bph.12324

17. Leonard CS, Kukkonen JP (2014) Orexin/hypocretin receptor signalling: a functional perspective. Br J Pharmacol 171(2):294–313. doi:10.1111/bph.12296
18. Bayer L, Eggermann E, Saint-Mleux B, Machard D, Jones BE, Muhlethaler M, Serafin M (2002) Selective action of orexin (hypocretin) on nonspecific thalamocortical projection neurons. J Neurosci 22(18):7835–7839
19. Hwang LL, Chen CT, Dun NJ (2001) Mechanisms of orexin-induced depolarizations in rat dorsal motor nucleus of vagus neurones in vitro. J Physiol 537(Pt 2):511–520
20. Ivanov A, Aston-Jones G (2000) Hypocretin/orexin depolarizes and decreases potassium conductance in locus coeruleus neurons. Neuroreport 11(8):1755–1758
21. Enyedi P, Czirjak G (2010) Molecular background of leak K+ currents: two-pore domain potassium channels. Physiol Rev 90(2):559–605. doi:10.1152/physrev.00029.2009
22. Logothetis DE, Mahajan R, Adney SK, Ha J, Kawano T, Meng XY, Cui M (2015) Unifying mechanism of controlling Kir3 channel activity by G proteins and phosphoinositides. Int Rev Neurobiol 123:1–26. doi:10.1016/bs.irn.2015.05.013
23. Doroshenko P, Renaud LP (2009) Acid-sensitive TASK-like K+ conductances contribute to resting membrane potential and to orexin-induced membrane depolarization in rat thalamic paraventricular nucleus neurons. Neuroscience 158(4):1560–1570. doi:10.1016/j.neuroscience.2008.12.008
24. Hoang QV, Bajic D, Yanagisawa M, Nakajima S, Nakajima Y (2003) Effects of orexin (hypocretin) on GIRK channels. J Neurophysiol 90(2):693–702
25. Hoang QV, Zhao P, Nakajima S, Nakajima Y (2004) Orexin (hypocretin) effects on constitutively active inward rectifier K+ channels in cultured nucleus basalis neurons. J Neurophysiol 92(6):3183–3191. doi:10.1152/jn.01222.2003
26. Brown RE, Sergeeva O, Eriksson KS, Haas HL (2001) Orexin A excites serotonergic neurons in the dorsal raphe nucleus of the rat. Neuropharmacology 40(3):457–459
27. Burlet S, Tyler CJ, Leonard CS (2002) Direct and indirect excitation of laterodorsal tegmental neurons by Hypocretin/Orexin peptides: implications for wakefulness and narcolepsy. J Neurosci 22(7):2862–2872
28. van den Pol AN, Ghosh PK, Liu RJ, Li Y, Aghajanian GK, Gao XB (2002) Hypocretin (orexin) enhances neuron activity and cell synchrony in developing mouse GFP-expressing locus coeruleus. J Physiol 541(Pt 1):169–185
29. Yang B, Ferguson AV (2002) Orexin-A depolarizes dissociated rat area postrema neurons through activation of a nonselective cationic conductance. J Neurosci 22(15):6303–6308
30. Burdakov D, Liss B, Ashcroft FM (2003) Orexin excites GABAergic neurons of the arcuate nucleus by activating the sodium--calcium exchanger. J Neurosci 23(12):4951–4957
31. Eriksson KS, Sergeeva O, Brown RE, Haas HL (2001) Orexin/hypocretin excites the histaminergic neurons of the tuberomammillary nucleus. J Neurosci 21(23):9273–9279
32. Wu M, Zhang Z, Leranth C, Xu C, van den Pol AN, Alreja M (2002) Hypocretin increases impulse flow in the septohippocampal GABAergic pathway: implications for arousal via a mechanism of hippocampal disinhibition. J Neurosci 22(17):7754–7765
33. Lytton J (2007) Na+/Ca2+ exchangers: three mammalian gene families control Ca2+ transport. Biochem J 406(3):365–382. doi:10.1042/BJ20070619
34. Kukkonen JP (2011) A ménage à trois made in heaven: G-protein-coupled receptors, lipids and TRP channels. Cell Calcium 50(1):9–26. doi:10.1016/j.ceca.2011.04.005
35. Nilius B, Szallasi A (2014) Transient receptor potential channels as drug targets: from the science of basic research to the art of medicine. Pharmacol Rev 66(3):676–814. doi:10.1124/pr.113.008268
36. Kolaj M, Zhang L, Renaud LP (2014) Novel coupling between TRPC-like and KNa channels modulates low threshold spike-induced afterpotentials in rat thalamic midline neurons. Neuropharmacology 86:88–96. doi:10.1016/j.neuropharm.2014.06.023
37. Kukkonen JP (2014) Lipid signaling cascades of orexin/hypocretin receptors. Biochimie 96:158–165. doi:10.1016/j.biochi.2013.06.015

38. Acuna-Goycolea C, van den Pol AN (2009) Neuroendocrine proopiomelanocortin neurons are excited by hypocretin/orexin. J Neurosci 29(5):1503–1513. doi:10.1523/JNEUROSCI. 5147-08.2009

39. Aracri P, Banfi D, Pasini ME, Amadeo A, Becchetti A (2015) Hypocretin (orexin) regulates glutamate input to fast-spiking interneurons in layer V of the Fr2 region of the murine prefrontal cortex. Cereb Cortex 25(5):1330–1347. doi:10.1093/cercor/bht326

40. Borgland SL, Chang SJ, Bowers MS, Thompson JL, Vittoz N, Floresco SB, Chou J, Chen BT, Bonci A (2009) Orexin A/hypocretin-1 selectively promotes motivation for positive reinforcers. J Neurosci 29(36):11215–11225. doi:10.1523/JNEUROSCI.6096-08.2009

41. Borgland SL, Storm E, Bonci A (2008) Orexin B/hypocretin 2 increases glutamatergic transmission to ventral tegmental area neurons. Eur J Neurosci 28(8):1545–1556. doi:10. 1111/j.1460-9568.2008.06397.x

42. Davis SF, Williams KW, Xu W, Glatzer NR, Smith BN (2003) Selective enhancement of synaptic inhibition by hypocretin (orexin) in rat vagal motor neurons: implications for autonomic regulation. J Neurosci 23(9):3844–3854

43. Dergacheva O, Bateman R, Byrne P, Mendelowitz D (2012) Orexinergic modulation of GABAergic neurotransmission to cardiac vagal neurons in the brain stem nucleus ambiguus changes during development. Neuroscience 209:12–20. doi:10.1016/j.neuroscience.2012.02. 020

44. Li Y, Gao XB, Sakurai T, van den Pol AN (2002) Hypocretin/Orexin excites hypocretin neurons via a local glutamate neuron-A potential mechanism for orchestrating the hypothalamic arousal system. Neuron 36(6):1169–1181

45. Li Y, Xu Y, van den Pol AN (2013) Reversed synaptic effects of hypocretin and NPY mediated by excitatory GABA-dependent synaptic activity in developing MCH neurons. J Neurophysiol 109(6):1571–1578. doi:10.1152/jn.00522.2012

46. Ono K, Kai A, Honda E, Inenaga K (2008) Hypocretin-1/orexin-A activates subfornical organ neurons of rats. Neuroreport 19(1):69–73. doi:10.1097/WNR.0b013e3282f32d64

47. Palus K, Chrobok L, Lewandowski MH (2015) Orexins/hypocretins modulate the activity of NPY-positive and -negative neurons in the rat intergeniculate leaflet via OX1 and OX2 receptors. Neuroscience 300:370–380. doi:10.1016/j.neuroscience.2015.05.039

48. Smith BN, Davis SF, Van Den Pol AN, Xu W (2002) Selective enhancement of excitatory synaptic activity in the rat nucleus tractus solitarius by hypocretin 2. Neuroscience 115 (3):707–714

49. van den Pol AN, Gao XB, Obrietan K, Kilduff TS, Belousov AB (1998) Presynaptic and postsynaptic actions and modulation of neuroendocrine neurons by a new hypothalamic peptide, hypocretin/orexin. J Neurosci 18(19):7962–7971

50. Selbach O, Bohla C, Barbara A, Doreulee N, Eriksson KS, Sergeeva OA, Haas HL (2010) Orexins/hypocretins control bistability of hippocampal long-term synaptic plasticity through co-activation of multiple kinases. Acta Physiol (Oxf) 198(3):277–285. doi:10.1111/j.1748-1716.2009.02021.x

51. Selbach O, Doreulee N, Bohla C, Eriksson KS, Sergeeva OA, Poelchen W, Brown RE, Haas HL (2004) Orexins/hypocretins cause sharp wave- and theta-related synaptic plasticity in the hippocampus via glutamatergic, gabaergic, noradrenergic, and cholinergic signaling. Neuroscience 127(2):519–528

52. Yang L, Zou B, Xiong X, Pascual C, Xie J, Malik A, Xie J, Sakurai T, Xie XS (2013) Hypocretin/orexin neurons contribute to hippocampus-dependent social memory and synaptic plasticity in mice. J Neurosci 33(12):5275–5284. doi:10.1523/JNEUROSCI.3200-12.2013

53. Huang H, Ghosh P, van den Pol AN (2006) Prefrontal cortex-projecting glutamatergic thalamic paraventricular nucleus-excited by hypocretin: a feedforward circuit that may enhance cognitive arousal. J Neurophysiol 95(3):1656–1668. doi:10.1152/jn.00927.2005

54. Kim J, Nakajima K, Oomura Y, Wayner MJ, Sasaki K (2009) Electrophysiological effects of orexins/hypocretins on pedunculopontine tegmental neurons in rats: an in vitro study. Peptides 30(2):191–209. doi:10.1016/j.peptides.2008.09.023

55. Kim J, Nakajima K, Oomura Y, Wayner MJ, Sasaki K (2009) Orexin-A and ghrelin depolarize the same pedunculopontine tegmental neurons in rats: an in vitro study. Peptides 30(7):1328–1335. doi:10.1016/j.peptides.2009.03.015
56. Kolaj M, Doroshenko P, Yan Cao X, Coderre E, Renaud LP (2007) Orexin-induced modulation of state-dependent intrinsic properties in thalamic paraventricular nucleus neurons attenuates action potential patterning and frequency. Neuroscience 147(4):1066–1075. doi:10.1016/j.neuroscience.2007.05.018
57. Mukai K, Kim J, Nakajima K, Oomura Y, Wayner MJ, Sasaki K (2009) Electrophysiological effects of orexin/hypocretin on nucleus accumbens shell neurons in rats: an in vitro study. Peptides 30(8):1487–1496. doi:10.1016/j.peptides.2009.04.018
58. Murai Y, Akaike T (2005) Orexins cause depolarization via nonselective cationic and K+ channels in isolated locus coeruleus neurons. Neurosci Res 51(1):55–65. doi:10.1016/j. neures.2004.09.005
59. Wu M, Zaborszky L, Hajszan T, van den Pol AN, Alreja M (2004) Hypocretin/orexin innervation and excitation of identified septohippocampal cholinergic neurons. J Neurosci 24(14):3527–3536
60. Yan J, He C, Xia JX, Zhang D, Hu ZA (2012) Orexin-A excites pyramidal neurons in layer 2/3 of the rat prefrontal cortex. Neurosci Lett 520(1):92–97. doi:10.1016/j.neulet.2012.05. 038
61. Yang B, Ferguson AV (2003) Orexin-A depolarizes nucleus tractus solitarius neurons through effects on nonselective cationic and K+ conductances. J Neurophysiol 89 (4):2167–2175. doi:10.1152/jn.01088.2002
62. Yang B, Samson WK, Ferguson AV (2003) Excitatory effects of orexin-A on nucleus tractus solitarius neurons are mediated by phospholipase C and protein kinase C. J Neurosci 23 (15):6215–6222
63. Zhang J, Li B, Yu L, He YC, Li HZ, Zhu JN, Wang JJ (2011) A role for orexin in central vestibular motor control. Neuron 69(4):793–804. doi:10.1016/j.neuron.2011.01.026
64. Yamanaka A, Tabuchi S, Tsunematsu T, Fukazawa Y, Tominaga M (2010) Orexin directly excites orexin neurons through orexin 2 receptor. J Neurosci 30(38):12642–12652. doi:10. 1523/JNEUROSCI.2120-10.2010
65. Zhou WL, Gao XB, Picciotto MR (2015) Acetylcholine acts through nicotinic receptors to enhance the firing rate of a subset of hypocretin neurons in the mouse hypothalamus through distinct presynaptic and postsynaptic mechanisms. eNeuro 2(1). doi:10.1523/ENEURO. 0052-14.2015
66. Xia J, Chen X, Song C, Ye J, Yu Z, Hu Z (2005) Postsynaptic excitation of prefrontal cortical pyramidal neurons by hypocretin-1/orexin A through the inhibition of potassium currents. J Neurosci Res 82(5):729–736. doi:10.1002/jnr.20667
67. Grabauskas G, Moises HC (2003) Gastrointestinal-projecting neurones in the dorsal motor nucleus of the vagus exhibit direct and viscerotopically organized sensitivity to orexin. J Physiol 549(Pt 1):37–56
68. Muroya S, Funahashi H, Yamanaka A, Kohno D, Uramura K, Nambu T, Shibahara M, Kuramochi M, Takigawa M, Yanagisawa M, Sakurai T, Shioda S, Yada T (2004) Orexins (hypocretins) directly interact with neuropeptide Y, POMC and glucose-responsive neurons to regulate Ca 2+ signaling in a reciprocal manner to leptin: orexigenic neuronal pathways in the mediobasal hypothalamus. Eur J Neurosci 19(6):1524–1534
69. Apergis-Schoute J, Iordanidou P, Faure C, Jego S, Schone C, Aitta-Aho T, Adamantidis A, Burdakov D (2015) Optogenetic evidence for inhibitory signaling from orexin to MCH neurons via local microcircuits. J Neurosci 35(14):5435–5441. doi:10.1523/jneurosci.5269- 14.2015
70. Kohlmeier KA, Watanabe S, Tyler CJ, Burlet S, Leonard CS (2008) Dual orexin actions on dorsal raphe and laterodorsal tegmentum neurons: noisy cation current activation and selective enhancement of Ca2+ transients mediated by L-type calcium channels. J Neurophysiol 100(4):2265–2281. doi:10.1152/jn.01388.2007

71. Liu F, Weng SJ, Yang XL, Zhong YM (2015) Orexin-A potentiates L-type calcium/barium currents in rat retinal ganglion cells. Neuroscience 305:225–237. doi:10.1016/j.neuroscience. 2015.08.008

72. Wu WN, Wu PF, Zhou J, Guan XL, Zhang Z, Yang YJ, Long LH, Xie N, Chen JG, Wang F (2013) Orexin-A activates hypothalamic AMP-activated protein kinase signaling through a Ca(2)(+)-dependent mechanism involving voltage-gated L-type calcium channel. Mol Pharmacol 84(6):876–887. doi:10.1124/mol.113.086744

73. Putula J, Pihlajamaa T, Kukkonen JP (2014) Calcium affects OX1 orexin (hypocretin) receptor responses by modifying both orexin binding and the signal transduction machinery. Br J Pharmacol 171(24):5816–5828. doi:10.1111/bph.12883

74. Bernard R, Lydic R, Baghdoyan HA (2002) Hypocretin-1 activates G proteins in arousal-related brainstem nuclei of rat. Neuroreport 13(4):447–450

75. Bernard R, Lydic R, Baghdoyan HA (2003) Hypocretin-1 causes G protein activation and increases ACh release in rat pons. Eur J Neurosci 18(7):1775–1785. doi:10.1046/j.1460-9568.2003.02905.x

76. Karteris E, Machado RJ, Chen J, Zervou S, Hillhouse EW, Randeva HS (2005) Food deprivation differentially modulates orexin receptor expression and signalling in the rat hypothalamus and adrenal cortex. Am J Physiol Endocrinol Metab 288(6):E1089–E1100. doi:10.1152/ajpendo.00351.2004

77. Haj-Dahmane S, Shen RY (2005) The wake-promoting peptide orexin-B inhibits glutamatergic transmission to dorsal raphe nucleus serotonin neurons through retrograde endocannabinoid signaling. J Neurosci 25(4):896–905. doi:10.1523/JNEUROSCI.3258-04. 2005

78. Ho YC, Lee HJ, Tung LW, Liao YY, Fu SY, Teng SF, Liao HT, Mackie K, Chiou LC (2011) Activation of orexin 1 receptors in the periaqueductal gray of male rats leads to antinociception via retrograde endocannabinoid (2-arachidonoylglycerol)-induced disinhibition. J Neurosci 31(41):14600–14610. doi:10.1523/JNEUROSCI.2671-11.2011

79. Kano M, Ohno-Shosaku T, Hashimotodani Y, Uchigashima M, Watanabe M (2009) Endocannabinoid-mediated control of synaptic transmission. Physiol Rev 89(1):309–380. doi:10.1152/physrev.00019.2008

80. Malendowicz LK, Tortorella C, Nussdorfer GG (1999) Orexins stimulate corticosterone secretion of rat adrenocortical cells, through the activation of the adenylate cyclase-dependent signaling cascade. J Steroid Biochem Mol Biol 70(4-6):185–188

81. Mazzocchi G, Malendowicz LK, Gottardo L, Aragona F, Nussdorfer GG (2001) Orexin A stimulates cortisol secretion from human adrenocortical cells through activation of the adenylate cyclase-dependent signaling cascade. J Clin Endocrinol Metab 86(2):778–782

82. Randeva HS, Karteris E, Grammatopoulos D, Hillhouse EW (2001) Expression of orexin-A and functional orexin type 2 receptors in the human adult adrenals: implications for adrenal function and energy homeostasis. J Clin Endocrinol Metab 86(10):4808–4813

83. Sunahara RK, Taussig R (2002) Isoforms of mammalian adenylyl cyclase: multiplicities of signaling. Mol Interv 2(3):168–184. doi:10.1124/mi.2.3.168

84. Ramanjaneya M, Conner AC, Chen J, Stanfield PR, Randeva HS (2008) Orexins stimulate steroidogenic acute regulatory protein expression through multiple signaling pathways in human adrenal H295R cells. Endocrinology 149(8):4106–4115. doi:10.1210/en.2007-1739

85. Wenzel J, Grabinski N, Knopp CA, Dendorfer A, Ramanjaneya M, Randeva HS, Ehrhart-Bornstein M, Dominiak P, Johren O (2009) Hypocretin/orexin increases the expression of steroidogenic enzymes in human adrenocortical NCI H295R cells. Am J Physiol Regul Integr Comp Physiol 297(5):R1601–R1609. doi:10.1152/ajpregu.91034.2008

86. Chen XW, Huang W, Yan JA, Fan HX, Guo N, Lu J, Xiu Y, Gu JL, Zhang CX, Ruan HZ, Hu ZA, Yu ZP, Zhou Z (2008) Reinvestigation of the effect of orexin A on catecholamine release from adrenal chromaffin cells. Neurosci Lett 436(2):181–184. doi:10.1016/j.neulet.2008.03. 014

87. Mazzocchi G, Malendowicz LK, Aragona F, Rebuffat P, Gottardo L, Nussdorfer GG (2001) Human pheochromocytomas express orexin receptor type 2 gene and display an in vitro secretory response to orexins A and B. J Clin Endocrinol Metab 86(10):4818–4821

88. Nakabayashi M, Suzuki T, Takahashi K, Totsune K, Muramatsu Y, Kaneko C, Date F, Takeyama J, Darnel AD, Moriya T, Sasano H (2003) Orexin-A expression in human peripheral tissues. Mol Cell Endocrinol 205(1-2):43–50

89. Chen C, Xu R (2003) The in vitro regulation of growth hormone secretion by orexins. Endocrine 22(1):57–66

90. Samson WK, Taylor MM (2001) Hypocretin/orexin suppresses corticotroph responsiveness in vitro. Am J Physiol Regul Integr Comp Physiol 281(4):R1140–R1145

91. Göncz E, Strowski MZ, Grotzinger C, Nowak KW, Kaczmarek P, Sassek M, Mergler S, El-Zayat BF, Theodoropoulou M, Stalla GK, Wiedenmann B, Plockinger U (2008) Orexin-A inhibits glucagon secretion and gene expression through a Foxo1-dependent pathway. Endocrinology 149(4):1618–1626. doi:10.1210/en.2007-1257

92. Wymann MP, Zvelebil M, Laffargue M (2003) Phosphoinositide 3-kinase signalling--which way to target? Trends Pharmacol Sci 24(7):366–376. doi:10.1016/S0165-6147(03)00163-9

93. Karteris E, Chen J, Randeva HS (2004) Expression of human prepro-orexin and signaling characteristics of orexin receptors in the male reproductive system. J Clin Endocrinol Metab 89(4):1957–1962. doi:10.1210/jc.2003-031778

94. Barreiro ML, Pineda R, Gaytan F, Archanco M, Burrell MA, Castellano JM, Hakovirta H, Nurmio M, Pinilla L, Aguilar E, Toppari J, Dieguez C, Tena-Sempere M (2005) Pattern of orexin expression and direct biological actions of orexin-a in rat testis. Endocrinology 146 (12):5164–5175. doi:10.1210/en.2005-0455

95. Barreiro ML, Pineda R, Navarro VM, Lopez M, Suominen JS, Pinilla L, Senaris R, Toppari J, Aguilar E, Dieguez C, Tena-Sempere M (2004) Orexin 1 receptor messenger ribonucleic acid expression and stimulation of testosterone secretion by orexin-A in rat testis. Endocrinology 145(5):2297–2306

96. Jöhren O, Neidert SJ, Kummer M, Dendorfer A, Dominiak P (2001) Prepro-orexin and orexin receptor mRNAs are differentially expressed in peripheral tissues of male and female rats. Endocrinology 142(8):3324–3331

97. Sakurai T, Amemiya A, Ishii M, Matsuzaki I, Chemelli RM, Tanaka H, Williams SC, Richardson JA, Kozlowski GP, Wilson S, Arch JR, Buckingham RE, Haynes AC, Carr SA, Annan RS, McNulty DE, Liu WS, Terrett JA, Elshourbagy NA, Bergsma DJ, Yanagisawa M (1998) Orexins and orexin receptors: a family of hypothalamic neuropeptides and G protein-coupled receptors that regulate feeding behavior. Cell 92(4):573–585. doi:10.1016/S0092-8674(00)80949-6

98. Tafuri S, Muto RL, Pavone LM, Valiante S, Costagliola A, Staiano N, Vittoria A (2010) Novel localization of orexin A in the tubular cytotypes of the rat testis. Regul Pept 164 (2-3):53–57. doi:10.1016/j.regpep.2010.06.011

99. Kirchgessner AL, Liu M (1999) Orexin synthesis and response in the gut. Neuron 24 (4):941–951

100. Matsuo K, Kaibara M, Uezono Y, Hayashi H, Taniyama K, Nakane Y (2002) Involvement of cholinergic neurons in orexin-induced contraction of guinea pig ileum. Eur J Pharmacol 452 (1):105–109

101. Squecco R, Garella R, Luciani G, Francini F, Baccari MC (2011) Muscular effects of orexin A on the mouse duodenum: mechanical and electrophysiological studies. J Physiol 589 (Pt 21):5231–5246. doi:10.1113/jphysiol.2011.214940

102. Rouet-Benzineb P, Rouyer-Fessard C, Jarry A, Avondo V, Pouzet C, Yanagisawa M, Laboisse C, Laburthe M, Voisin T (2004) Orexins acting at native OX(1) receptor in colon cancer and neuroblastoma cells or at recombinant OX(1) receptor suppress cell growth by inducing apoptosis. J Biol Chem 279(44):45875–45886. doi:10.1074/jbc.M404136200

103. Voisin T, El Firar A, Fasseu M, Rouyer-Fessard C, Descatoire V, Walker F, Paradis V, Bedossa P, Henin D, Lehy T, Laburthe M (2011) Aberrant expression of OX1 receptors for

orexins in colon cancers and liver metastases: an openable gate to apoptosis. Cancer Res 71 (9):3341–3351. doi:10.1158/0008-5472.CAN-10-3473

104. Voisin T, Firar AE, Avondo V, Laburthe M (2006) Orexin-induced apoptosis: the key role of the seven-transmembrane domain orexin type 2 receptor. Endocrinology 147 (10):4977–4984. doi:10.1210/en.2006-0201

105. Sellayah D, Bharaj P, Sikder D (2011) Orexin is required for brown adipose tissue development, differentiation, and function. Cell Metab 14(4):478–490. doi:10.1016/j.cmet.2011.08.010

106. Sellayah D, Sikder D (2012) Orexin receptor-1 mediates brown fat developmental differentiation. Adipocyte 1(1):58–63. doi:10.4161/adip.18965

107. Digby JE, Chen J, Tang JY, Lehnert H, Matthews RN, Randeva HS (2006) Orexin receptor expression in human adipose tissue: effects of orexin-A and orexin-B. J Endocrinol 191 (1):129–136. doi:10.1677/joe.1.06886

108. Skrzypski M, Le TT, Kaczmarek P, Pruszynska-Oszmalek E, Pietrzak P, Szczepankiewicz D, Kolodziejski PA, Sassek M, Arafat A, Wiedenmann B, Nowak KW, Strowski MZ (2011) Orexin A stimulates glucose uptake, lipid accumulation and adiponectin secretion from 3T3-L1 adipocytes and isolated primary rat adipocytes. Diabetologia 54(7):1841–1852. doi:10.1007/s00125-011-2152-2

109. Skrzypski M, Kaczmarek P, Le TT, Wojciechowicz T, Pruszynska-Oszmalek E, Szczepankiewicz D, Sassek M, Arafat A, Wiedenmann B, Nowak KW, Strowski MZ (2012) Effects of orexin A on proliferation, survival, apoptosis and differentiation of 3T3-L1 preadipocytes into mature adipocytes. FEBS Lett 586(23):4157–4164. doi:10.1016/j.febslet.2012.10.013

110. Zwirska-Korczala K, Adamczyk-Sowa M, Sowa P, Pilc K, Suchanek R, Pierzchala K, Namyslowski G, Misiolek M, Sodowski K, Kato I, Kuwahara A, Zabielski R (2007) Role of leptin, ghrelin, angiotensin II and orexins in 3T3 L1 preadipocyte cells proliferation and oxidative metabolism. J Physiol Pharmacol 58(Suppl 1):53–64

111. Kim MK, Park HJ, Kim SR, Choi YK, Bae SK, Bae MK (2015) Involvement of heme oxygenase-1 in orexin-a-induced angiogenesis in vascular endothelial cells. Korean J Physiol Pharmacol 19(4):327–334. doi:10.4196/kjpp.2015.19.4.327

112. Johansson L, Ekholm ME, Kukkonen JP (2008) Multiple phospholipase activation by OX (1) orexin/hypocretin receptors. Cell Mol Life Sci 65(12):1948–1956. doi:10.1007/s00018-008-8206-z

113. Lund PE, Shariatmadari R, Uustare A, Detheux M, Parmentier M, Kukkonen JP, Åkerman KEO (2000) The orexin OX1 receptor activates a novel Ca2+ influx pathway necessary for coupling to phospholipase C. J Biol Chem 275(40):30806–30812. doi:10.1074/jbc.M002603200

114. Kukkonen JP (2016) OX orexin/hypocretin receptor signal transduction in recombinant Chinese hamster ovary cells. Cell Signal 28(2):51–60. doi:10.1016/j.cellsig.2015.11.009

115. Turunen PM, Jäntti MH, Kukkonen JP (2012) OX1 orexin/hypocretin receptor signaling via arachidonic acid and endocannabinoid release. Mol Pharmacol 82(2):156–167. doi:10.1124/mol.112.078063

116. Jäntti MH, Putula J, Somerharju P, Frohman MA, Kukkonen JP (2012) OX1 orexin/hypocretin receptor activation of phospholipase D. Br J Pharmacol 165(4b):1109–1123. doi:10.1111/j.1476-5381.2011.01565.x

117. Kukkonen JP (2016) G-protein-dependency of orexin/hypocretin receptor signalling in recombinant Chinese hamster ovary cells. Biochem Biophys Res Commun 476:379–385. doi:10.1016/j.bbrc.2016.05.130

118. Holmqvist T, Johansson L, Östman M, Ammoun S, Åkerman KE, Kukkonen JP (2005) OX1 orexin receptors couple to adenylyl cyclase regulation via multiple mechanisms. J Biol Chem 280(8):6570–6579. doi:10.1074/jbc.M407397200

119. Ammoun S, Johansson L, Ekholm ME, Holmqvist T, Danis AS, Korhonen L, Sergeeva OA, Haas HL, Åkerman KE, Kukkonen JP (2006) OX1 orexin receptors activate extracellular

signal-regulated kinase (ERK) in CHO cells via multiple mechanisms: the role of Ca2+ influx in OX1 receptor signaling. Mol Endocrinol 20(1):80–99. doi:10.1210/me.2004-0389

120. Ammoun S, Lindholm D, Wootz H, Åkerman KE, Kukkonen JP (2006) G-protein-coupled OX1 orexin/hcrtr-1 hypocretin receptors induce caspase-dependent and -independent cell death through p38 mitogen-/stress-activated protein kinase. J Biol Chem 281(2):834–842. doi:10.1074/jbc.M508603200

121. Ekholm ME, Johansson L, Kukkonen JP (2007) IP3-independent signalling of OX1 orexin/hypocretin receptors to Ca2+ influx and ERK. Biochem Biophys Res Commun 353 (2):475–480. doi:10.1016/j.bbrc.2006.12.045

122. Larsson KP, Peltonen HM, Bart G, Louhivuori LM, Penttonen A, Antikainen M, Kukkonen JP, Åkerman KE (2005) Orexin-A-induced Ca2+ entry: evidence for involvement of TRPC channels and protein kinase C regulation. J Biol Chem 280(3):1771–1781. doi:10.1074/jbc.M406073200

123. Johansson L, Ekholm ME, Kukkonen JP (2007) Regulation of OX(1) orexin/hypocretin receptor-coupling to phospholipase C by Ca(2+) influx. Br J Pharmacol 150(1):97–104. doi:10.1038/sj.bjp.0706959

124. Turunen PM, Ekholm ME, Somerharju P, Kukkonen JP (2010) Arachidonic acid release mediated by OX1 orexin receptors. Br J Pharmacol 159(1):212–221. doi:10.1111/j.1476-5381.2009.00535.x

125. El Firar A, Voisin T, Rouyer-Fessard C, Ostuni MA, Couvineau A, Laburthe M (2009) Discovery of a functional immunoreceptor tyrosine-based switch motif in a 7-transmembrane-spanning receptor: role in the orexin receptor OX1R-driven apoptosis. FASEB J 23 (12):4069–4080. doi:10.1096/fj.09-131367

126. Putula J, Kukkonen JP (2012) Mapping of the binding sites for the OX(1) orexin receptor antagonist, SB-334867, using orexin/hypocretin receptor chimaeras. Neurosci Lett 506 (1):111–115. doi:10.1016/j.neulet.2011.10.061

127. Putula J, Turunen PM, Jäntti MH, Ekholm ME, Kukkonen JP (2011) Agonist ligand discrimination by the two orexin receptors depends on the expression system. Neurosci Lett 494 (1):57–60. doi:10.1016/j.neulet.2011.02.055

128. Peltonen HM, Magga JM, Bart G, Turunen PM, Antikainen MS, Kukkonen JP, Åkerman KE (2009) Involvement of TRPC3 channels in calcium oscillations mediated by OX(1) orexin receptors. Biochem Biophys Res Commun 385(3):408–412. doi:10.1016/j.bbrc.2009.05.077

129. Tang J, Chen J, Ramanjaneya M, Punn A, Conner AC, Randeva HS (2008) The signalling profile of recombinant human orexin-2 receptor. Cell Signal 20(9):1651–1661. doi:10.1016/j.cellsig.2008.05.010

130. Holmqvist T, Åkerman KEO, Kukkonen JP (2002) Orexin signaling in recombinant neuron-like cells. FEBS Lett 526:11–14. doi:10.1016/S0014-5793(02)03101-0

131. Zhu Y, Miwa Y, Yamanaka A, Yada T, Shibahara M, Abe Y, Sakurai T, Goto K (2003) Orexin receptor type-1 couples exclusively to pertussis toxin-insensitive G-proteins, while orexin receptor type-2 couples to both pertussis toxin-sensitive and -insensitive G-proteins. J Pharmacol Sci 92(3):259–266

132. Louhivuori LM, Jansson L, Nordstrom T, Bart G, Nasman J, Akerman KE (2010) Selective interference with TRPC3/6 channels disrupts OX1 receptor signalling via NCX and reveals a distinct calcium influx pathway. Cell Calcium 48(2-3):114–123. doi:10.1016/j.ceca.2010.07.005

133. Näsman J, Bart G, Larsson K, Louhivuori L, Peltonen H, Åkerman KE (2006) The orexin OX1 receptor regulates Ca2+ entry via diacylglycerol-activated channels in differentiated neuroblastoma cells. J Neurosci 26(42):10658–10666. doi:10.1523/JNEUROSCI.2609-06.2006

134. Wang Z, Liu S, Kakizaki M, Hirose Y, Ishikawa Y, Funato H, Yanagisawa M, Yu Y, Liu Q (2014) Orexin/hypocretin activates mTOR complex 1 (mTORC1) via an Erk/Akt-independent and calcium-stimulated lysosome v-ATPase pathway. J Biol Chem 289 (46):31950–31959. doi:10.1074/jbc.M114.600015

135. Gorojankina T, Grebert D, Salesse R, Tanfin Z, Caillol M (2007) Study of orexins signal transduction pathways in rat olfactory mucosa and in olfactory sensory neurons-derived cell line Odora: multiple orexin signalling pathways. Regul Pept 141(1-3):73–85. doi:10.1016/j. regpep.2006.12.012

136. Voisin T, El Firar A, Rouyer-Fessard C, Gratio V, Laburthe M (2008) A hallmark of immunoreceptor, the tyrosine-based inhibitory motif ITIM, is present in the G protein-coupled receptor OX1R for orexins and drives apoptosis: a novel mechanism. FASEB J 22 (6):1993–2002. doi:10.1096/fj.07-098723

137. Hungs M, Fan J, Lin L, Lin X, Maki RA, Mignot E (2001) Identification and functional analysis of mutations in the hypocretin (orexin) genes of narcoleptic canines. Genome Res 11 (4):531–539. doi:10.1101/gr.161001

138. Chen J, Randeva HS (2004) Genomic organization of mouse orexin receptors: characterization of two novel tissue-specific splice variants. Mol Endocrinol 18(11):2790–2804. doi:10. 1210/me.2004-0167

139. Wang C, Pan Y, Zhang R, Bai B, Chen J, Randeva HS (2014) Heterodimerization of Mouse Orexin type 2 receptor variants and the effects on signal transduction. Biochim Biophys Acta 1843(3):652–663. doi:10.1016/j.bbamcr.2013.12.010

140. Maggio R, Innamorati G, Parenti M (2007) G protein-coupled receptor oligomerization provides the framework for signal discrimination. J Neurochem 103(5):1741–1752. doi:10. 1111/j.1471-4159.2007.04896.x

141. Milligan G (2009) G protein-coupled receptor hetero-dimerization: contribution to pharmacology and function. Br J Pharmacol 158(1):5–14. doi:10.1111/j.1476-5381.2009.00169.x

142. Jäntti MH, Mandrika I, Kukkonen JP (2014) Human orexin/hypocretin receptors form constitutive homo- and heteromeric complexes with each other and with human CB1 cannabinoid receptors. Biochem Biophys Res Commun 445(2):486–490. doi:10.1016/j. bbrc.2014.02.026

143. Xu TR, Ward RJ, Pediani JD, Milligan G (2011) The orexin OX1 receptor exists predominantly as a homodimer in the basal state: potential regulation of receptor organization by both agonist and antagonist ligands. Biochem J 439(1):171–183. doi:10.1042/BJ20110230

144. Ward RJ, Pediani JD, Milligan G (2011) Hetero-multimerization of the cannabinoid CB1 receptor and the orexin OX1 receptor generates a unique complex in which both protomers are regulated by orexin A. J Biol Chem 286(43):37414–37428. doi:10.1074/jbc.M111. 287649

145. Jäntti MH, Putula J, Turunen PM, Näsman J, Reijonen S, Kukkonen JP (2013) Autocrine endocannabinoid signaling potentiates orexin receptor signaling upon CB1 cannabinoid-OX1 orexin receptor coexpression. Mol Pharmacol 83(3):621–632. doi:10.1124/mol.112.080523

146. Robinson JD, McDonald PH (2015) The orexin 1 receptor modulates kappa opioid receptor function via a JNK-dependent mechanism. Cell Signal 27(7):1449–1456. doi:10.1016/j. cellsig.2015.03.026

147. Chen J, Zhang R, Chen X, Wang C, Cai X, Liu H, Jiang Y, Liu C, Bai B (2015) Heterodimerization of human orexin receptor 1 and kappa opioid receptor promotes protein kinase A/cAMP-response element binding protein signaling via a Galphas-mediated mechanism. Cell Signal 27(7):1426–1438. doi:10.1016/j.cellsig.2015.03.027

148. Davies J, Chen J, Pink R, Carter D, Saunders N, Sotiriadis G, Bai B, Pan Y, Howlett D, Payne A, Randeva H, Karteris E (2015) Orexin receptors exert a neuroprotective effect in Alzheimer's disease (AD) via heterodimerization with GPR103. Sci Rep 5:12584. doi:10. 1038/srep12584

149. Navarro G, Quiroz C, Moreno-Delgado D, Sierakowiak A, McDowell K, Moreno E, Rea W, Cai NS, Aguinaga D, Howell LA, Hausch F, Cortes A, Mallol J, Casado V, Lluis C, Canela EI, Ferre S, McCormick PJ (2015) Orexin-corticotropin-releasing factor receptor heteromers in the ventral tegmental area as targets for cocaine. J Neurosci 35(17):6639–6653. doi:10. 1523/JNEUROSCI.4364-14.2015

150. Scarselli M, Annibale P, McCormick PJ, Kolachalam S, Aringhieri S, Radenovic A, Corsini GU, Maggio R (2015) Revealing GPCR oligomerization at the single-molecule level through a nanoscopic lens: methods, dynamics and biological function. FEBS J. doi:10.1111/febs. 13577

151. Cockcroft S (2009) Phosphatidic acid regulation of phosphatidylinositol 4-phosphate 5-kinases. Biochim Biophys Acta 1791(9):905–912. doi:10.1016/j.bbalip.2009.03.007

152. Bjornstrom K, Turina D, Strid T, Sundqvist T, Eintrei C (2014) Orexin A inhibits propofol-induced neurite retraction by a phospholipase D/protein kinase Cepsilon-dependent mechanism in neurons. PLoS One 9(5), e97129. doi:10.1371/journal.pone.0097129

153. Ahn S, Shenoy SK, Wei H, Lefkowitz RJ (2004) Differential kinetic and spatial patterns of beta-arrestin and G protein-mediated ERK activation by the angiotensin II receptor. J Biol Chem 279(34):35518–35525. doi:10.1074/jbc.M405878200

154. Daaka Y, Luttrell LM, Lefkowitz RJ (1997) Switching of the coupling of the beta2-adrenergic receptor to different G proteins by protein kinase A. Nature 390(6655):88–91

155. Kukkonen JP, Näsman J, Åkerman KEO (2001) Modelling of promiscuous receptor-Gi/Gs-protein coupling and effector response. Trends Pharmacol Sci 22(12):616–622

156. Newman-Tancredi A, Cussac D, Marini L, Millan MJ (2002) Antibody capture assay reveals bell-shaped concentration-response isotherms for h5-HT(1A) receptor-mediated Galpha (i3) activation: conformational selection by high-efficacy agonists, and relationship to traf ficking of receptor signaling. Mol Pharmacol 62(3):590–601

157. Sun Y, Huang J, Xiang Y, Bastepe M, Juppner H, Kobilka BK, Zhang JJ, Huang XY (2007) Dosage-dependent switch from G protein-coupled to G protein-independent signaling by a GPCR. EMBO J 26(1):53–64. doi:10.1038/sj.emboj.7601502

158. Zhang L, Zhao H, Qiu Y, Loh HH, Law PY (2009) Src phosphorylation of micro-receptor is responsible for the receptor switching from an inhibitory to a stimulatory signal. J Biol Chem 284(4):1990–2000. doi:10.1074/jbc.M807971200

159. Chen J, Randeva HS (2010) Genomic organization and regulation of the human orexin (hypocretin) receptor 2 gene: identification of alternative promoters. Biochem J 427 (3):377–390. doi:10.1042/BJ20091755

160. Kalogiannis M, Hsu E, Willie JT, Chemelli RM, Kisanuki YY, Yanagisawa M, Leonard CS (2011) Cholinergic modulation of narcoleptic attacks in double orexin receptor knockout mice. PLoS One 6(4), e18697. doi:10.1371/journal.pone.0018697

161. Lin L, Faraco J, Li R, Kadotani H, Rogers W, Lin X, Qiu X, de Jong PJ, Nishino S, Mignot E (1999) The sleep disorder canine narcolepsy is caused by a mutation in the hypocretin (orexin) receptor 2 gene. Cell 98(3):365–376. doi:10.1016/S0092-8674(00)81965-0

162. Willie JT, Chemelli RM, Sinton CM, Tokita S, Williams SC, Kisanuki YY, Marcus JN, Lee C, Elmquist JK, Kohlmeier KA, Leonard CS, Richardson JA, Hammer RE, Yanagisawa M (2003) Distinct narcolepsy syndromes in Orexin receptor-2 and Orexin null mice: molecular genetic dissection of non-REM and REM sleep regulatory processes. Neuron 38 (5):715–730

163. Asahi S, Egashira S, Matsuda M, Iwaasa H, Kanatani A, Ohkubo M, Ihara M, Morishima H (2003) Development of an orexin-2 receptor selective agonist, [Ala(11), D-Leu(15)]orexin-B. Bioorg Med Chem Lett 13(1):111–113

164. Kukkonen JP (2013) Physiology of the orexinergic/hypocretinergic system: a revisit in 2012. Am J Physiol Cell Physiol 301(1):C2–C32. doi:10.1152/ajpcell.00227.2012

165. Karhu L, Turku A, Xhaard H (2015) Modeling of the OX1R-orexin-A complex suggests two alternative binding modes. BMC Struct Biol 15:9. doi:10.1186/s12900-015-0036-2

166. Lang M, Soll RM, Durrenberger F, Dautzenberg FM, Beck-Sickinger AG (2004) Structure-activity studies of orexin A and orexin B at the human orexin 1 and orexin 2 receptors led to orexin 2 receptor selective and orexin 1 receptor preferring ligands. J Med Chem 47 (5):1153–1160

167. McElhinny CJ Jr, Lewin AH, Mascarella SW, Runyon S, Brieaddy L, Carroll FI (2012) Hydrolytic instability of the important orexin 1 receptor antagonist SB-334867: possible

confounding effects on in vivo and in vitro studies. Bioorg Med Chem Lett 22 (21):6661–6664. doi:10.1016/j.bmcl.2012.08.109
168. Kukkonen JP (2016) G-protein inhibition profile of the reported Gq/11 inhibitor UBO-QIC. Biochem Biophys Res Commun 1(1):101–107. doi:10.1016/j.bbrc.2015.11.078
169. Schrage R, Schmitz AL, Gaffal E, Annala S, Kehraus S, Wenzel D, Bullesbach KM, Bald T, Inoue A, Shinjo Y, Galandrin S, Shridhar N, Hesse M, Grundmann M, Merten N, Charpentier TH, Martz M, Butcher AJ, Slodczyk T, Armando S, Effern M, Namkung Y, Jenkins L, Horn V, Stossel A, Dargatz H, Tietze D, Imhof D, Gales C, Drewke C, Muller CE, Holzel M, Milligan G, Tobin AB, Gomeza J, Dohlman HG, Sondek J, Harden TK, Bouvier M, Laporte SA, Aoki J, Fleischmann BK, Mohr K, Konig GM, Tuting T, Kostenis E (2015) The experimental power of FR900359 to study Gq-regulated biological processes. Nat Commun 6:10156. doi:10.1038/ncomms10156
170. Kukkonen JP (2012) Recent progress in orexin/hypocretin physiology and pharmacology. Biomol Concepts 3(5):447–463. doi:10.1515/bmc-2012-0013
171. Yin J, Babaoglu K, Brautigam CA, Clark L, Shao Z, Scheuermann TH, Harrell CM, Gotter AL, Roecker AJ, Winrow CJ, Renger JJ, Coleman PJ, Rosenbaum DM (2016) Structure and ligand-binding mechanism of the human OX and OX orexin receptors. Nat Struct Mol Biol. doi:10.1038/nsmb.3183
172. Yin J, Mobarec JC, Kolb P, Rosenbaum DM (2015) Crystal structure of the human OX2 orexin receptor bound to the insomnia drug suvorexant. Nature 519(7542):247–250. doi:10.1038/nature14035

Orexin/Hypocretin and Organizing Principles for a Diversity of Wake-Promoting Neurons in the Brain

Cornelia Schöne and Denis Burdakov

Abstract An enigmatic feature of behavioural state control is the rich diversity of wake-promoting neural systems. This diversity has been rationalized as 'robustness via redundancy', wherein wakefulness control is not critically dependent on one type of neuron or molecule. Studies of the brain orexin/hypocretin system challenge this view by demonstrating that wakefulness control fails upon loss of this neurotransmitter system. Since orexin neurons signal arousal need, and excite other wake-promoting neurons, their actions illuminate nonredundant principles of arousal control. Here, we suggest such principles by reviewing the orexin system from a collective viewpoint of biology, physics and engineering. Orexin peptides excite other arousal-promoting neurons (noradrenaline, histamine, serotonin, acetylcholine neurons), either by activating mixed-cation conductances or by inhibiting potassium conductances. Ohm's law predicts that these opposite conductance changes will produce opposite effects on sensitivity of neuronal excitability to current inputs, thus enabling orexin to differentially control input-output gain of its target networks. Orexin neurons also produce other transmitters, including glutamate. When orexin cells fire, glutamate-mediated downstream excitation displays temporal decay, but orexin-mediated excitation escalates, as if orexin transmission enabled arousal controllers to compute a time integral of arousal need. Since the anatomical and functional architecture of the orexin system contains negative feedback loops (e.g. orexin $\rightarrow$ histamine $\rightarrow$ noradrenaline/serotonin—orexin), such computations may stabilize wakefulness via integral feedback, a basic engineering strategy for set point control in uncertain environments. Such dynamic

C. Schöne
Department of Neurology, University of Bern, Bern University Hospital, 3010 Bern, Switzerland

D. Burdakov (✉)
The Francis Crick Institute, Mill Hill Laboratory, London NW7 1AA, UK
e-mail: denis.burdakov@crick.ac.uk

© Springer International Publishing AG 2016
Curr Topics Behav Neurosci (2017) 33: 51–74
DOI 10.1007/7854_2016_45
Published Online: 8 November 2016

behavioural control requires several distinct wake-promoting modules, which perform nonredundant transformations of arousal signals and are connected in feedback loops.

Keywords Arousal • Brain state • Control theory • Hypocretin • Hypothalamus • Neurons • Orexin

Contents

1 Postsynaptic Actions of Orexin/Hypocretin Mediate Arousal Control

Efficient behaviour requires optimal adjustment of behavioural state (i.e. arousal, wakefulness, activity, energy expenditure) to perturbations in the environment. The environmental perturbations come in many diverse types, which differ greatly in their speed and predictability, necessitating an evolution of behavioural control systems that can deal with this perturbation diversity. A fundamental example of a slow and predictable perturbation in the environment is the ca. 24 h day-night cycle on Earth. The hypothalamic suprachiasmatic nucleus (SCN) adjusts behaviour to this cycle by emitting sinusoidal (ca. 24 h period) neural signals that schedule behavioural activity for either the day (for diurnal animals) or the night (for nocturnal animals), depending on specific survival advantages afforded to different animals by being active during light or dark. These slow daily rhythms of behavioural activity collapse upon description of the SCN [1].

However, wakefulness also needs to be controlled on a much more rapid and unpredictable timescale than that controlled by the SCN. Most of us take this rapid wakefulness adjustment for granted, assuming that we will not fall asleep in the middle of laughing or talking. This is not so for patients suffering from the sleep-wake disorder narcolepsy, which affects about 1:2000 people and where sleep and paralysis suddenly and uncontrollably intrude into normal wakefulness [2, 3]. Most cases of human narcolepsy are associated with reduced levels of orexin/hypocretin peptides in the CSF and lack of central orexin/hypocretin-producing neurons in the brain [4–7]. Loss of orexin/hypocretin peptides in humans, dogs, mice and rats impairs arousal control, resulting in abnormally frequent and rapid loss of consciousness ('sleep attacks'). It seems that without the orexin/hypocretin system, wakefulness is prone to instability in the face of rapid perturbations in the environment, and processes enabled by orexin/hypocretin cells keep this instability under control. Orexin/hypocretin and the SCN systems are thus two hypothalamic systems that are essential for appropriate matching of behavioural state to the environment on rapid and slow timescales, respectively.

The orexin/hypocretin cells act as controllers rather than critical generators of wakefulness: with them, the average amount of arousal (sleep and waking) does not change, but the ability to control arousal based on salient environmental set point is impaired [8, 9]. Coordination of many other behaviours, such as reward-seeking, also critically relies on orexin/hypocretin neurons [10–12], but since these behaviours are wakefulness dependent, it is often unclear to what extent these effects are secondary to wakefulness control. Since the discovery of orexin/hypocretin neurons, a key question has been why they are so critical for arousal control, considering the numerous other arousal-controlling neurons in the brain (e.g. noradrenaline, histamine, acetylcholine, serotonin cells). Here, we review the biological properties of brain orexin/hypocretin circuits related to wakefulness/arousal, with particular emphasis on postsynaptic actions of orexin/hypocretin peptides and from a viewpoint of basic principles of signal processing and dynamic set point control. For more comprehensive overviews of orexin/hypocretin physiology, we refer the reader elsewhere [10, 13].

In this article, we will take the view that orexin/hypocretin neurons signal arousal need (or, in control system language, arousal error – see below). We define arousal need as a need to counteract actual or potential dangers such as low energy levels, high CO_2 levels, or potentially threatening sensory stimuli (e.g. sudden sounds, presence of another animal). Orexin neurons sense all these signals (Fig. 1), and thus their activity represents a sum of diverse 'arousal demands' (e.g. they are inhibited by glucose but excited by H_3O^-/CO_2) [8, 14–18]. Orexin/hypocretin neurons are also inhibited by at least some of the other wakefulness-promoting transmitters such as serotonin and noradrenaline, i.e. transmitters that may represent the actual level of arousal [19–21] (discussed below). Thus, orexin/hypocretin cell output may represent an 'arousal error' (actual arousal minus required arousal), thereby signalling how much arousal should be increased. In the absence of these orexin/hypocretin signals, arousal is no longer appropriately coupled to internal and external environment, which is an alternative way of describing sleep-wake instability.

The wake-sleep instability seen upon loss of orexin/hypocretin-producing neurons is recapitulated by the loss of orexin/hypocretin type 2 receptors or of orexin/hypocretin peptides [22–24]. This suggests that postsynaptic actions of orexin/hypocretin peptides on orexin/hypocretin type 2 G-protein-coupled receptors are responsible for arousal control mediated by orexin/hypocretin cells. This also suggests that orexin type 2 receptor-independent actions of orexin/hypocretin cells, such as those mediated by their other transmitters (glutamate, dynorphin, Narp) or by orexin/hypocretin via type 1 receptors, are insufficient to achieve

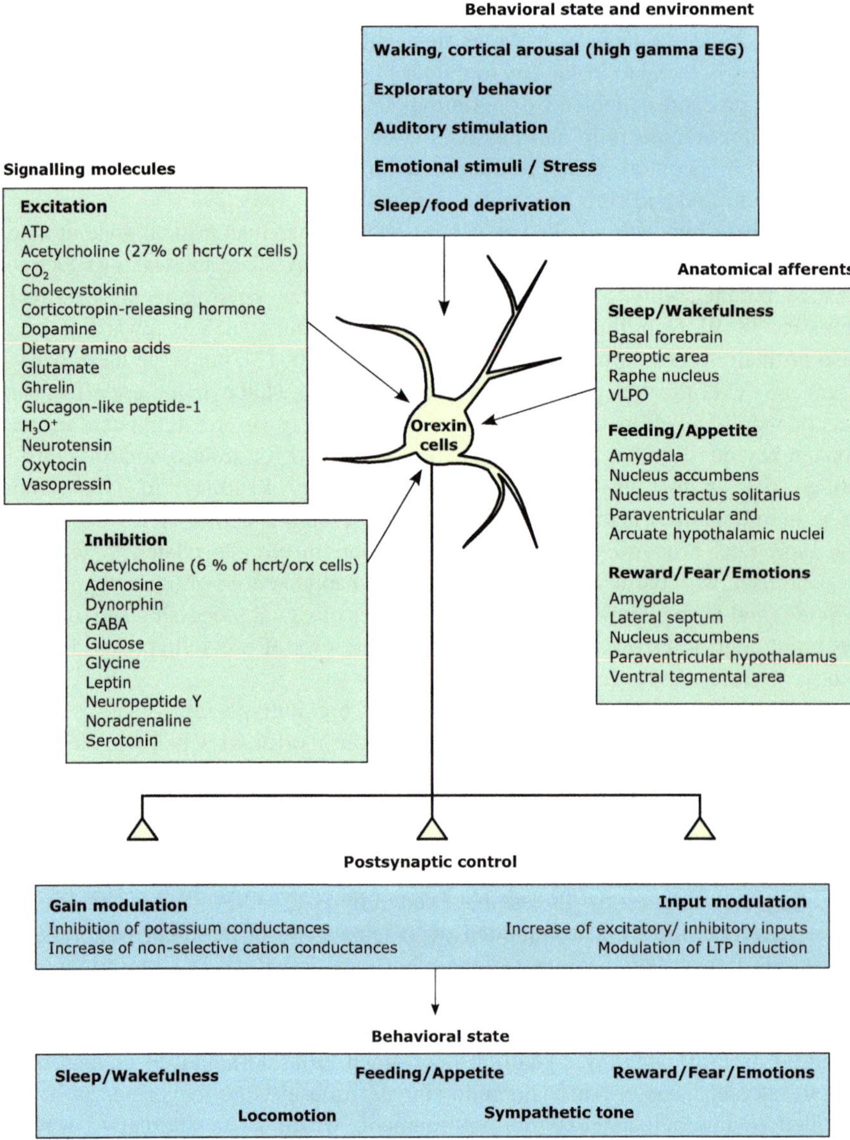

Fig. 1 Input and outputs of orexin/hypocretin cells. Orexin/hypocretin neurons are activated during high vigilance states associated with high gamma EEG, requiring increased arousal such as exploratory behaviour or upon sensory or emotional stimulation. A multitude of excitatory and inhibitory substances modulates orexin/hypocretin cell activity. These include hormones, neuropeptides and small molecule transmitters as well as homeostatic signals. Orexin/hypocretin neurons receive direct inputs from brain areas involved in sleep/wake control, appetite control and reward. Orexin/hypocretin neurons integrate this information via the release of orexin peptides, affecting postsynaptic gain and synaptic drive in target neurons. Orexin/hypocretin activity may thereby gate relevant information based on environmental and homeostatic needs [14, 16, 124–129]

proper arousal control without orexin/hypocretin action of type 2 orexin/hypocretin receptors. The latter point may seem surprising, considering that some arousal-promoting neurons, such as noradrenaline neurons of the LC which are important for orexin/hypocretin-induced wakefulness [25], are excited by orexin/hypocretin via type 1 rather than type 2 OX receptors [26]. In contrast, other arousal-promoting neurons, such as histamine cells of the tuberomammillary hypothalamus, are excited by orexin/hypocretin via type 2 receptors [26, 27]. Below, we catalogue this biological complexity of postsynaptic actions of orexin/hypocretin cells in more detail and comment on some functional biophysical implications of this complexity. Then, we propose a framework that simplifies and generalizes the diversity of postsynaptic orexin/hypocretin actions into a control systems model that accounts for general dynamic features of orexin/hypocretin-dependent arousal and offers an organizing principle for the puzzling diversity of wakefulness-promoting neurons in the brain.

2 Sites and Biophysics of Postsynaptic Actions of Orexin/Hypocretin Neurons

From their location in the lateral hypothalamus, orexin/hypocretin cells project axons to the entire brain [28, 29] (Fig. 2). The anatomical distribution of these projections largely mirrors that of two G-protein-coupled receptors for orexin [30]. Increased firing rate of orexin/hypocretin neurons produces awakening [31], and most of the brain's classical arousal-related systems are innervated by orexin/hypocretin axons and excited by orexin/hypocretin peptides (see Table 1, which lists many key findings alongside corresponding references [19, 26, 27, 32–106]). Orexin/hypocretin peptides also modulate neuronal activity in brain areas related to eating, emotion, autonomic function and motor control (Table 1). Here, we only list (Table 1) but do not discuss the latter actions of orexin/hypocretin in detail, since this has been covered extensively in recent publications (e.g. [10, 107]). We would just like to note that combined activation of arousal and reward systems may ensure that a heightened arousal accompanies reward-seeking, thereby increasing probability of reward discovery and of avoiding danger while exploring for rewards. Also, arousal and exploration may require motivational signals, since these behaviours are not intrinsically rewarding but are energy expending and potentially dangerous. Orexin/hypocretin may provide this motivation [10, 108], at least until reward consumption beings [109].

From a biophysical perspective on signal processing, the excitatory/depolarizing actions of orexin/hypocretin on central neurons (Table 1) can be divided into those increasing membrane conductance (e.g. activation of non-selective cation currents) and those decreasing it (e.g. inhibition of K^+ currents). This has profound implications for input processing capabilities of orexin/hypocretin-modulated neurons. The ability of a current input (I) to change membrane potential (V) is inversely related to

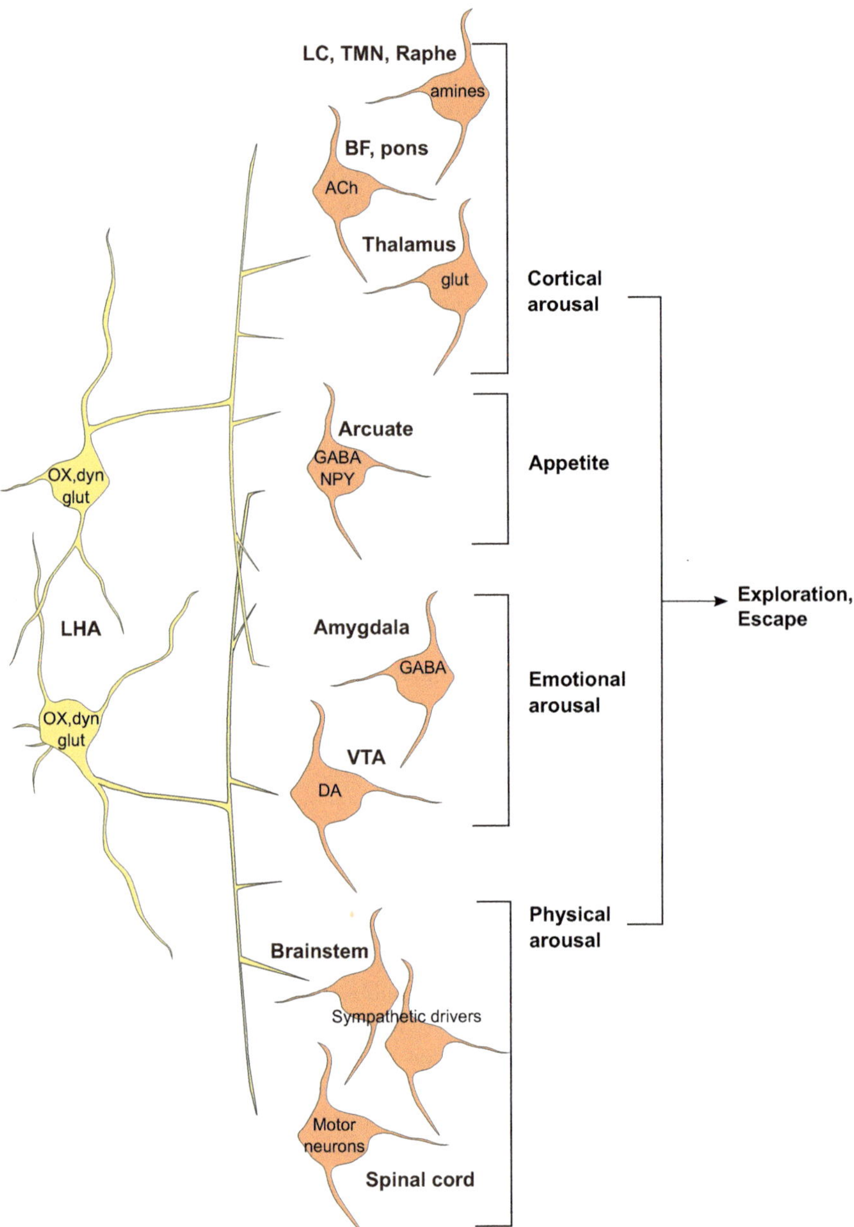

Fig. 2 Main anatomical targets for orexin/hypocretin control of cortical arousal, appetite, emotional arousal and physical arousal. Together, orexin/hypocretin control of these targets would facilitate exploratory or escape behaviour

Table 1 Postsynaptic effects of orexin/hypocretin neurons on sleep-/wakefulness-/arousal-related targets

Brain area	Nucleus	Targets	Receptor/ligand	Mechanism of action	References
Cerebral cortex	PFC		OR1/orexin	Excitation: presynaptic excitation of boutons from excitatory thalamocortical projecting neurons in layer V	[40–42]
	Visual cortex	Layer 6b pyramidal	OR2/orexin	Excitation: TTX insensitive	[43]
		Other layers		Increase synaptic inputs, improved spike-timing of layer VIa responses to rhomboid nucleus	[43–44]
Hippo-campus	DG		OR1/orexin	OR1 mediates induction of LTP in dentate gyrus of freely moving rats, Orexin A enhances LTP in dentate gyrus of urethane anaesthetized rats	[45–46]
	CA1		OR1/orexin A	Orexin A but not B induced LTP in Schaffer collateral – CA1 synapses in adult, while it induced LTD in juvenile mouse hippocampal sections	[47]
Cerebral nuclei	SI, MPO	Cholinergic neurons	OR2/orexin	Excitation: inhibition of constitutively active inward rectifier K+ current; potentiation of evoked EPSCs	[32–34]
			k-opioid/Dyn	Inhibition: activation of inward rectifier potassium current, inhibition of voltage sensitive calcium current, presynaptic depression of glutamatergic inputs	[35]
		Wake-active GABAergic	OR2/orexin	Excitation: unknown mechanism, potentiation of evoked EPSCs	[33]
			Dyn	No effect	[33]
		Sleep-active GABAergic	Orexin	No effect	[33]
			k-opioid/Dyn	Inhibition: subset of non-cholinergic neurons, suppression of evoked EPSCs	[33]
	Medial septum, DBB	Septo-hippo-campal cholinergic	OR2/orexin	Excitation: two ionic mechanisms with 80% overlap: inhibition of an inwardly rectifying K+ conductance and activation of an Na+/Ca2+ exchanger; Inhibition: increased GABA release (spike dependent, via activation of interneuron)	[36, 37]
		Septo-hippocampal GABAergic	OR2/orexin	Excitation, activation of sodium–calcium exchanger; Inhibition by increasing local, spike dependent GABA release	[37]
	NA	Accumbens shell	OR2/orexin	Excitation: increase of nonselective cationic conductance and decrease of K(+) conductance; Inhibition: decrease of NMDA currents and increased GABA currents in dissociated cells	[38, 39]
	BNST		OR1/orexins	Excitation: subset of neurons; Inhibition: OR1 dependent depression of excitatory transmission	[49, 50]

(continued)

Table 1 (continued)

Brain area	Nucleus	Targets	Receptor/ligand	Mechanism of action	References
Thalamus	MTN		OR2/orexin	Excitation: closure of potassium channels and activation of nonselective cation channels	[48, 51–53]
	IGL	NPY+ and − neurons	OR1/2/orexin	Excitation (58% of neurons): direct; Inhibition: increased inhibitory synaptic inputs	[54]
	VGN	NO+ and − neurons	OR1/2/orexin	Excitation	[55]
	CLN, PFN, MDT, TRN		OR1/2/orexin	Excitation of subset of neurons: inhibition of K+ current	[56]
	LP, LD P, LGN			No effect	[56]
Hypo-thalamus	TMN	Histamine	OR2/orexin	Excitation: activation of sodium–calcium exchanger, inhibition of G-protein coupled inward rectifier channel in cell culture; Inhibition: presynaptic increase GABA inputs	[27, 57, 59, 60, 61]
			AMPA/Glut	Excitation	[61]
	POA	GABAergic	Orexin	No effect	[32]
	SCN		OR1/2/orexins	Two subsets of cells showing either direct excitation or increase in GABAA mediated synaptic inputs during the day, and during the night direct inhibition via activation of leak-potassium conductance	[62]
	LHA	MCH	OR2/orexin	Excitation: direct inward current, inhibition of K+ current, increase excitatory synaptic inputs Inhibition: activate inhibitory GABAA-mediated synaptic currents	[76]
		Orexin	OR2/orexin	Excitation: increase of excitatory synaptic inputs	[73]
	Arc	GABA neurons	OR2/orexin	Excitation: activation of sodium–calcium exchanger	[78]
		NPY /AGRP neurons	OR1/2/orexin	Excitation: activation of inward current, phospholipase C dependent increase in intracellular calcium	[79, 77, 80]
		POMC neurons	OR2/orexin	Inhibition: decrease intracellular calcium, increase of inhibitory synaptic tone, decrease excitatory synaptic tone	[42, 79, 81]
	PVN	Type 1 and type 2 neurons	OR2/orexin B	Excitation: 50–80% of neurons	[82]
	VMH	Glucose-responsive neurons	OR1/orexins	Inhibition: decrease intracellular calcium	[79]

Cerebellum	CIN		OR2/orexins	Excitation: 67% of neurons	[105]
Midbrain	PPT		OR1/orexins	Excitation: decrease of potassium current, increase of nonselective cationic current	[65]
	DR	Serotonergic	OR2/orexins	Excitation: activation of sodium–potassium nonselective cation channels, protein kinase C dependent enhancement of Ca2+ transients mediated by L-type Ca2+ channels	[63, 68, 72, 73, 88]
				Inhibition: enhanced spontaneous IPSCs, suppression of evoked EPSCs via retrograde endocannabinoid release and stimulation of postsynaptic orexin receptors (activation of PLC and DAG lipase signalling pathways)	[73, 74]
		GABA neurons	OR1/2/orexins	TTX insensitive excitation	[42, 73]
	VTA	Dopamine neurons	OR1/2/orexins	Excitation: increase of intracellular calcium via phosphatidylcholine-specific phospholipase C- and protein kinase C- signalling pathways to activate of L- and N-type Ca2+ channels	[42, 83, 84, 86]
			OR2/orexin B	Excitation: protein kinase C dependent postsynaptic potentiation of NMDA receptors; Inhibition: reduction of spike-timing induced LTP	[85]
		Non-dopamine neurons	Orexins	Excitation	[86]
	RMR		OR1/orexins		[42, 89]
	SN-Pars reticulata	GABAergic	OR2/orexins	Excitation: protein kinase A dependent	[104]
	SN-Pars compacta	Dopaminergic	Orexins	No effect	[104]
Pons	LDT	Cholinergic neurons	OR1/orexins	Excitation: activation of noisy cation channels, activation of a protein kinase C (PKC)-dependent enhancement of Ca2+ transients mediated by L-type Ca2+ channels, enhanced spontaneous EPSCs	[26, 64, 88]
		Non-cholinergic	OR1/2/orexins	Excitation: activation of inward current and membrane noise, enhanced spontaneous EPSCs but not IPSC	[26, 42, 64]
	LC	Noradrenergic	OR1/orexins	Excitation: depolarization and reduction in the slow component of the afterhyperpolarization (AHP)	[66]
		LC cells	OR1/orexins	Excitation: augmentation of nonselective cation conductance, inhibition of sustained potassium conductance, inhibition of G-protein coupled inward rectifier channel in cell culture	[59, 67–69, 101]
			Glut	Glutamate co-expression in orexin positive terminals	[71]
	TGMN	Motor neurons	OR1/orexins	Excitation: NMDA dependent excitation	[42, 102]

(continued)

Table 1 (continued)

Brain area	Nucleus	Targets	Receptor/ligand	Mechanism of action	References
Medulla	ReTN	Acid sensing neurons	OR1/orexins	Excitation	[89–91]
	Pre-Bötz		OR1/2/orexins	Increase in diaphragm electromyographic activity	[90, 92]
	NAc		OR1/2/orexins	Excitation: 90% of NTS cells through inhibition of sustained K+ current, activation of nonselective cationic conductance, hypocretin 2 increased mEPSCs in dorsal NTS neurons	[42, 93, 94]
	NAm	Cardiac vagal neurons	OR2/orexins	In postnatal day 5 rats: inhibition: enhanced GABAergic postsynaptic currents In postnatal day 20 and 30 rats: excitation: inhibition of GABAergic postsynaptic currents	[42, 106]
	RVMM		OR1/orexin A	Excitation: majority of neurons, possible indirect mechanisms; Inhibition: subset of neurons with irregular firing pattern that lack fast afterhyperpolarization, possible indirect action	[95]
	RVLM		OR1<2/orexinz A	Excitation: subset of neurons	[90, 96]
	LVN		OR1/2/orexins	Excitation: activation of sodium–calcium exchanger, block of inward rectifier K+ current	[99]
	IVN		OR1/2/orexins	Direct excitation	[103]
Spinal cord		Sympathetic preganglion neurons	OX1/2/orexins	Excitation, inhibition of potassium conductance via protein kinase A-dependent pathway	[97, 98]
		Phrenic/hypoglossal motor neurons	OR1/2/Orexins	Excitation, persistent in 1 μM tetrodotoxin	[89, 90, 99, 100]

Abbreviations: *Arc* arcuate nucleus, *BNST* bed nucleus stria terminalis, *CeM* central medial amygdala, *CIN* cerebellar interpositus nucleus, *CLN* centrolateral nucleus of the thalamus, *DBB* diagonal band of broca, *DR* dorsal raphe, *IGL* intergeniculate leaflet, *Dyn* dynorphin, *Glut* glutamate, *IVN* inferior vestibular nucleus, *LC* locus ceruleus, *LDT* laterodorsal tegmental nucleus, *LGN* lateral geniculate nuclei, *LHA* lateral hypothalamic area, *LP* lateral posterior geniculate nucleus, *LD* laterodorsal geniculate nucleus, *LVN* lateral vestibular nucleus, *MD* mediodorsal, *MPO* magnocellular preoptic nucleus, *MTN* midline thalamic nuclei, *NAc* nucleus accumbens, *NAm* nucleus ambiguus, *NTS* nucleus tractus solitarius, *OR1/2* orexin receptor 1/2, *P* posterior geniculate nuclei, *PFN* parafascicular nucleus, *PFC* prefrontal cortex, *PPT* pedunculo pontine tegmental nucleus, *Pre-Bötz* pre-Bötzinger complex, *PVN* paraventricular nucleus of the hypothalamus, *ReTN* retro trapezoid nucleus, *RVLM* rostral ventro-lateral medulla, *RVMM* rostral ventro-medial medulla, *RMR* rostral medullary raphe, *SCN* suprachiasmatic nucleus, *SI* substantia innominata, *SN* substantia nigra, *TGMN* trigeminal motor nucleus, *TMN* tuberomammillary nucleus, *TRN* reticular nucleus, *VGN* ventrolateral geniculate nucleus, *VMH* ventromedial hypothalamus, *VTA* ventral tegmental area

membrane conductance (g). This dependence is described by Ohm's law, $V = I/g$. The neuronal firing output depends on the membrane potential (it is increased by depolarization, [110, 111]). Therefore, conductance-increasing actions of orexin/hypocretin will not only depolarize and electrically excite the target neuron but also reduce the sensitivity of the neuron's firing to other inputs. This would effectively lock the neuron in a high-output state, which could be useful for overriding other inputs in times of danger. In turn, conductance-reducing actions of orexin/hypocretin will not only depolarize and excite the neuron but also increase its sensitivity to other inputs. This would enable the neuron to be readily modulated by other inputs (both stimulatory and inhibitory), thus allowing other inputs to either augment or cancel the orexin/hypocretin-induced excitation. Overall, these conductance-related actions of orexin/hypocretin can be viewed as not simply excitatory but also 'gain modulating'. The ability to modulate the input-output gain is an important feature of neural computation [112]. Therefore, the functional/behavioural implications of gain-modulating postsynaptic actions of orexin/hypocretin are an important question for future investigations.

Although the direct postsynaptic actions of orexin/hypocretin are usually excitatory, there are some exceptions. Per1 neurons of the hypothalamic suprachiasmatic nucleus that function as the brain's master circadian clock are inhibited by orexin/hypocretin via activation of leak-like K^+ channels, as well as presynaptically by increasing GABA release [62]. Signal transduction pathways linking orexin/hypocretin receptors to the inhibitory channels remain undefined. The inhibition of mouse Per1 neurons by orexin/hypocretin may enable the circadian clock signals to be overridden by arousal need signalled by orexin/hypocretin cells.

The above summary and classification of postsynaptic actions of orexin/hypocretin highlight the diversity and some general themes of brain-wide orexin/hypocretin signalling. However, such descriptions do not reveal critical components of orexin/hypocretin actions nor why orexins/hypocretins are vital for brain function stability. To achieve the latter insights, it is necessary to understand the functionally critical modules mediating orexin/hypocretin action and the overall control system architecture implemented by these modules. We address this in the next section, continuing to focus on wakefulness control.

3 Which Orexin/Hypocretin-Regulated Sites Are Critical for Wake Stability in the Normal Brain?

The diversity of orexin/hypocretin-excited neurons throughout the brain raises the question of the relative roles of different orexin targets in preventing the narcoleptic instability of wakefulness. An optimal way to deconstruct these natural roles would be to examine the effects on wakefulness stability of targeted, specific and reversible inactivation of orexin receptors in molecularly defined neurons in adult mice. Such technically demanding experiments have not yet been accomplished. The relevant studies performed to date used other approaches, such as global receptor deletion followed by targeted receptor restoration [113, 114] or experimental stimulation of orexin/hypocretin cells concurrent with experimental silencing of specific downstream targets [25]. These approaches have caveats as far as the natural roles of orexin/ hypocretin targets in wake stability are concerned. For example, given the feedback loops in wakefulness circuits (see below), the role of an orexin/hypocretin receptor site in wakefulness control when all other orexin/hypocretin sites are genetically deleted is not the same as its role in the natural brain. In turn, experimental stimulation of orexin/ hypocretin cells does not reproduce their natural firing patterns, and the ability of orexin/hypocretin cells to stimulate awakening is not an assay of wakefulness stability. Nevertheless, the existing studies provide fundamental information about causal links between specific neurons and wakefulness, as well as proof-of-concept information relevant to narcolepsy treatment. Therefore, we briefly comment on some of them here (for more in-depth discussions of current literature on this topic, see [2, 115]).

Carter et al. examined the mechanism of orexin-mediated wakefulness by optogenetically stimulating orexin/hypocretin neurons while concurrently optogenetically silencing one of their downstream effectors, the orexin type 1 receptor expressing noradrenaline neurons of the locus coeruleus (LC) [25]. Note that this does not address the question of which orexin/hypocretin targets are critical for orexin/hypocretin-dependent wakefulness stability. They found that when the LC noradrenaline neurons were inactivated, stimulation of orexin/hypocretin neurons no longer produced awakening from sleep. This seminal finding establishes the noradrenaline neurons as critical generators for the orexin/hypocretin-dependent stimulation of wakefulness. However, it remains unclear how these generators are controlled in order to maintain stable wakefulness, especially since the orexin type 2 receptor (not the type 1 expressed by the LC noradrenaline neurons) is essential for the wake stability. In the next sections, we will propose a unifying framework that can reconcile the wake-generator function of the noradrenaline neurons with wake-controller functions of orexin type 2 receptor neurons.

Mochizuki et al. globally deleted orexin type 2 receptors in mice, producing a narcoleptic instability of wakefulness [113]. They then used a viral expression strategy to restore these receptors locally in the tuberomammillary hypothalamus, an area rich in histamine neurons that normally express high levels of orexin type 2 receptors. This local manipulation rescued the wakefulness instability (but interestingly, not sleep instability that also results from loss of orexin/hypocretin function). This suggests that the histamine neurons could be critical for wake-controllers that signal to wake-generators (such as noradrenaline cells) to adjust their signals properly.

Hasegawa et al. knocked out both types of orexin/hypocretin receptors in mice and then reintroduced both of them at specific brain sites by viral delivery under a non-specific promoter. They found that such receptor overexpression in the LC restored the normal duration and number of wakefulness episodes [114]. It is not clear whether this is the normal function of the orexin signals to the locus coeruleus or an outcome of overexpression of orexin receptors that are not normally there. In contrast, the dual orexin/hypocretin receptor overexpression in the tuberomammillary hypothalamus did not restore the normal duration and number of wakefulness episodes. This shows that orexin/hypocretin signalling in the tuberomammillary hypothalamus is insufficient for normal wakefulness when all orexin receptors are missing from the locus coeruleus. Together with the data of Mochizuki et al., this can be interpreted to suggest that the tuberomammillary hypothalamus requires an orexin-sensitive downstream wakefulness generator in order to control wakefulness. For effective wakefulness control, orexin/hypocretin may need to alter the activity of both wakefulness regulators and generators in the brain, with different kinetics and via different receptors (see Fig. 3, discussed below).

A Error-based integral feedback scheme

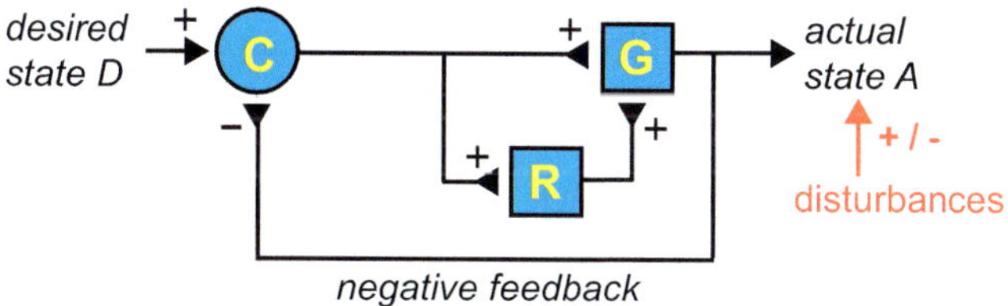

Operations that help $D \rightarrow A$:
C - summation
R - integration
G - amplification

B Possible mapping to orexin/hypocretin biology

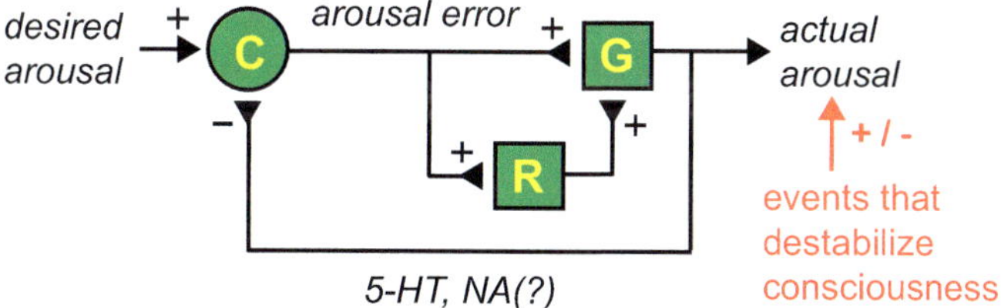

Possible neural implementation:
C - orexin neurons sensitive to arousal need
R - orexin type-2 receptor neurons (e.g. HA?)
G - orexin type-1 receptor neurons (e.g. NA?)

Fig. 3 Brain arousal systems as control modules in a feedback loop. (**a**) A generalized control system architecture (integral feedback loop) for tracking a desired set point (D) despite unpredictable disturbances. After [120]. (**b**) Possible implementation of A by a diversity of wake-promoting neurons in the brain (from more detail, see [18])

4 Mapping Orexin/Hypocretin Biology onto Control Operations

The above-discussed biological measurements define functionally important components of orexin/hypocretin systems and the general signs (plus or minus) for interactions between these components. This knowledge is fundamental, but alone is insufficient to account for control operations performed by orexin/hypocretin to achieve stable wakefulness. To clarify what we mean by control operations, a brief formal definition of tracking and stability is warranted. From a general evolutionary perspective, a highly desirable attribute of arousal control is set point tracking, i.e. the ability to adjust a set point to relevant inputs while rejecting disturbances. A good tracking system will follow salient inputs while resisting disturbances. Disturbance resistance is the ability to protect a set point from irrelevant disturbance, e.g. noise in brain/body internal signals, external events not requiring arousal responses, etc. A system capable of disturbance-resistant tracking can be considered 'robust yet flexible'. Note that this robust flexibility has to exist in the real world, i.e. where neither noise/disturbance nor important inputs are completely predictable, i.e. the control system has to be uncertainty proof. This need to deal with uncertainty imposes important requirements (and thus constraints) on system architecture (see below). For more detailed discussions of control principles as applied to orexin/hypocretin networks, see [18].

Can the actions of orexin/hypocretin be considered to implement such robust-yet-flexible arousal? We believe the answer is yes, since without orexin/hypocretin, arousal becomes both flexible and less robust. For example, when orexin/hypocretin is knocked out, mice cannot respond to potentially dangerous intrusions by properly increasing blood pressure [116], and they cannot properly adapt to a fall in their energy levels by increasing locomotion [8]. Thus, a vital flexibility of arousal is lost without orexin/hypocretin. In terms of robustness, it is well known that without orexin/hypocretin arousal can dip to inappropriately low levels (unconsciousness) upon disturbances such as laughter in humans or sight of delicious food in animals [2, 117, 118]. Without orexin/hypocretin, there is no appropriate tracking/adjustment of arousal state to internal and external state.

If orexin/hypocretin actions implement arousal tracking, the understanding of arousal control will increase by viewing orexin/hypocretin system from general perspectives of robust-yet-flexible control systems. Such systems generally must contain autocorrecting feedback loops, since neither the world nor system performance can be precisely predicted [119, 120]. As a minimum, such a feedback loop circuit must contain at least three operationally different elements in order to be robust yet flexible, which we here call a comparator, a controller and a generator (Fig. 3). This error-based feedback system is a canonical engineering strategy to track a set point despite noise [119, 120]. Note that although artificial, stimulation of each of these elements would increase the final output of the system (e.g. arousal). However, this 'test' does not mean that the elements are redundant: their functions and dynamics are fundamentally distinct.

These distinct functions of the three components in this autocorrecting system architecture (Fig. 3) have been discussed in detail in control engineering literature [119, 121] and recently in arousal control literature [18, 58]. A summary of the latter discussions is that orexin/hypocretin neurons display functional hallmarks of comparators, some orexin type 2 receptor neurons (histamine cells) exhibit functional signatures of controllers, while some orexin type 1 receptor neurons (noradrenaline cells) have operational features of generators (for detailed arguments, see [18]). A particularly curious feature of some orexin type 2 receptor cells is that they appear to transmit a signal resembling a temporal integral of orexin/hypocretin neuron activity (Fig. 4) [18, 58]. This integration may enable them to function as integral controllers, engineering signals that are theoretically necessary and sufficient for robust-and-flexible control mediated by orexin/hypocretin in general and its type 2 receptors in particular [18]. Therefore, from an operational perspective, integral feedback is an important candidate mechanism for how orexin/hypocretin maintains appropriate behavioural state.

5 Explanatory and Predictive Value of Viewing Orexin/Hypocretin Actions as Control Computations

What is the scientific value, for orexin/hypocretin biology and clinical applications, of control engineering theories such as those shown in Fig. 3?

First, an important corollary is that these control schemes assign a clear operational reason for the hitherto puzzling diversity of seemingly redundant wake-promoting neurons in the brain. If brain wakefulness control was operating via integral control or a related feedback scheme, there would have to be several operationally nonredundant neural types (comparators, controllers, generators) cooperating together. Note that these neurons are nonredundant in the sense of operations they perform, for example in this case, addition, integration and amplification, respectively. However, the comparator, controller and generator neurons are redundant in the sense that they all promote wakefulness if separately stimulated (this follows mathematically from the scheme in Fig. 3). The latter 'redundancy', however, is a by-product of experimental manipulation – it could be useful clinically for achieving rapid arousal, but it does not mean that the wakefulness control architecture is redundant. Such considerations are attractive because they settle a long-standing enigma in the field – the diversity of wake-promoting neurons – and emphasize that wakefulness control/stability/flexibility/robustness is a separate process from wakefulness stimulation, which allows the terms redundant and nonredundant to be applied more precisely. Thus, a control view adds clarity and explanatory power to understanding how the complex biology of arousal control relates to the need for the brain to operate flexibly yet robustly in uncertain environments.

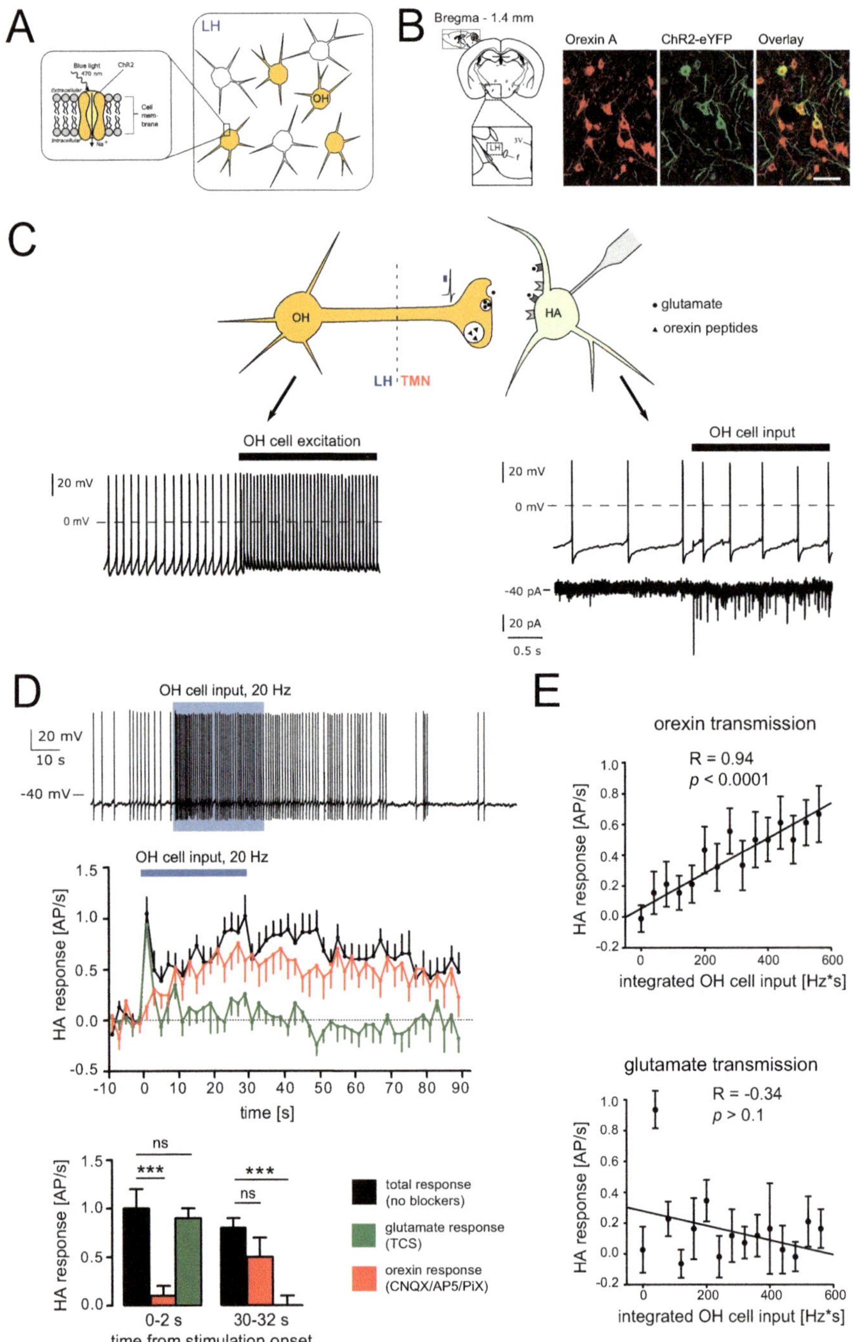

Fig. 4 Orexin/hypocretin and glutamate co-transmission enables fast and sustained control of histamine neurons. (**a**) Targeted expression of light-sensitive ion channel ChR2 enables selective control of excitatory membrane currents in orexin/hypocretin cells. (**b**) Cre-dependent expression of virally delivered ChR2-eYFP in orexin-cre mice allows selective expression in lateral

Second, control schemes such as those shown in Fig. 3 are formal mathematical theories that produce precise experimentally testable predictions about the temporal dynamics of distinct neurons. Such predictions of dynamics can be directly compared with real biological dynamics (measuring of temporal patterns of neuronal activity) and are thus essential for falsifying any theories of dynamic brain function. For example, when mathematically simulated, comparator, regulator and generator neurons produce different temporal signatures of activity in response to an input [18], and this can be experimentally tested. Furthermore, a mathematical control scheme such as that in Fig. 3 allows a proof-of-concept examination of whether a particular experimentally discovered neural operation is necessary to account for wakefulness stability. For example, if integration by orexin type 2 receptor cells is taken out of the model and replaced by a different computation (amplification), it can be mathematically demonstrated that both robustness and flexibility of the system are lost [18]. In contrast, more conventional (in biology) descriptions of arousal-implicated orexin/hypocretin biology that we gave earlier in this chapter (Fig. 1 and Table 1) are not mathematical theories and do not produce useful predictions and wakefulness dynamics. Thus, control engineering theories are useful for biology because they generate clearer predictions to guide experiments aimed at brain dynamics.

6 Overview, Omissions and Future Perspectives

In summary, we have reviewed postsynaptic orexin/hypocretin actions relating to arousal and presented a control theoretical view of these actions. This view theoretically accounts both for how orexin/hypocretin generates robust-yet-flexible arousal and for why multiple nonredundant types of arousal-promoting neurons exist in the brain.

We have omitted from this brief article many publications on the topic that have potential bearing on our interpretations. However, to the best of our knowledge, there are currently no experimental observations that invalidate our general argument. For example, under some behavioural manipulations, the actions of

---

Fig. 4 (continued) hypothalamic orexin-positive neurons. Data are from [61]. (**c**) *Left*: Light-activated action potential firing in ChR2-eYFP-expressing orexin/hypocretin cells recorded using whole-cell patch clamp. *Right*: Whole-cell recording of histamine neurons shows increased glutamate inputs (*bottom*) and action potential firing (*top*) upon light stimulation of adjacent orexin/hypocretin fibres. Adapted from [130], Fig. 2. (**d**) *Top*: Prolonged stimulation of orexin/hypocretin fibres for 30 s at 20 Hz produces fast and sustained increase in histamine cell firing. *Bottom*: Blockade of CNQX-sensitive glutamate currents blocks the fast rise in histamine firing, while a slow, long-lasting component remains. In contrary, orexin 2 receptor blockade (TCS) abolishes the slow component, leaving the fast component unaltered. Data from [58]. (**e**) Orexin-mediated increase in histamine firing integrates orexin/hypocretin activity (*top*), while glutamate-mediated increase in histamine firing does not (*bottom*). Data from [58]

noradrenaline on orexin/hypocretin cells in vitro have been reported to switch from inhibition to excitation [21, 122]. However, this does not mean that under such circumstances the negative feedback of arousal signals to orexin/hypocretin neurons does not exist; for example, it can be signalled by 5-HT. It would also be important to determine whether fast transmitters expressed by arousal-promoting neurons (e.g. GABA) affect orexin/hypocretin neurons. It has also been reported that some orexin/hypocretin neurons sense orexin/hypocretin themselves via orexin type 2 receptors [123], although this is controversial [19]. This opens up the possibility that orexin/hypocretin population may perform a dual function of regulators and comparators/error generators). Furthermore, we did not discuss the possible reasons for why different arousal-regulating neurons synthesize and use different transmitters, considering that activating or inhibiting downstream targets could be accomplished via glutamate, GABA and their many receptors. We speculate that transmitter diversity evolved to facilitate parallel signalling in a tight space [62], but it is beyond the scope of this review to discuss this in detail.

Overall, we feel that the results of our analysis provide some evidence that control-based logic is used by orexin/hypocretin system to dynamically control arousal. Furthermore, the integral feedback model provides a framework for studying wakefulness stability in both animal models and patients with narcolepsy. This model predicts transient responses of orexin/hypocretin neurons and more sustained responses of their downstream effector neurons to a change in arousal need [58]. These specific predictions can be tested in animal models with tools such as cell type-specific neural recordings, and this testing may aid the development of biomimetic medical robotics for patients with wakefulness disabilities.

Acknowledgements This work was funded by The Francis Crick Institute, which receives its core funding from Cancer Research UK, the UK Medical Research Council, and the Wellcome Trust.

References

1. Saper CB, Scammell TE, Lu J (2005) Hypothalamic regulation of sleep and circadian rhythms. Nature 437(7063):1257–1263
2. Scammell TE (2015) Narcolepsy. N Engl J Med 373(27):2654–2662
3. Dauvilliers Y, Arnulf I, Mignot E (2007) Narcolepsy with cataplexy. Lancet 369 (9560):499–511
4. Peyron C et al (2000) A mutation in a case of early onset narcolepsy and a generalized absence of hypocretin peptides in human narcoleptic brains. Nat Med 6(9):991–997
5. Ripley B et al (2001) CSF hypocretin/orexin levels in narcolepsy and other neurological conditions. Neurology 57(12):2253–2258
6. Nishino S et al (2001) Low cerebrospinal fluid hypocretin (orexin) and altered energy homeostasis in human narcolepsy. Ann Neurol 50(3):381–388
7. Thannickal TC et al (2000) Reduced number of hypocretin neurons in human narcolepsy. Neuron 27(3):469–474

8. Yamanaka A et al (2003) Hypothalamic orexin neurons regulate arousal according to energy balance in mice. Neuron 38(5):701–713
9. Hara J et al (2001) Genetic ablation of orexin neurons in mice results in narcolepsy, hypophagia, and obesity. Neuron 30(2):345–354
10. Mahler SV et al (2014) Motivational activation: a unifying hypothesis of orexin/hypocretin function. Nat Neurosci 17(10):1298–1303
11. Harris GC, Wimmer M, Aston-Jones G (2005) A role for lateral hypothalamic orexin neurons in reward seeking. Nature 437(7058):556–559
12. Boutrel B et al (2005) Role for hypocretin in mediating stress-induced reinstatement of cocaine-seeking behavior. Proc Natl Acad Sci U S A 102(52):19168–19173
13. Sakurai T (2014) The role of orexin in motivated behaviours. Nat Rev Neurosci 15 (11):719–731
14. Williams RH et al (2008) Adaptive sugar sensors in hypothalamic feeding circuits. Proc Natl Acad Sci U S A 105(33):11975–11980
15. Williams RH et al (2007) Control of hypothalamic orexin neurons by acid and CO_2. Proc Natl Acad Sci U S A 104(25):10685–10690
16. Karnani MM et al (2011) Activation of central orexin/hypocretin neurons by dietary amino acids. Neuron 72(4):616–629
17. Mileykovskiy BY, Kiyashchenko LI, Siegel JM (2005) Behavioral correlates of activity in identified hypocretin/orexin neurons. Neuron 46(5):787–798
18. Kosse C, Burdakov D (2014) A unifying computational framework for stability and flexibility of arousal. Front Syst Neurosci 8:192
19. Li Y et al (2002) Hypocretin/orexin excites hypocretin neurons via a local glutamate neuron—A potential mechanism for orchestrating the hypothalamic arousal system. Neuron 36 (6):1169–1181
20. Yamanaka A et al (2003) Regulation of orexin neurons by the monoaminergic and cholinergic systems. Biochem Biophys Res Commun 303(1):120–129
21. Grivel J et al (2005) The wake-promoting hypocretin/orexin neurons change their response to noradrenaline after sleep deprivation. J Neurosci 25(16):4127–4130
22. Lin L et al (1999) The sleep disorder canine narcolepsy is caused by a mutation in the hypocretin (orexin) receptor 2 gene. Cell 98(3):365–376
23. Chemelli RM et al (1999) Narcolepsy in orexin knockout mice: molecular genetics of sleep regulation. Cell 98(4):437–451
24. Willie JT et al (2003) Distinct narcolepsy syndromes in orexin receptor-2 and orexin null mice: molecular genetic dissection of Non-REM and REM sleep regulatory processes. Neuron 38(5):715–730
25. Carter ME et al (2012) Mechanism for hypocretin-mediated sleep-to-wake transitions. Proc Natl Acad Sci U S A 109(39):E2635–E2644
26. Mieda M et al (2011) Differential roles of orexin receptor-1 and -2 in the regulation of non-REM and REM sleep. J Neurosci 31(17):6518–6526
27. Eriksson KS et al (2001) Orexin/hypocretin excites the histaminergic neurons of the tuberomammillary nucleus. J Neurosci 21(23):9273–9279
28. Peyron C et al (1998) Neurons containing hypocretin (orexin) project to multiple neuronal systems. J Neurosci 18(23):9996–10015
29. Ciriello J, Caverson MM (2014) Hypothalamic orexin-A (hypocretin-1) neuronal projections to the vestibular complex and cerebellum in the rat. Brain Res 1579:20–34
30. Sakurai T (2007) The neural circuit of orexin (hypocretin): maintaining sleep and wakefulness. Nat Rev Neurosci 8(3):171–181
31. Adamantidis AR et al (2007) Neural substrates of awakening probed with optogenetic control of hypocretin neurons. Nature 450(7168):420–424
32. Eggermann E et al (2001) Orexins/hypocretins excite basal forebrain cholinergic neurones. Neuroscience 108(2):177–181

33. Arrigoni E, Mochizuki T, Scammell TE (2010) Activation of the basal forebrain by the orexin/hypocretin neurones. Acta Physiol (Oxf) 198(3):223–235
34. Hoang QV et al (2014) Orexin (hypocretin) effects on constitutively active inward rectifier K + channels in cultured nucleus basalis neurons. J Neurophysiol 92(6):3183–3191
35. Ferrari LL et al (2015) Dynorphin inhibits basal forebrain cholinergic neurons by pre- and post-synaptic mechanisms. J Physiol 594(4):1069–1085
36. Wu M (2004) Hypocretin/orexin innervation and excitation of identified septohippocampal cholinergic neurons. J Neurosci 24(14):3527–3536
37. Wu M et al (2002) Hypocretin increases impulse flow in the septohippocampal GABAergic pathway: implications for arousal via a mechanism of hippocampal disinhibition. J Neurosci 22(17):7754–7765
38. Mukai K et al (2009) Electrophysiological effects of orexin/hypocretin on nucleus accumbens shell neurons in rats: an in vitro study. Peptides 30(8):1487–1496
39. Martin G et al (2002) Interaction of the hypocretins with neurotransmitters in the nucleus accumbens. Regul Pept 104(1-3):111–117
40. Lambe EK, Aghajanian GK (2003) Hypocretin (orexin) induces calcium transients in single spines postsynaptic to identified thalamocortical boutons in prefrontal slice. Neuron 40 (1):139–150
41. Lambe EK et al (2005) Hypocretin and nicotine excite the same thalamocortical synapses in prefrontal cortex: correlation with improved attention in rat. J Neurosci 25(21):5225–5229
42. Marcus JN et al (2001) Differential expression of orexin receptors 1 and 2 in the rat brain. J Comp Neurol 435(1):6–25
43. Bayer L et al (2004) Exclusive postsynaptic action of hypocretin-orexin on sublayer 6b cortical neurons. J Neurosci 24(30):6760–6764
44. Hay YA et al (2015) Orexin-dependent activation of layer VIb enhances cortical network activity and integration of non-specific thalamocortical inputs. Brain Struct Funct 220:3497–3512
45. Akbari E et al (2011) Orexin-1 receptor mediates long-term potentiation in the dentate gyrus area of freely moving rats. Behav Brain Res 216(1):375–380
46. Wayner MJ et al (2004) Orexin-A (hypocretin-1) and leptin enhance LTP in the dentate gyrus of rats in vivo. Peptides 25(6):991–996
47. Selbach O et al (2010) Orexins/hypocretins control bistability of hippocampal long-term synaptic plasticity through co-activation of multiple kinases. Acta Physiol (Oxf) 198 (3):277–285
48. Bayer L et al (2002) Selective action of orexin (hypocretin) on nonspecific thalamocortical projection neurons. J Neurosci 22(18):7835–7839
49. Lungwitz EA et al (2012) Orexin-A induces anxiety-like behavior through interactions with glutamatergic receptors in the bed nucleus of the stria terminalis of rats. Physiol Behav 107 (5):726–732
50. Conrad KL et al (2012) Yohimbine depresses excitatory transmission in BNST and impairs extinction of cocaine place preference through orexin-dependent, norepinephrine-independent processes. Neuropsychopharmacology 37(10):2253–2266
51. Ishibashi M et al (2005) Effects of orexins/hypocretins on neuronal activity in the paraventricular nucleus of the thalamus in rats in vitro. Peptides 26(3):471–481
52. Huang H, Ghosh P, Pol AN (2006) Prefrontal cortex-projecting glutamatergic thalamic paraventricular nucleus-excited by hypocretin: a feedforward circuit that may enhance cognitive arousal. J Neurophysiol 95(3):1656–1668
53. Kolaj M et al (2007) Orexin-induced modulation of state-dependent intrinsic properties in thalamic paraventricular nucleus neurons attenuates action potential patterning and frequency. Neuroscience 147(4):1066–1075
54. Palus K, Chrobok L, Lewandowski MH (2015) Orexins/hypocretins modulate the activity of NPY-positive and -negative neurons in the rat intergeniculate leaflet via OX1 and OX2 receptors. Neuroscience 300:370–380

55. Chrobok L, Palus K, Lewandowski MH (2015) Orexins excite ventrolateral geniculate nucleus neurons predominantly via OX2 receptors. Neuropharmacology 103:236–246
56. Govindaiah G, Cox CL (2006) Modulation of thalamic neuron excitability by orexins. Neuropharmacology 51(3):414–425
57. Yamanaka A et al (2002) Orexins activate histaminergic neurons via the orexin 2 receptor. Biochem Biophys Res Commun 290(4):1237–1245
58. Schöne C et al (2014) Coreleased orexin and glutamate evoke nonredundant spike outputs and computations in histamine neurons. Cell Rep 7(3):697–704
59. Hoang QV et al (2003) Effects of orexin (hypocretin) on GIRK channels. J Neurophysiol 90 (2):693–702
60. Eriksson KS et al (2004) Orexin (hypocretin)/dynorphin neurons control GABAergic inputs to tuberomammillary neurons. Eur J Neurosci 19(5):1278–1284
61. Schöne C et al (2012) Optogenetic probing of fast glutamatergic transmission from hypocretin/orexin to histamine neurons in situ. J Neurosci 32(36):12437–12443
62. Belle MD et al (2014) Acute suppressive and long-term phase modulation actions of orexin on the mammalian circadian clock. J Neurosci 34(10):3607–3621
63. Kohlmeier KA, Inoue T, Leonard CS (2004) Hypocretin/orexin peptide signaling in the ascending arousal system: elevation of intracellular calcium in the mouse dorsal raphe and laterodorsal tegmentum. J Neurophysiol 92(1):221–235
64. Burlet S, Tyler CJ, Leonard CS (2002) Direct and indirect excitation of laterodorsal tegmental neurons by hypocretin/orexin peptides: implications for wakefulness and narcolepsy. J Neurosci 22(7):2862–2872
65. Kim J et al (2009) Orexin-A and ghrelin depolarize the same pedunculopontine tegmental neurons in rats: an in vitro study. Peptides 30(7):1328–1335
66. Horvath TL et al (1999) Hypocretin (orexin) activation and synaptic innervation of the locus coeruleus noradrenergic system. J Comp Neurol 415(2):145–159
67. Soffin EM et al (2002) SB-334867-A antagonises orexin mediated excitation in the locus coeruleus. Neuropharmacology 42(1):127–133
68. Soffin EM et al (2004) Pharmacological characterisation of the orexin receptor subtype mediating postsynaptic excitation in the rat dorsal raphe nucleus. Neuropharmacology 46 (8):1168–1176
69. Murai Y, Akaike T (2005) Orexins cause depolarization via nonselective cationic and K+ channels in isolated locus coeruleus neurons. Neurosci Res 51(1):55–65
70. Kreibich A et al (2008) Presynaptic inhibition of diverse afferents to the locus ceruleus by kappa-opiate receptors: a novel mechanism for regulating the central norepinephrine system. J Neurosci 28(25):6516–6525
71. Henny P et al (2010) Immunohistochemical evidence for synaptic release of glutamate from orexin terminals in the locus coeruleus. Neuroscience 169(3):1150–1157
72. Brown RE et al (2002) Convergent excitation of dorsal raphe serotonin neurons by multiple arousal systems (orexin/hypocretin, histamine and noradrenaline). J Neurosci 22 (20):8850–8859
73. Liu RJ, Pol AN, Aghajanian GK (2002) Hypocretins (orexins) regulate serotonin neurons in the dorsal raphe nucleus by excitatory direct and inhibitory indirect actions. J Neurosci 22 (21):9453–9464
74. Haj-Dahmane S, Shen RY (2005) The wake-promoting peptide orexin-B inhibits glutamatergic transmission to dorsal raphe nucleus serotonin neurons through retrograde endocannabinoid signaling. J Neurosci 25(4):896–905
75. Bisetti A et al (2006) Excitatory action of hypocretin/orexin on neurons of the central medial amygdala. Neuroscience 142(4):999–1004
76. Pol AN et al (2004) Physiological properties of hypothalamic MCH neurons identified with selective expression of reporter gene after recombinant virus infection. Neuron 42 (4):635–652

77. Li Y, Pol AN (2006) Differential target-dependent actions of coexpressed inhibitory dynorphin and excitatory hypocretin/orexin neuropeptides. J Neurosci 26(50):13037–13047
78. Burdakov D, Liss B, Ashcroft FM (2003) Orexin excites GABAergic neurons of the arcuate nucleus by activating the sodium–calcium exchanger. J Neurosci 23(12):4951–4957
79. Muroya S et al (2004) Orexins (hypocretins) directly interact with neuropeptide Y, POMC and glucose-responsive neurons to regulate Ca 2+ signaling in a reciprocal manner to leptin: orexigenic neuronal pathways in the mediobasal hypothalamus. Eur J Neurosci 19 (6):1524–1534
80. Top M et al (2004) Orexigen-sensitive NPY/AgRP pacemaker neurons in the hypothalamic arcuate nucleus. Nat Neurosci 7(5):493–494
81. Ma X et al (2007) Electrical inhibition of identified anorexigenic POMC neurons by orexin/ hypocretin. J Neurosci 27(7):1529–1533
82. Shirasaka T et al (2001) Orexin depolarizes rat hypothalamic paraventricular nucleus neurons. Am J Physiol Regul Integr Comp Physiol 281(4):R1114–R1118
83. Muschamp JW et al (2014) Hypocretin (orexin) facilitates reward by attenuating the antireward effects of its cotransmitter dynorphin in ventral tegmental area. Proc Natl Acad Sci U S A 111(16):E1648–E1655
84. Uramura K et al (2001) Orexin-a activates phospholipase C- and protein kinase C-mediated Ca2+ signaling in dopamine neurons of the ventral tegmental area. Neuroreport 12 (9):1885–1889
85. Borgland SL, Storm E, Bonci A (2008) Orexin B/hypocretin 2 increases glutamatergic transmission to ventral tegmental area neurons. Eur J Neurosci 28(8):1545–1556
86. Korotkova TM et al (2003) Excitation of ventral tegmental area dopaminergic and nondopaminergic neurons by orexins/hypocretins. J Neurosci 23(1):7–11
87. Li A, Nattie E (2014) Orexin, cardio-respiratory function, and hypertension. Front Neurosci 8:22–22
88. Kohlmeier KA et al (2008) Dual orexin actions on dorsal raphe and laterodorsal tegmentum neurons: noisy cation current activation and selective enhancement of Ca2+ transients mediated by L-type calcium channels. J Neurophysiol 100(4):2265–2281
89. Dias MB, Li A, Nattie EE (2009) Antagonism of orexin receptor-1 in the retrotrapezoid nucleus inhibits the ventilatory response to hypercapnia predominantly in wakefulness. J Physiol 587(Pt 9):2059–2067
90. Shahid IZ, Rahman AA, Pilowsky PM (2012) Orexin and central regulation of cardiorespiratory system. Vitam Horm 89:159–184
91. Lazarenko RM et al (2011) Orexin A activates retrotrapezoid neurons in mice. Respir Physiol Neurobiol 175(2):283–287
92. Young JK et al (2005) Orexin stimulates breathing via medullary and spinal pathways. J Appl Physiol (1985) 98(4):1387–1395
93. Smith BN et al (2002) Selective enhancement of excitatory synaptic activity in the rat nucleus tractus solitarius by hypocretin 2. Neuroscience 115(3):707–714
94. Yang B, Ferguson AV (2003) Orexin-A depolarizes nucleus tractus solitarius neurons through effects on nonselective cationic and K+ conductances. J Neurophysiol 89 (4):2167–2175
95. Azhdari-Zarmehri H, Semnanian S, Fathollahi Y (2015) Orexin-a modulates firing of rat rostral ventromedial medulla neurons: an in vitro study. Cell J 17(1):163–170
96. Huang SC et al (2010) Orexins depolarize rostral ventrolateral medulla neurons and increase arterial pressure and heart rate in rats mainly via orexin 2 receptors. J Pharmacol Exp Ther 334(2):522–529
97. Antunes VR et al (2001) Orexins/hypocretins excite rat sympathetic preganglionic neurons in vivo and in vitro. Am J Physiol Regul Integr Comp Physiol 281(6):R1801–R1807
98. Top M et al (2003) Orexins induce increased excitability and synchronisation of rat sympathetic preganglionic neurones. J Physiol 549(3):809–821

99. Zhang GH et al (2014) Orexin A activates hypoglossal motoneurons and enhances genioglossus muscle activity in rats. Br J Pharmacol 171(18):4233–4246
100. Corcoran A, Richerson G, Harris M (2010) Modulation of respiratory activity by hypocretin-1 (orexin A) in situ and in vitro. Adv Exp Med Biol 669:109–113
101. Ivanov A, Aston-Jones G (2000) Hypocretin/orexin depolarizes and decreases potassium conductance in locus coeruleus neurons. Neuroreport 11(8):1755–1758
102. Peever JH, Lai Y-Y, Siegel JM (2003) Excitatory effects of hypocretin-1 (orexin-A) in the trigeminal motor nucleus are reversed by NMDA antagonism. J Neurophysiol 89(5):2591–2600
103. Yu L et al (2015) Orexin excites rat inferior vestibular nuclear neurons via co-activation of OX1 and OX 2 receptors. J Neural Transm (Vienna) 122(6):747–755
104. Korotkova TM et al (2002) Selective excitation of GABAergic neurons in the substantia nigra of the rat by orexin/hypocretin in vitro. Regul Pept 104(1-3):83–89
105. Yu L et al (2010) Orexins excite neurons of the rat cerebellar nucleus interpositus via orexin 2 receptors in vitro. Cerebellum 9(1):88–95
106. Dergacheva O et al (2012) Orexinergic modulation of GABAergic neurotransmission to cardiac vagal neurons in the brain stem nucleus ambiguus changes during development. Neuroscience 209:12–20
107. Lecea L et al (2006) Addiction and arousal: alternative roles of hypothalamic peptides. J Neurosci 26(41):10372–10375
108. Patyal R, Woo EY, Borgland SL (2012) Local hypocretin-1 modulates terminal dopamine concentration in the nucleus accumbens shell. Front Behav Neurosci 6:82
109. Gonzalez JA et al (2016) Inhibitory interplay between orexin neurons and eating. Curr Biol 26(18):2486–2491
110. Hodgkin AL, Huxley AF (1952) A quantitative description of membrane current and its application to conduction and excitation in nerve. J Physiol 117(4):500–544
111. Koch C (1999) Biophysics of computation. Oxford Universtiy Press, New York
112. Salinas E, Thier P (2000) Gain modulation: a major computational principle of the central nervous system. Neuron 27(1):15–21
113. Mochizuki T et al (2011) Orexin receptor 2 expression in the posterior hypothalamus rescues sleepiness in narcoleptic mice. Proc Natl Acad Sci U S A 108(11):4471–4476
114. Hasegawa E et al (2014) Orexin neurons suppress narcolepsy via 2 distinct efferent pathways. J Clin Invest 124(2):604–616
115. Sakurai T (2013) Orexin deficiency and narcolepsy. Curr Opin Neurobiol 23(5):760–766
116. Kayaba Y et al (2003) Attenuated defense response and low basal blood pressure in orexin knockout mice. Am J Physiol Regul Integr Comp Physiol 285(3):R581–R593
117. Meletti S et al (2015) The brain correlates of laugh and cataplexy in childhood narcolepsy. J Neurosci 35(33):11583–11594
118. Burgess CR et al (2013) Amygdala lesions reduce cataplexy in orexin knock-out mice. J Neurosci 33(23):9734–9742
119. DiStefano J Stubberud A, Williams I (2012) Feedback and control systems, 2nd ed. McGrawHill
120. Csete ME, Doyle JC (2002) Reverse engineering of biological complexity. Science 295:1664–1669
121. Aström K, Hagglund T (1995) PID controllers: theory, design, and tuning. Intrument Society of America (ISA). ISBN-10: 1556175167
122. Uschakov A et al (2011) Sleep-deprivation regulates alpha-2 adrenergic responses of rat hypocretin/orexin neurons. PLoS One 6(2):e16672
123. Yamanaka A et al (2010) Orexin directly excites orexin neurons through orexin 2 receptor. J Neurosci 30(38):12642–12652
124. Ohno K, Sakurai T (2008) Orexin neuronal circuitry: role in the regulation of sleep and wakefulness. Front Neuroendocrinol 29(1):70–87

125. Lu J et al (2006) A putative flip-flop switch for control of REM sleep. Nature 441 (7093):589–594
126. Sakurai T et al (2005) Input of orexin/hypocretin neurons revealed by a genetically encoded tracer in mice. Neuron 46(2):297–308
127. Yoshida K et al (2006) Afferents to the orexin neurons of the rat brain. J Comp Neurol 494 (5):845–861
128. Tsujino N et al (2005) Cholecystokinin activates orexin/hypocretin neurons through the cholecystokinin A receptor. J Neurosci 25(32):7459–7469
129. Gonzalez JA et al (2016) Awake dynamics and brain-wide direct inputs of hypothalamic MCH and orexin networks. Nat Commun 7:11395
130. Schöne C, Burdakov D (2012) Glutamate and GABA as rapid effectors of hypothalamic "peptidergic" neurons. Front Behav Neurosci 6:81

The Hypocretin/Orexin Neuronal Networks in Zebrafish

Idan Elbaz, Talia Levitas-Djerbi, and Lior Appelbaum

Abstract The hypothalamic Hypocretin/Orexin (Hcrt) neurons secrete two Hcrt neuropeptides. These neurons and peptides play a major role in the regulation of feeding, sleep wake cycle, reward-seeking, addiction, and stress. Loss of Hcrt neurons causes the sleep disorder narcolepsy. The zebrafish has become an attractive model to study the Hcrt neuronal network because it is a transparent vertebrate that enables simple genetic manipulation, imaging of the structure and function of neuronal circuits in live animals, and high-throughput monitoring of behavioral performance during both day and night. The zebrafish Hcrt network comprises ~16–60 neurons, which similar to mammals, are located in the hypothalamus and widely innervate the brain and spinal cord, and regulate various fundamental behaviors such as feeding, sleep, and wakefulness. Here we review how the zebrafish contributes to the study of the Hcrt neuronal system molecularly, anatomically, physiologically, and pathologically.

Keywords Behavior · Hypocretin · Narcolepsy · Orexin · Sleep · Zebrafish

Contents

The authors "Idan Elbaz" and "Talia Levitas-Djerbi" contributed equally to this work.

I. Elbaz, T. Levitas-Djerbi, and L. Appelbaum (✉)
The Mina & Everard Goodman Faculty of Life Sciences and The Leslie and Susan Gonda Multidisciplinary Brain Research Center, Bar-Ilan University, Ramat Gan 5290002, Israel
e-mail: lior.appelbaum@biu.ac.il

© Springer International Publishing AG 2016
Curr Topics Behav Neurosci (2017) 33: 75–92
DOI 10.1007/7854_2016_59
Published Online: 18 December 2016

1   Introduction

In all vertebrates, cell populations located in specific hypothalamic nuclei regulate various vital functions such as energy homeostasis, reproduction, and circadian rhythms. Specifically, the lateral hypothalamus area (LHA), from which neuronal populations project widely in the brain, regulates key behaviors including sleep/ wake cycles, feeding, and reward [1–4]. Two decades ago, the LHA-expressed neuropeptides called hypocretins/orexins (Hcrt-1/Orexin-A and Hcrt-2/Orexin-B) were identified [5, 6]. These peptides are cleaved from a common precursor, prepro-Hcrt, encoded by the *hcrt* gene. This gene consists of two exons, and is translated into a 131 and 130 amino acid (aa) protein in humans and rodents, respectively [5, 6]. Hcrt neurons are localized in the LHA and the adjacent periformical area (PFA) in animal and human brains, and project extensively throughout the central nervous system (CNS) [5–12]. In humans, there are between 50,000 and 80,000 Hcrt neurons [12] while in rodents there are a few thousand [13, 14]. The Hcrt neuro-peptides bind to two G protein-coupled Hcrt receptors, HcrtR-1 and HcrtR-2, which have distinct expression patterns in the brain and different binding affinities for Hcrt-1 and Hcrt-2 [6, 15].

The role of the Hcrt system is diverse. Hcrt neurons were first identified as appetite regulators [6] and as sleep/wakefulness regulators in mammals [5, 7]. Confirming their role in sleep regulation, Hcrt knockout mice, Hcrt neuron-ablated mice, and HcrtR-2 deficient dogs showed highly similar phenotypes to human narcolepsy, including fragmented sleep/wake cycles and cataplexy-like behavior [7, 16–18]. Human narco-lepsy is a relatively common sleep disorder characterized by excessive daytime sleep-iness, fragmented sleep during the night, premature transitions to REM sleep, and cataplexy (brief loss of muscle tone triggered by emotional stimuli). The link between a deficient Hcrt system and human narcolepsy was confirmed when an *hcrt* mutation was associated with narcolepsy in a single case, and when an undetectable level of Hcrt-1 in the cerebrospinal fluid (CSF) of narcolepsy patients was found [19]. Furthermore, postmortem studies in narcoleptic patients showed an 80–100% reduction of *hcrt* mRNA-producing neurons or Hcrt immunoreactivity [12, 19, 20]. Altogether, studies in humans and mammalian models have significantly advanced our understanding of the neuroanatomy and function of the Hcrt system. Nevertheless, evolutionarily conserved and simpler experimental models could complement and enhance mammalian data. Here, we review how research of the Hcrt system in zebrafish (*Danio rerio*) has complemented mammalian studies and, importantly, revealed new molecular and cellular mechanistic insights.

2 The Zebrafish as a Model to Study Hcrt Neurons

Fishes have conserved organization and function of the brain, including the hypothalamic–pituitary axis. More specifically, the zebrafish has emerged over the past two decades as an attractive vertebrate model to study genetics, developmental biology, and neurobiology. Zebrafish are easy to maintain, produce a large number of embryos, have rapid external development, and are transparent [21–24]. Furthermore, hundreds of independent animals can be screened simultaneously, making this model well suited for large-scale genetic and behavioral assays [25–27]. Another important feature is the ease of genetic manipulation to establish transgenic zebrafish lines and to overexpress or knockout selected genes using highly efficient techniques such as the tol2 transposon [28] and the clustered regularly interspaced short palindromic repeats (CRISPR/Cas9) systems [29, 30]. Finally, zebrafish have the ability to absorb compounds through the water, making them suitable and advantageous for high-throughput pharmacological screening [31, 32].

In fish, the *hcrt* gene was first identified in two pufferfish species [33], and later in other teleost fish such as goldfish and zebrafish [34–38]. The primary structure of the Hcrt precursor is conserved in fish [15, 34, 36]. The zebrafish *hcrt* gene and Hcrt protein were first identified by Kaslin and colleagues [36], and the 5′ DNA fragment, including the first exon and functional promotor, was later cloned by Faraco and colleagues [34]. In zebrafish, the *hcrt* gene consists of two exons, which, similar to mammals, encodes for two Hcrt neuropeptides. The zebrafish Hcrt-1 and Hcrt-2 peptides are 47 and 28 aa residues long, respectively, and they both show the highest sequence homology with their pufferfish (*Fugu rubripes*) counterparts. In the zebrafish brain, only one hypothalamic cell cluster was found to contain both *hcrt* mRNA and the Hcrt protein [34, 36]. The onset of *hcrt* expression is at 22 h post fertilization (hpf), when it is detected bilaterally in cell clusters of the developing hypothalamus [34]. Unlike in humans, there are only ~16–60 Hcrt-expressing neurons in larvae and adult zebrafish, respectively [34, 36, 39, 40]. This Hcrt-expressing nucleus sends projections to the telencephalon, diencephalon, mesencephalon, and rhombencephalon, toward the noradrenic, dopaminergic, serotonergic, cholinergic, and histaminergic nuclei ([36, 41–44], Fig. 1). Only one Hcrt receptor (Hcrtr) has been identified in zebrafish, and it corresponds structurally to the mammalian HcrtR-2 [15, 40, 45]. The relative simplicity of the zebrafish Hcrt system, which includes a low number of Hcrt neurons and only one Hcrt receptor, makes this vertebrate an ideal model to study the diverse functions of Hcrt's neuronal networks.

3 The Heterogeneous Molecular Profile of Hcrt Neurons

Hcrt neurons regulate diverse processes that include the sleep/wake cycle, feeding, stress response, energy homeostasis, pain, emotion, and reward [4–6, 11, 13]. The marker that defines these neurons is the Hcrt protein, which has a key role in mediating these neuronal functions and interactions. However, it is evident that

Hcrt neurons express additional key proteins that may also regulate these diverse processes. In order to study the molecular identity of Hcrt neurons, a number of studies have been conducted in mammals [46–48]. These studies resulted in a list of genes that are expressed in Hcrt neurons. However, the complexity of the opaque mammalian brain, which includes a few thousand Hcrt cells that are relatively rare and are intermingled with other hypothalamic neurons, makes isolation of the pure Hcrt neuronal population challenging. The zebrafish Hcrt neuronal network [34, 36, 39, 40] enables the study of the molecular signature of Hcrt neurons in a relatively simple, yet evolutionarily conserved system. To assess whether Hcrt cells are excitatory or inhibitory, double in situ hybridization (ISH) of *hcrt* with vesicular glutamate transporter genes *vglut1*, *vglut2a*, *vglut2b* (markers of excitatory glutamatergic neurons), and with the glutamate decarboxylase gene *gad67* (marker of inhibitory GABAergic neurons) was performed in adult brain sections. These experiments showed that, as in mammals [49], the Hcrt neurons are glutamatergic [39]. This result further supports the idea of a conserved role for these neurons among vertebrates.

In order to examine the Hcrt neuron specification, a screen for regulatory factors was conducted in the early stages of zebrafish development [26 h post-fertilization (hpf)], [50]. Microarray gene-expression analysis revealed that similar to mammals [47], the LIM homeobox transcription factor Lhx9, which is widely expressed in the brain, including in the HCRT neurons, can induce the specification of Hcrt neurons [50]. In a recent study, 7-day post-fertilization (dpf) transgenic larvae that express EGFP under the *hcrt* promoter were used to isolate the entire Hcrt neuronal population using fluorescence-activated cell sorting (FACS). Whole transcriptome RNA sequencing (RNA-seq) and extensive anatomical validations were used to identify hundreds of novel Hcrt-neuron-specific transcripts and annotated genes [51]. The functional roles of the genes were diverse and included regulation of metabolism, sleep, synaptogenesis, and synaptic plasticity. Some candidate genes, including *kcnh4a*, *hmx3*, *lhx9*, and *dennd1b*, were found to be expressed in most Hcrt neurons, while other genes were expressed in a subset of Hcrt neurons. Particularly *kcnh4a*, which encodes for a voltage-dependent potassium channel, was located in close proximity to the *hcrt* gene in the zebrafish and mammalian genomes. The genomic location and analysis of transcription factor binding sites in zebrafish and mammals suggested a shared transcription regulation for *hcrt* and *kcnh4a*. *kcnh4a* mutant (*kcnh4a*−/−) fish demonstrated reduced sleep time and a reduction in the number and length of sleep episodes specifically during the night [51]. Similarly, in mouse and fly models, loss of voltage-dependent potassium channels alters sleep architecture [52, 53].

Transcriptome profiling in zebrafish [51] demonstrated that there is, as expected, a battery of genes that are expressed in all or subsets of Hcrt neurons, suggesting that the Hcrt neurons are heterogeneous and that various genes regulate their diverse functions. Supporting this notion, several additional studies showed partial co-localization of various genes with the *hcrt* transcript. For example, *ecto-nucleoside triphosphate diphosphohydrolase 3* (*entpd3*) and *p2rx8* are expressed in a small subset of Hcrt neurons in larvae. These findings suggest that Hcrt neurons mediate purinergic signaling by cell surface P2X receptors and Entpd3, which control the extracellular concentration and accessibility of ATP [54]. As adenosine signaling was linked with sleep regulation in zebrafish [55], these results suggest that some Hcrt neurons may regulate metabolism and the sleep/wake cycle by sensing purine nucleotides. An additional gene that was found to be expressed in a subset of Hcrt neurons is the neuronal activity-regulated pentraxin 2b (*nptx2b*), which encodes for a protein that is implicated in AMPA receptor clustering. Overexpression of *nptx2b* specifically in Hcrt cells abolishes structural synaptic rhythmicity in Hcrt axons and reduces the sensitivity of the larvae to melatonin [56]. Moreover, study of the Leptin-neurotensin (Nts)-Hcrt neuronal circuit demonstrates that a portion of both larvae and adult Hcrt neurons express the Nts receptor [57], suggesting that Nts neurons regulate the function of Hcrt neurons in zebrafish, as is the case in mammals [58]. Altogether, the studies in mammals and zebrafish suggest that the functional diversity of Hcrt neurons can be regulated by their diverse molecular actors. Accordingly, the hypocretinergic system does not necessarily act as one homogenous unit, and different neurons could be responsible for the regulation of different processes.

4 The Hcrt Neuron Anatomy and Neuronal Networks

In accordance with the versatile roles of Hcrt neuropeptides, the Hcrt neurons innervate several areas in the brain (Fig. 1). In mammals, the Hcrt axons regulate centers that control wakefulness, including the paraventricular nucleus (PVN), the arcuate nucleus, dorsal raphe (containing serotonergic neurons), tuberomammillary nucleus (TMN, containing histaminergic neurons), and locus coeruleus (LC, containing noradrenergic neurons); as well as centers that regulate sleep, including the ventrolateral preoptic area [VLPO, containing γ-aminobutyric acid (GABA) neurons], laterodorsal tegmental nucleus, and pedunculopontine tegmental nucleus [LDT/PPT containing acetylcholine (ACh) neurons]. Furthermore, the Hcrt neurons regulate the reward system including the ventral tegmental area (VTA, containing dopamine neurons) and serotonin [5-hydroxy tryptamine (5-HT)] containing neurons in the dorsal raphe nuclei (DR). In addition, the Hcrt neurons mediate the stress response by innervating the limbic system, which includes the amygdala and the bed nucleus of the stria terminalis (BST). Moreover, Hcrt neurons mediate metabolic regulation through the humoral system using the hormones leptin and ghrelin [4–6, 11, 13].

The zebrafish toolkit provides unique capabilities to study the Hcrt neuronal networks. This model enables high-resolution in vivo imaging and the ability to optogenetically manipulate and monitor neuronal activity in any neuron of interest in the entire brain of live larvae [59, 60]. An additional advantage of this model is the ability to perform continuous live imaging of single neurons and synapses in the whole brain [23, 61]. Using two photon imaging and fluorescent synaptic markers, it was shown that the circadian clock and sleep-dependent homeostatic processes regulate rhythmic structural synaptic plasticity in the Hcrt axons [56]. In zebrafish, Hcrt axons project widely throughout the brain including the hypothalamus,

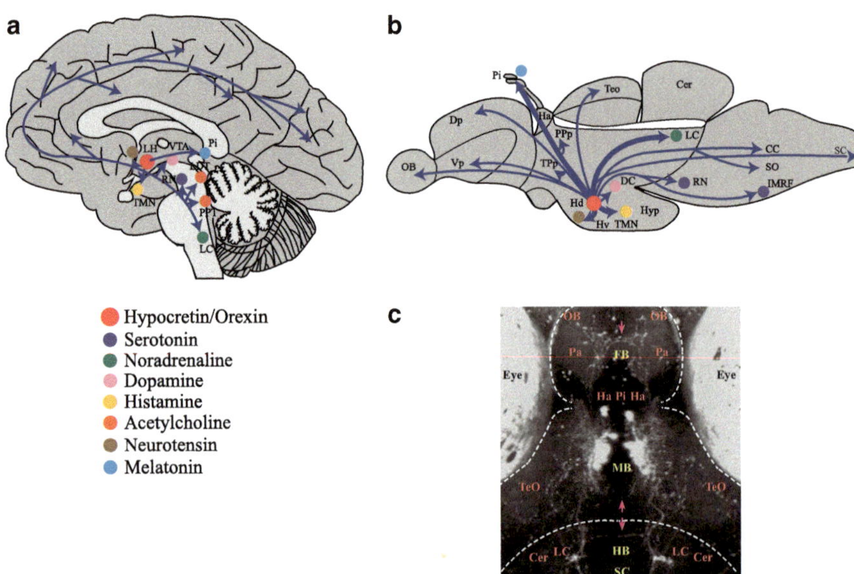

Fig. 1 Comparative illustration of Hcrt neuronal circuitry in the human and zebrafish brain. The approximate location of a number of neuronal centers in the human (**a**) and zebrafish (**b**) brain with emphasis on Hcrt neuron innervations. *Abbreviations*: *LH* lateral hypothalamus, *VTA* ventral tegmental area, *PPT* pedunculopontine, *LDT* laterodorsal tegmental nuclei, *TMN* tuberomammillary nucleus, *RN* raphe nuclei, *LC* locus coeruleus, *OB* olfactory bulb, *Dp* dorsal telencephalic area, *Dv* ventral telencephalic area, *Pi* pineal gland, *Ha* habenula, *PPp* parvocellular preoptic nycleus, posterior part, *TPp* periventricular nucleus of posterior tuberculum, *Hv* ventral zone of periventricular hypothalamus, *Hd* dorsal zone of the periventricular hypothalamus, *TeO* tectum opticum, *Cer* cerebellum, *Hyp* hypothalamus, *DC* dopaminergic diencephalic cluster, *IMRF* intermediate reticular formation, *CC* crista cerebellaris, *SO* secondary octaval population. Note that human and zebrafish brains are not depicted to scale. *Thin arrows* indicate Hcrt neuronal network interactions. *Thick arrows* indicate Hcrt neuronal network interactions studied in zebrafish. (**c**) Dorsal view (head points to top) of a 6 dpf transgenic larvae expressing TagRFP under the *hcrt* promoter, using the UAS/GAL4 system. Maximum projection of confocal imaging, showing the Hcrt cell bodies in the midbrain (MB) and their processes in the hindbrain (HB) and spinal cord (SC) and in the forebrain (FB). *White areas* on both sides of the larva correspond to autofluorescence of the eyes. *Pink arrows* indicate commissural projections

forebrain, habenula-pineal gland complex, hindbrain, and spinal cord (SC) [39] (Fig. 1). The ligand expression and neuronal projection correspond with the expression pattern of the single Hcrtr in the larval and adult zebrafish nervous systems. The *hcrtr* mRNA expression is apparent in wide areas of the brain including the telencephalon, hypothalamus, hypophysis, posterior tuberculum, ventral rhombomere, and SC [40, 45]. Interestingly, *hcrtr* expression is increased in Hcrt neuron-ablated larvae, suggesting that loss of these neurons may trigger compensatory enhancement of *hcrtr* mRNA levels [62]. The anatomical and functional properties of four Hcrt neuronal circuits were studied in more detail in zebrafish. In agreement with findings in rodents [63], in zebrafish, the Hcrt neurons innervate LC noradrenergic neurons, and optogenetic stimulation of Hcrt neurons activates the LC, suggesting a direct and functional inter-neuronal connection [42]. In addition, a direct interaction was found between the Hcrt axons and the habenula-pineal gland complex where *hcrtr* is expressed [39]. Furthermore, as is the case in mammals, Hcrt neurons also innervate and are innervated by the histaminergic system in zebrafish [41, 64]. In the posterior hypothalamus, Hcrt axons innervate the dendrites of histaminergic neurons [44]. Additionally, similar to mammals, the neurotensinergic system interacts with the Hcrt neurons in zebrafish [57]. In mammals, Nts is involved in the regulation of locomotor activity, feeding, sleep, arousal, and the reward system [65–69]. In the zebrafish brain, Nts neurons are located primarily in the preoptic area and the hypothalamus, and project towards a number of Hcrt neurons, which express the Nts receptor [57]. More anatomical and functional studies using high resolution imaging of both Hcrt fibers and *hcrtr*-expressing neurons, as well as cloning of the *hcrtr* promoter are required in order to better characterize the Hcrt neuronal networks in zebrafish.

5 The Role of Hcrt Neurons in Regulating Locomotor Activity and State-Transitions

Like humans, larvae and adult zebrafish are diurnal animals and demonstrate rhythmic locomotor activity that peaks during the day. At 4 dpf, the larvae start to exhibit a stable diurnal rhythm of locomotor activity [70]. An important hallmark of a circadian clock-driven rhythm is that it persists under constant photic conditions. Indeed, the locomotor activity of zebrafish larvae is rhythmic under constant dim light (dim LL) [39, 71]. Similarly, adult zebrafish exhibit rhythmic locomotor activity and increase activity during the subjective day under constant dark (DD) conditions [72].

The role of Hcrt neurons in mediating locomotor activity was studied in a number of reports. Global over-expression of the Hcrt neuropeptide in the whole body increases locomotor activity during both the day and night under LD and DD conditions in 5–7 dpf larvae without altering the rhythmicity of locomotor activity [40]. Overexpression of *nptx2b*, an AMPA receptor clustering regulator, specifically in Hcrt neurons, did not affect the rhythmic locomotor activity of 5 dpf larvae. Notably, the *Tg(hcrt:nptx2b)* larvae were more resistant to the sleep-promoting effect of melatonin and showed higher locomotor activity, presumably because the Nptx2b protein increases the neuronal activity of Hcrt neurons [56]. A study that monitored bioluminescence of the Ca^{2+}-sensitive photoprotein GFP-Aequorin in Hcrt neurons of freely behaving larvae showed that Hcrt-neuronal activity was associated with periods of increased locomotor activity. Furthermore, under DD conditions, the larvae demonstrated spontaneous Hcrt-neuron activity primarily during subjective night/day transitions [73]. In this line, ablation of the Hcrt neurons altered the larvae behavioral response to alternating light/dark and sound stimuli [62]. These findings suggest that Hcrt neurons mediate the behavioral response to external stimuli and state-transitions.

In addition to the ligand, the role of the Hcrt system in regulating locomotor activity was also studied using an *hcrt* receptor mutant (*hcrtr−/−*). At 5 dpf, *hcrtr−/−* larvae showed normal rhythmic locomotor activity under dim LL. Exposure of the *hcrtr−/−* larvae to a subthreshold dose of melatonin (1 µM) showed that loss of *hcrtr* increased the sensitivity to melatonin because the *hcrtr−/−* fish were less active [39]. Altogether, the data suggests that the Hcrt system mediates locomotor activity particularly during state-transitions; however, the intrinsic clock maintains rhythmic locomotor activity even in the absence of the Hcrt neurons.

6   The Role of Hcrt Neurons in Regulating the Sleep/Wake Cycle

Sleep state is associated with a period of quiescence and decreased level of sensory awareness to external stimuli. In addition to behavioral criteria, in mammals, sleep and brain activities are monitored by electroencephalograph (EEG). However, in non-mammalian animals the use of EEG to define sleep stages is challenging. Therefore, in non-mammalian animals, sleep is characterized by the following behavioral criteria: (1) quick reversibility to wakefulness (the ability to distinguish sleep from coma or hibernation), (2) homeostatic control (sleep deprivation will be followed by sleep rebound), (3) a period of immobility associated with a species-specific posture, (4) circadian regulation, and (5) increased arousal threshold to external stimuli (low level of sensory awareness to the environment). These behavioral criteria were also used to define sleep state in zebrafish larvae. A minimum of 1 min of immobility is associated with elevated arousal threshold in 5–7 dpf larvae [40, 62].

The role of the Hcrt system in regulating the sleep/wake cycle in zebrafish was studied using various genetic and neurobiological techniques. Global overexpression of *hcrt* in larvae promoted a high level of arousal and inhibited sleep during the day and night [40]. Supporting this observation, ablation of Hcrt neurons [62, 74] and administration of a *hcrtr* antagonist [75] mildly increased sleep time during the day. Furthermore, sleep was reduced specifically during the night in *kcnh4a* mutant (*kcnh4a−/−*) larvae. This gene is expressed in all Hcrt neurons and encodes for a voltage-gated potassium channel; thus, Hcrt neurons may be hyperactive in *kcnh4a−/−* larvae and inhibit sleep. In line with these studies, specific pharmacological activation of Hcrt neurons reduces sleep time during the night [74]. These data suggest that the Hcrt system in zebrafish regulates the sleep/wake cycle, particularly sleep/wake transitions. Supporting this notion, ablation of Hcrt neurons resulted in an increase in the number of sleep/wake transitions [62, 74] and sleep fragmentation was demonstrated in *hcrtr−/−* adult zebrafish [45]. Furthermore, *hcrt* mutant (*hcrt−/−*) larvae demonstrated increased sleep/wake transitions during the day [75] and specific activation of Hcrt neurons resulted in a lower number of sleep/wake transitions during the night [74]. This idea is further strengthened by the role of neuronal networks that are targeted by and interact with Hcrt neurons. In zebrafish, Hcrt neurons innervate and activate the arousal-mediating norepinephrine neurons [42] and the melatonergic sleep-prompting cells in the pineal gland [39]. These anatomical and functional interactions support the role of Hcrt in regulating sleep/wake transitions and in consolidating both sleep and wakefulness in zebrafish. Similarly, based on mammalian data, it was proposed that the Hcrt neurons are part of the "flip-flop switch," in which they play a crucial role in balancing between wake and sleep states [4]. In line with this proposal, in vivo optogenetic experiments in rodents have demonstrated that Hcrt neuronal activity is most important in sleep-to-wake transitions [76]. Indeed, in narcoleptic human patients and mammalian models for narcolepsy, the total amount of sleep over the day/night cycle is normal but sleep and wakefulness are fragmented [7, 77–80]. Altogether, these studies suggest that in zebrafish, as in mammals, the Hcrt neuronal network regulates behavioral state transitions, such as between sleep and wake.

7 The Role of Hcrt in the Regulation of Feeding

Hcrt neurons are located in areas of the brain which are implicated in the regulation of feeding behavior and energy homeostasis, and central administration of Hcrt peptides stimulates food intake. Additionally, *hcrt* expression was upregulated in rats fasted for 48 h compared with fed control rats [6]. Hcrt-sensitive feeding sites were identified by injection of the peptides into different brain locations. Hcrt-1 was shown to stimulate feeding in the LHA and PFA in rats [81, 82]. Furthermore, Hcrt neurons were found to be glucose sensitive, and it was suggested that they play an important role in energy homeostasis [83]. In goldfish, similar to rodent models, ICV administration of Hcrt-1 stimulated food intake. Expression of prepro-Hcrt mRNA in the brain changed according to feeding status [38]; fasting increased the levels of both Hcrt mRNA and the number of Hcrt immunoreactive neurons. In contrast, the number of Hcrt immunoreactive neurons was decreased in glucose injected fish [38, 84, 85]. In contrast to sleep regulation, the involvement of Hcrt in the feeding behavior of zebrafish has not been extensively studied. One study tested the hypothesis that caloric restriction alters physical activity through its effects on Hcrt levels. Consistent with data in mammals and goldfish, results indicated that while short term fasting had no significant effect on prepro-Hcrt mRNA in the zebrafish brain, long term fasting (14 days) was associated with a significant increase in prepro-Hcrt levels [86]. Supporting this notion, prepro-Hcrt levels were higher in fish fasted for 7 days than those in fish that had been fed normally. Moreover, food intake was significantly increased by ICV administration of synthetic human Hcrt-1, which shares 32% amino acid sequence identity with its zebrafish counterpart. Co-injection of the Hcrt receptor antagonist completely suppressed the action of human Hcrt-1, confirming that the Hcrt-1 induced action in zebrafish is mediated through an endogenous Hcrt receptor [87]. However, in a satiety study performed in *hcrtr−/−* adult fish under both normal conditions and after food restriction, no phenotype was detected [45]. Thus, the effect of Hcrt on feeding in adult fish should be further explored using manipulation of the ligand or acute manipulation of the neurons; however, the limited data in larvae suggests that similar to mammals, the Hcrt system induces food intake in zebrafish.

Another feeding-regulating hormone is leptin, a satiety hormone that regulates energy balance by suppressing food intake [88]. In mammals, leptin was shown to inhibit Hcrt neurons both directly and indirectly, via Nts neurons [58]. In zebrafish, Nts neurons co-express the leptin receptor, and a portion of Hcrt neurons express the Nts receptor, thus establishing structural leptin-Nts-Hcrt circuitry [57]. These results suggest that in zebrafish, as in mammals, the hypothalamic Hcrt neuronal circuits have an important role in feeding behavior and energy expenditure.

8 The Role of Hcrt in Stress Regulation

The interactions between Hcrt and the stress systems are well conserved in mammals and zebrafish. In recent years, the use of zebrafish as a model of stress has become increasingly common. The main marker for the stress response in zebrafish, as in humans, is cortisol hormone levels. The genes involved in the hypothalamus–pituitary–interrenal (HPI) axis [equivalent to the mammalian hypothalamic–pituitary–adrenal (HPA) axis] are expressed very early in development. Larval cortisol levels begin to rise at hatch, while an elevated whole-body cortisol concentration after exposure to stressors is observed at 4 dpf [89]. In mammals, corticotropin-releasing factor (CRF) is released from PVN, activating the HPA stress axis and causing an increase in levels of adrenocorticotropin hormone (ACTH) and cortisol. Similarly, in fish, hypothalamic CRF is released from the preoptic area (PO), an analogue of the mammalian PVN, into the anterior lobe of the pituitary gland where it binds to its receptor, located in ACTH secreted corticotropic cells. ACTH binds to the melanocortin 2 receptor and triggers the release of cortisol from the interrenal gland [89].

The effect of the Hcrt neurons and peptide on the modulation of the stress system has been studied in mammals and zebrafish. In rodents, ICV Hcrt administration stimulates the release of stress hormones, causing elevation of ACTH and corticosterone levels, and enhances CRF-mediated anxiety behavior [90–94]. An antagonist for HcrtR-1 was shown to attenuate stressor-induced increases in ACTH secretion [95, 96]. Selective optogenetic activation of Hcrt neurons in mice was sufficient to drive stress responses, including HPA axis activation [97]. Furthermore, the leptin hormone was shown to attenuate Hcrt-mediated HPA axis activation through a network of leptin receptor (LepRb)-expressing inhibitory neurons [97]. These results in mammals demonstrate that Hcrt neurons have an important role in the modulation of stress and anxiety-related behaviors.

In zebrafish, whole body cortisol levels are elevated after exposure to stressors such as acute net handling, chronic crowding, and exposure to other anxiogenic factors [98, 99]. Zebrafish often form dominant-subordinate relationships and display agonistic or anxiety-like behavior [100, 101]. Both dominant and subordinate fish showed higher whole-body cortisol concentrations than controls [102]. In socially dominant zebrafish, *hcrt* levels increased compared to subordinate and control zebrafish. This provides more evidence of a stimulatory role for Hcrt in the HPI axis. Furthermore, there was a positive correlation of mRNA expression levels between Hcrt and the HPI stress axis related genes glucocorticoid receptor (*nr3c1*), mineralocorticoid receptor (*nr3c2*), and *crf* [102]. Moreover, acute and chronic stress of high intensity, consisting of optical, husbandry, and social stressors, resulted in increased levels of Hcrt mRNA [103]. This function of Hcrt in the stress response may be through the cholinergic and/or aminergic systems [36], or through CRF action on Hcrt neurons [95]. These results demonstrate that Hcrt is a modulator of the neural circuitry of stress, and may be used as a biomarker to define the regulation of stress and allostatic load in both zebrafish and mammals.

9 Disorders and Pharmacology

The hypocretinergic system has been implicated in several neurological disorders. The hallmark Hcrt neuron-related disorder is narcolepsy-cataplexy, which is caused by elimination of Hcrt-producing neurons in humans [4, 19, 104] and mutation of *hcrtr* in dogs [17]. This pathological condition has been proposed to be caused by an auto-immune process [105, 106]. Therefore, the Hcrt system is primarily implicated in the mechanisms that regulate sleep disorders; however, accumulating data point to the involvement of the Hcrt system in neurodegenerative diseases as well. Loss of Hcrt neurons has been identified in both Parkinson's and Alzheimer's disease, and is correlated with the progress of the disease [107–109]. A disruption of the hypocretinergic system has also been identified in other brain deficiencies including post-traumatic injury, dementia with Lewy bodies, and Prader–Willi syndrome. Since disruptions of the sleep/wake cycle are associated with many additional psychiatric diseases, such as depression, bipolar disorder, and schizophrenia, the Hcrt neuropeptide has been suggested as a potential biomarker. For example, in depression, insomnia can precede the disease and thus Hcrt can be used as an early indicator [110]. Therefore, the Hcrt system could be an effective target for drug development and treatment of sleep and other neurological disorders. Thus, further study of the role of Hcrt neurons in these diseases may provide new insights into our understanding of these and other interacting neuronal systems.

The zebrafish provides a powerful tool to understand the mechanisms of disease and for drug discovery because of their small size, transparency, neuroanatomical and physiological homology to mammals, robust phenotypes, and ease of genetic manipulation that enable the establishment of disease models. A major advantage of the zebrafish model in pharmacological studies is that they are able to absorb a wide range of chemicals from the medium in which they swim [31]. A recent bioinformatic study in zebrafish identified *hcrt* as an important candidate for targeting drug development with application to Alzheimer's disease clinical trials

[111]. In another study, an *hcrt−/−* zebrafish was generated in order to determine whether profiling of zebrafish behavior could be used to classify sleep–wake modifiers corresponding to their mode of action [75]. Indeed, the drug Modafinil was delivered to larval zebrafish and was shown to reduce sleep in a dose-dependent manner, consistent with data in mammals [112]. In mammals, Modafinil administration results in a pattern that activates arousal-related cell groups, including Hcrt cells [113]. Additionally, Modafinil administration results in elevated histamine release via activation of Hcrt cells [114]. In zebrafish, the interaction between the Hcrt and histamine systems requires further study; however, it has been suggested that the wake inducing effect of Modafinil is mediated by the dopaminergic and norepinephrinergic systems [112, 114]. Supporting this idea, in zebrafish, these systems are innervated by Hcrt neurons [36, 41, 42].

While the mechanisms of action of certain chemicals on the Hcrt system as well as the involvement of Hcrt in different diseases are still being characterized in mammals, zebrafish offer a unique opportunity to define a basic form of behavioral regulation for both simple and complex behaviors, which are comparable with human behavior at a functional level [115, 116]. Due to the small size of larval zebrafish and the ease of administration of chemicals, these drug- or disease-related behaviors can be easily assessed using a high-throughput video tracking system, allowing observation of abnormal phenotypes [117]. This makes the zebrafish an important and unique model for the study of the mechanisms involved in Hcrt-related disorders, as a potential treatment, or even as an early biomarker for a range of neurodegenerative and psychiatric diseases.

10 Conclusions and Future Perspective

In zebrafish, similar to mammals, the hypocretinergic system is conserved and comprises the afferent and efferent connections to the main neuromodulatory nuclei. These neurons regulate behavioral state transitions and consolidate sleep. Loss of Hcrt neurons in zebrafish results in narcolepsy-like symptoms, including wake and sleep fragmentation, and they are also involved in feeding and stress regulation.

The effect of Hcrt neurons on the sleep/wake cycle is the main mechanism studied in zebrafish. Clearly, there is a need to further study how the Hcrt system regulates additional processes including feeding, stress, reward, fear, memory, and addiction. In addition, since the zebrafish is an ideal neurodevelopmental model, further studies should focus on basic questions regarding Hcrt neuron development, such as why the ligand is expressed prior to formation of neurites and synapses. Furthermore, the anatomical and functional interaction between Hcrt neurons and other key neuronal systems such as the dopaminergic, GABAergic, serotonergic, and histaminergic nuclei require further research in zebrafish. The zebrafish toolbox comprises vast opportunities to further reveal Hcrt neuron development and neuron connectomics using molecular, genetic, pharmacological, and optical methods. For example, the CRISPR method can be used to establish mutant fish lines for genes that are specifically expressed in Hcrt neurons or in neurons that anatomically interact with Hcrt neurons [42, 51]. Furthermore, zebrafish are the only vertebrate that enable optically monitored brain-wide activity at a single-cell resolution during behavior [118–120]. In addition, the zebrafish enables high-throughput behavioral and pharmacological studies that allow screening of thousands of neuroactive chemical compounds in a relatively short period of time [121, 122]. These methods, together with activation or suppression of neurons using optogenetics for specific circuits in parallel to behavior monitoring, can contribute to the understanding of the role of Hcrt neurons in controlling fundamental processes. Special emphasis should be on drug research [31] for the treatment of narcolepsy, neurodegenerative and psychiatric diseases, and to place the zebrafish as a first-line for further studies in mammals. These studies are likely to elucidate the role of the hypocretinergic system and may provide treatment for various brain disorders.

Acknowledgments This work was supported by the Israel Science Foundation (grant no. 690/15), the Legacy Heritage Biomedical Program of the Israel Science Foundation (grant no. 992/14), and by the US-Israel Binational Science Foundation (BSF, grant no. 2011335). IE is supported by the Nehemia Levtzion scholarship from the Council for Higher Education, Israel.

References

1. Burt J, Alberto CO, Parsons MP, Hirasawa M (2011) Local network regulation of orexin neurons in the lateral hypothalamus. Am J Physiol Regul Integr Comp Physiol 301:R572–R580
2. DiLeone RJ, Georgescu D, Nestler EJ (2003) Lateral hypothalamic neuropeptides in reward and drug addiction. Life Sci 73:759–768
3. Saper CB (2006) Staying awake for dinner: hypothalamic integration of sleep, feeding, and circadian rhythms. Prog Brain Res 153:243–252
4. Saper CB, Scammell TE, Lu J (2005) Hypothalamic regulation of sleep and circadian rhythms. Nature 437:1257–1263
5. de Lecea L, Kilduff TS, Peyron C, Gao X, Foye PE, Danielson PE, Fukuhara C, Battenberg EL, Gautvik VT, Bartlett FS, et al. (1998) The hypocretins: hypothalamus-specific peptides with neuroexcitatory activity. Proc Natl Acad Sci U S A 95:322–327
6. Sakurai T, Amemiya A, Ishii M, Matsuzaki I, Chemelli RM, Tanaka H, Williams SC, Richardson JA, Kozlowski GP, Wilson S, et al. (1998) Orexins and orexin receptors: a family of hypothalamic neuropeptides and G protein-coupled receptors that regulate feeding behavior. Cell 92:573–585
7. Chemelli RM, Willie JT, Sinton CM, Elmquist JK, Scammell T, Lee C, Richardson JA, Williams SC, Xiong Y, Kisanuki Y, et al. (1999) Narcolepsy in orexin knockout mice: molecular genetics of sleep regulation. Cell 98:437–451
8. Date Y, Ueta Y, Yamashita H, Yamaguchi H, Matsukura S, Kangawa K, Sakurai T, Yanagisawa M, Nakazato M (1999) Orexins, orexigenic hypothalamic peptides, interact with autonomic, neuroendocrine and neuroregulatory systems. Proc Natl Acad Sci U S A 96:748–753

9. Elias CF, Saper CB, Maratos-Flier E, Tritos NA, Lee C, Kelly J, Tatro JB, Hoffman GE, Ollmann MM, Barsh GS, et al. (1998) Chemically defined projections linking the mediobasal hypothalamus and the lateral hypothalamic area. J Comp Neurol 402:442–459

10. Nambu T, Sakurai T, Mizukami K, Hosoya Y, Yanagisawa M, Goto K (1999) Distribution of orexin neurons in the adult rat brain. Brain Res 827:243–260

11. Peyron C, Tighe DK, van den Pol AN, de Lecea L, Heller HC, Sutcliffe JG, Kilduff TS (1998) Neurons containing hypocretin (orexin) project to multiple neuronal systems. J Neurosci 18:9996–10015

12. Thannickal TC, Moore RY, Nienhuis R, Ramanathan L, Gulyani S, Aldrich M, Cornford M, Siegel JM (2000) Reduced number of hypocretin neurons in human narcolepsy. Neuron 27:469–474

13. Sakurai T (2014) The role of orexin in motivated behaviours. Nat Rev Neurosci 15:719–731

14. Tabuchi S, Tsunematsu T, Black SW, Tominaga M, Maruyama M, Takagi K, Minokoshi Y, Sakurai T, Kilduff TS, Yamanaka A (2014) Conditional ablation of orexin/hypocretin neurons: a new mouse model for the study of narcolepsy and orexin system function. J Neurosci 34:6495–6509

15. Wong KKY, Ng SYL, Lee LTO, Ng HKH, Chow BKC (2011) Orexins and their receptors from fish to mammals: a comparative approach. Gen Comp Endocrinol 171:124–130

16. Hara J, Beuckmann CT, Nambu T, Willie JT, Chemelli RM, Sinton CM, Sugiyama F, Yagami K, Goto K, Yanagisawa M, et al. (2001) Genetic ablation of orexin neurons in mice results in narcolepsy, hypophagia, and obesity. Neuron 30:345–354

17. Lin L, Faraco J, Li R, Kadotani H, Rogers W, Lin X, Qiu X, de Jong PJ, Nishino S, Mignot E (1999) The sleep disorder canine narcolepsy is caused by a mutation in the hypocretin (orexin) receptor 2 gene. Cell 98:365–376

18. Mochizuki T, Crocker A, McCormack S, Yanagisawa M, Sakurai T, Scammell TE (2004) Behavioral state instability in orexin knock-out mice. J Neurosci 24:6291–6300

19. Peyron C, Faraco J, Rogers W, Ripley B, Overeem S, Charnay Y, Nevsimalova S, Aldrich M, Reynolds D, Albin R, et al. (2000) A mutation in a case of early onset narcolepsy and a generalized absence of hypocretin peptides in human narcoleptic brains. Nat Med 6:991–997

20. Nishino S, Ripley B, Overeem S, Lammers GJ, Mignot E (2000) Hypocretin (orexin) deficiency in human narcolepsy. Lancet 355:39–40

21. Fernandes AM, Fero K, Driever W, Burgess HA (2013) Enlightening the brain: linking deep brain photoreception with behavior and physiology. Bioessays 35:775–779

22. Varshney GK, Burgess SM (2014) Mutagenesis and phenotyping resources in zebrafish for studying development and human disease. Brief Funct Genomics 13:82–94

23. Wang G, Grone B, Colas D, Appelbaum L, Mourrain P (2011) Synaptic plasticity in sleep: learning, homeostasis and disease. Trends Neurosci 34:452–463

24. Wolman M, Granato M (2012) Behavioral genetics in larval zebrafish: learning from the young. Dev Neurobiol 72:366–372

25. Leung LC, Wang GX, Mourrain P (2013) Imaging zebrafish neural circuitry from whole brain to synapse. Front Neural Circuits 7:76

26. Maximino C, de Brito TM, da Silva Batista AW, Herculano AM, Morato S, Gouveia A (2010) Measuring anxiety in zebrafish: a critical review. Behav Brain Res 214:157–171

27. Norton W, Bally-Cuif L (2010) Adult zebrafish as a model organism for behavioural genetics. BMC Neurosci 11:90

28. Kawakami K (2007) Tol2: a versatile gene transfer vector in vertebrates. Genome Biol 8:S7

29. Auer TO, Del Bene F (2014) CRISPR/Cas9 and TALEN-mediated knock-in approaches in zebrafish. Methods 69:142–150

30. Kleinstiver BP, Prew MS, Tsai SQ, Topkar VV, Nguyen NT, Zheng Z, Gonzales APW, Li Z, Peterson RT, Yeh J-RJ, et al. (2015) Engineered CRISPR-Cas9 nucleases with altered PAM specificities. Nature 523:481–485

31. MacRae CA, Peterson RT (2015) Zebrafish as tools for drug discovery. Nat Rev Drug Discov 14:721–731

32. Rubinstein AL (2006) Zebrafish assays for drug toxicity screening. Expert Opin Drug Metab Toxicol 2:231–240
33. Alvarez CE, Sutcliffe JG (2002) Hypocretin is an early member of the incretin gene family. Neurosci Lett 324:169–172
34. Faraco JH, Appelbaum L, Marin W, Gaus SE, Mourrain P, Mignot E (2006) Regulation of hypocretin (orexin) expression in embryonic zebrafish. J Biol Chem 281:29753–29761
35. Huesa G, van den Pol AN, Finger TE (2005) Differential distribution of hypocretin (orexin) and melanin-concentrating hormone in the goldfish brain. J Comp Neurol 488:476–491
36. Kaslin J, Nystedt JM, Ostergård M, Peitsaro N, Panula P (2004) The orexin/hypocretin system in zebrafish is connected to the aminergic and cholinergic systems. J Neurosci 24:2678–2689
37. Matsuda K, Azuma M, Kang KS (2012) Orexin system in teleost fish. Vitam Horm 89:341–361
38. Nakamachi T, Matsuda K, Maruyama K, Miura T, Uchiyama M, Funahashi H, Sakurai T, Shioda S (2006) Regulation by orexin of feeding behaviour and locomotor activity in the goldfish. J Neuroendocrinol 18:290–297
39. Appelbaum L, Wang GX, Maro GS, Mori R, Tovin A, Marin W, Yokogawa T, Kawakami K, Smith SJ, Gothilf Y, et al. (2009) Sleep-wake regulation and hypocretinmelatonin interaction in zebrafish. Proc Natl Acad Sci U S A 106:21942–21947
40. Prober DA, Rihel J, Onah AA, Sung RJ, Schier AF (2006) Hypocretin/orexin overexpression induces an insomnia-like phenotype in zebrafish. J Neurosci 26:13400–13410
41. Panula P, Chen Y-C, Priyadarshini M, Kudo H, Semenova S, Sundvik M, Sallinen V (2010) The comparative neuroanatomy and neurochemistry of zebrafish CNS systems of relevance to human neuropsychiatric diseases. Neurobiol Dis 40:46–57
42. Singh C, Oikonomou G, Prober DA (2015) Norepinephrine is required to promote wakefulness and for hypocretin-induced arousal in zebrafish. eLife 4:e07000
43. Sundvik M, Panula P (2015) Interactions of the orexin/hypocretin neurones and the histaminergic system. Acta Physiol (Oxf) 213:321–333
44. Sundvik M, Kudo H, Toivonen P, Rozov S, Chen Y-C, Panula P (2011) The histaminergic system regulates wakefulness and orexin/hypocretin neuron development via histamine receptor H1 in zebrafish. FASEB J 25:4338–4347
45. Yokogawa T, Marin W, Faraco J, Pézeron G, Appelbaum L, Zhang J, Rosa F, Mourrain P, Mignot E (2007) Characterization of sleep in zebrafish and insomnia in hypocretin receptor mutants. PLoS Biol 5:e277
46. Cvetkovic-Lopes V, Bayer L, Dorsaz S, Maret S, Pradervand S, Dauvilliers Y, Lecendreux M, Lammers G-J, Donjacour CEHM, Du Pasquier RA, et al. (2010) Elevated Tribbles homolog 2-specific antibody levels in narcolepsy patients. J Clin Invest 120:713–719
47. Dalal J, Roh JH, Maloney SE, Akuffo A, Shah S, Yuan H, Wamsley B, Jones WB, de Guzman Strong C, Gray PA, et al. (2013) Translational profiling of hypocretin neurons identifies candidate molecules for sleep regulation. Genes Dev 27:565–578
48. Honda M, Arai T, Fukazawa M, Honda Y, Tsuchiya K, Salehi A, Akiyama H, Mignot E (2009) Absence of ubiquitinated inclusions in hypocretin neurons of patients with narcolepsy. Neurology 73:511–517
49. Rosin DL, Weston MC, Sevigny CP, Stornetta RL, Guyenet PG (2003) Hypothalamic orexin (hypocretin) neurons express vesicular glutamate transporters VGLUT1 or VGLUT2. J Comp Neurol 465:593–603
50. Liu J, Merkle FT, Gandhi AV, Gagnon JA, Woods IG, Chiu CN, Shimogori T, Schier AF, Prober DA (2015) Evolutionarily conserved regulation of hypocretin neuron specification by Lhx9. Development 142:1113–1124
51. Yelin-Bekerman L, Elbaz I, Diber A, Dahary D, Gibbs-Bar L, Alon S, Lerer-Goldshtein T, Appelbaum L (2015) Hypocretin neuron-specific transcriptome profiling identifies the sleep modulator Kcnh4a. Elife 4:e08638
52. Cirelli C, Bushey D, Hill S, Huber R, Kreber R, Ganetzky B, Tononi G (2005) Reduced sleep in Drosophila Shaker mutants. Nature 434:1087–1092

53. Douglas CL, Vyazovskiy V, Southard T, Chiu S-Y, Messing A, Tononi G, Cirelli C (2007) Sleep in Kcna2 knockout mice. BMC Biol 5:42

54. Appelbaum L, Skariah G, Mourrain P, Mignot E (2007) Comparative expression of p2x receptors and ecto-nucleoside triphosphate diphosphohydrolase 3 in hypocretin and sensory neurons in zebrafish. Brain Res 1174:66–75

55. Gandhi AV, Mosser EA, Oikonomou G, Prober DA (2015) Melatonin is required for the circadian regulation of sleep. Neuron 85:1193–1199

56. Appelbaum L, Wang G, Yokogawa T, Skariah GM, Smith SJ, Mourrain P, Mignot E (2010) Circadian and homeostatic regulation of structural synaptic plasticity in hypocretin neurons. Neuron 68:87–98

57. Levitas-Djerbi T, Yelin-Bekerman L, Lerer-Goldshtein T, Appelbaum L (2015) Hypothalamic leptin-neurotensin-hypocretin neuronal networks in zebrafish. J Comp Neurol 523:831–848

58. Leininger GM, Opland DM, Jo Y-H, Faouzi M, Christensen L, Cappellucci LA, Rhodes CJ, Gnegy ME, Becker JB, Pothos EN, et al. (2011) Leptin action via neurotensin neurons controls orexin, the mesolimbic dopamine system and energy balance. Cell Metab 14:313–323

59. Portugues R, Severi KE, Wyart C, Ahrens MB (2013) Optogenetics in a transparent animal: circuit function in the larval zebrafish. Curr Opin Neurobiol 23:119–126

60. Wyart C, Del Bene F (2011) Let there be light: zebrafish neurobiology and the optogenetic revolution. Rev Neurosci 22:121–130

61. Elbaz I, Foulkes NS, Gothilf Y, Appelbaum L (2013) Circadian clocks, rhythmic synaptic plasticity and the sleep-wake cycle in zebrafish. Front Neural Circuits 7:9

62. Elbaz I, Yelin-Bekerman L, Nicenboim J, Vatine G, Appelbaum L (2012) Genetic ablation of hypocretin neurons alters behavioral state transitions in zebrafish. J Neurosci 32:12961–12972

63. Carter ME, de Lecea L, Adamantidis A (2013) Functional wiring of hypocretin and LC-NE neurons: implications for arousal. Front Behav Neurosci 7:43

64. Sundvik M, Panula P (2012) Organization of the histaminergic system in adult zebrafish (*Danio rerio*) brain: neuron number, location, and cotransmitters. J Comp Neurol 520:3827–3845

65. Boules M, Li Z, Smith K, Fredrickson P, Richelson E (2013) Diverse roles of neurotensin agonists in the central nervous system. Front Endocrinol 4:36

66. Furutani N, Hondo M, Kageyama H, Tsujino N, Mieda M, Yanagisawa M, Shioda S, Sakurai T (2013) Neurotensin co-expressed in orexin-producing neurons in the lateral hypothalamus plays an important role in regulation of sleep/wakefulness states. PLoS One 8:e62391

67. Kleczkowska P, Lipkowski AW (2013) Neurotensin and neurotensin receptors: characteristic, structure-activity relationship and pain modulation – a review. Eur J Pharmacol 716:54–60

68. Mustain WC, Rychahou PG, Evers BM (2011) The role of neurotensin in physiologic and pathologic processes. Curr Opin Endocrinol Diabetes Obes 18:75–82

69. Vincent JP, Mazella J, Kitabgi P (1999) Neurotensin and neurotensin receptors. Trends Pharmacol Sci 20:302–309

70. Cahill GM, Hurd MW, Batchelor MM (1998) Circadian rhythmicity in the locomotor activity of larval zebrafish. Neuroreport 9:3445–3449

71. Tovin A, Alon S, Ben-Moshe Z, Mracek P, Vatine G, Foulkes NS, Jacob-Hirsch J, Rechavi G, Toyama R, Coon SL, et al. (2012) Systematic identification of rhythmic genes reveals camk1gb as a new element in the circadian clockwork. PLoS Genet 8:e1003116

72. Hurd MW, Debruyne J, Straume M, Cahill GM (1998) Circadian rhythms of locomotor activity in zebrafish. Physiol Behav 65:465–472

73. Naumann EA, Kampff AR, Prober DA, Schier AF, Engert F (2010) Monitoring neural activity with bioluminescence during natural behavior. Nat Neurosci 13:513–520

74. Chen S, Chiu CN, McArthur KL, Fetcho JR, Prober DA (2016) TRP channel mediated neuronal activation and ablation in freely behaving zebrafish. Nat Methods 13:147–150

75. Nishimura Y, Okabe S, Sasagawa S, Murakami S, Ashikawa Y, Yuge M, Kawaguchi K, Kawase R, Tanaka T (2015) Pharmacological profiling of zebrafish behavior using chemical and genetic classification of sleep-wake modifiers. Front Pharmacol 6:257

76. Adamantidis AR, Zhang F, Aravanis AM, Deisseroth K, de Lecea L (2007) Neural substrates of awakening probed with optogenetic control of hypocretin neurons. Nature 450:420–424
77. Hishikawa Y, Wakamatsu H, Furuya E, Sugita Y, Masaoka S (1976) Sleep satiation in narcoleptic patients. Electroencephalogr Clin Neurophysiol 41:1–18
78. Mitler MM, Dement WC (1977) Sleep studies on canine narcolepsy: pattern and cycle comparisons between affected and normal dogs. Electroencephalogr Clin Neurophysiol 43:691–699
79. Montplaisir J, Billiard M, Takahashi S, Bell IR, Guilleminault C, Dement WC (1978) Twenty-four-hour recording in REM-narcoleptics with special reference to nocturnal sleep disruption. Biol Psychiatry 13:73–89
80. Zhang S, Zeitzer JM, Sakurai T, Nishino S, Mignot E (2007) Sleep/wake fragmentation disrupts metabolism in a mouse model of narcolepsy. J Physiol 581:649–663
81. Kiwaki K, Kotz CM, Wang C, Lanningham-Foster L, Levine JA (2004) Orexin A (hypocretin 1) injected into hypothalamic paraventricular nucleus and spontaneous physical activity in rats. Am J Physiol Endocrinol Metab 286:E551–E559
82. Sweet DC, Levine AS, Billington CJ, Kotz CM (1999) Feeding response to central orexins. Brain Res 821:535–538
83. Williams RH, Alexopoulos H, Jensen LT, Fugger L, Burdakov D (2008) Adaptive sugar sensors in hypothalamic feeding circuits. Proc Natl Acad Sci U S A 105:11975–11980
84. Matsuda K, Kang KS, Sakashita A, Yahashi S, Vaudry H (2011) Behavioral effect of neuropeptides related to feeding regulation in fish. Ann N Y Acad Sci 1220:117–126
85. Volkoff H, Bjorklund JM, Peter RE (1999) Stimulation of feeding behavior and food consumption in the goldfish, *Carassius auratus*, by orexin-A and orexin-B. Brain Res 846:204–209
86. Novak CM, Jiang X, Wang C, Teske JA, Kotz CM, Levine JA (2005) Caloric restriction and physical activity in zebrafish (*Danio rerio*). Neurosci Lett 383:99–104
87. Yokobori E, Kojima K, Azuma M, Kang KS, Maejima S, Uchiyama M, Matsuda K (2011) Stimulatory effect of intracerebroventricular administration of orexin A on food intake in the zebrafish, *Danio rerio*. Peptides 32:1357–1362
88. Klok MD, Jakobsdottir S, Drent ML (2007) The role of leptin and ghrelin in the regulation of food intake and body weight in humans: a review. Obes Rev 8:21–34
89. Alsop D, Vijayan MM (2009) Molecular programming of the corticosteroid stress axis during zebrafish development. Comp Biochem Physiol A Mol Integr Physiol 153:49–54
90. Berridge CW, España RA, Vittoz NM (2010) Hypocretin/orexin in arousal and stress. Brain Res 1314:91–102
91. Johnson PL, Truitt W, Fitz SD, Minick PE, Dietrich A, Sanghani S, Träskman-Bendz L, Goddard AW, Brundin L, Shekhar A (2010) A key role for orexin in panic anxiety. Nat Med 16:111–115
92. Kuru M, Ueta Y, Serino R, Nakazato M, Yamamoto Y, Shibuya I, Yamashita H (2000) Centrally administered orexin/hypocretin activates HPA axis in rats. Neuroreport 11:1977–1980
93. Russell SH, Small CJ, Dakin CL, Abbott CR, Morgan DG, Ghatei MA, Bloom SR (2001) The central effects of orexin-A in the hypothalamic-pituitary-adrenal axis in vivo and in vitro in male rats. J Neuroendocrinol 13:561–566
94. Suzuki M, Beuckmann CT, Shikata K, Ogura H, Sawai T (2005) Orexin-A (hypocretin-1) is possibly involved in generation of anxiety-like behavior. Brain Res 1044:116–121
95. Johnson PL, Molosh A, Fitz SD, Truitt WA, Shekhar A (2012) Orexin, stress, and anxiety/panic states. Prog Brain Res 198:133–161
96. Samson WK, Bagley SL, Ferguson AV, White MM (2007) Hypocretin/orexin type 1 receptor in brain: role in cardiovascular control and the neuroendocrine response to immobilization stress. Am J Physiol Regul Integr Comp Physiol 292:R382–R387
97. Bonnavion P, Jackson AC, Carter ME, de Lecea L (2015) Antagonistic interplay between hypocretin and leptin in the lateral hypothalamus regulates stress responses. Nat Commun 6:6266
98. Ramsay JM, Feist GW, Varga ZM, Westerfield M, Kent ML, Schreck CB (2006) Whole-body cortisol is an indicator of crowding stress in adult zebrafish, *Danio rerio*. Aquaculture 258:565–574

99. Ramsay JM, Feist GW, Varga ZM, Westerfield M, Kent ML, Schreck CB (2009) Whole-body cortisol response of zebrafish to acute net handling stress. Aquaculture 297:157–162
100. Larson ET, O'Malley DM, Melloni RH (2006) Aggression and vasotocin are associated with dominant-subordinate relationships in zebrafish. Behav Brain Res 167:94–102
101. Spence R, Gerlach G, Lawrence C, Smith C (2008) The behaviour and ecology of the zebrafish, *Danio rerio*. Biol Rev Camb Philos Soc 83:13–34
102. Pavlidis M, Sundvik M, Chen Y-C, Panula P (2011) Adaptive changes in zebrafish brain in dominant-subordinate behavioral context. Behav Brain Res 225:529–537
103. Pavlidis M, Theodoridi A, Tsalafouta A (2015) Neuroendocrine regulation of the stress response in adult zebrafish, *Danio rerio*. Prog Neuropsychopharmacol Biol Psychiatry 60:121–131
104. Sakurai T (2007) The neural circuit of orexin (hypocretin): maintaining sleep and wakefulness. Nat Rev Neurosci 8:171–181
105. Han F (2012) Sleepiness that cannot be overcome: narcolepsy and cataplexy. Respirology 17:1157–1165
106. Mahlios J, De la Herrán-Arita AK, Mignot E (2013) The autoimmune basis of narcolepsy. Curr Opin Neurobiol 23:767–773
107. Fronczek R, Overeem S, Lee SYY, Hegeman IM, van Pelt J, van Duinen SG, Lammers GJ, Swaab DF (2007) Hypocretin (orexin) loss in Parkinson's disease. Brain J Neurol 130:1577–1585
108. Fronczek R, van Geest S, Frölich M, Overeem S, Roelandse FWC, Lammers GJ, Swaab DF (2012) Hypocretin (orexin) loss in Alzheimer's disease. Neurobiol Aging 33:1642–1650
109. Thannickal TC, Lai Y-Y, Siegel JM (2007) Hypocretin (orexin) cell loss in Parkinson's disease. Brain J Neurol 130:1586–1595
110. Nollet M, Leman S (2013) Role of orexin in the pathophysiology of depression: potential for pharmacological intervention. CNS Drugs 27:411–422
111. Holden T, Nguyen A, Lin E, Cheung E, Dehipawala S, Ye J, Tremberger G, Lieberman D, Cheung T (2013) Exploratory bioinformatics study of lncRNAs in Alzheimer's disease mRNA sequences with application to drug development. Comput Math Methods Med 2013:1–8
112. Sigurgeirsson B, Thorsteinsson H, Arnardóttir H, Jóhannesdóttir IT, Karlsson KA (2011) Effects of modafinil on sleep-wake cycles in larval zebrafish. Zebrafish 8:133–140
113. Scammell TE, Estabrooke IV, McCarthy MT, Chemelli RM, Yanagisawa M, Miller MS, Saper CB (2000) Hypothalamic arousal regions are activated during modafinilinduced wakefulness. J Neurosci 20:8620–8628
114. Ishizuka T, Murotani T, Yamatodani A (2010) Modanifil activates the histaminergic system through the orexinergic neurons. Neurosci Lett 483:193–196
115. Bruni G, Lakhani P, Kokel D (2014) Discovering novel neuroactive drugs through high-throughput behavior-based chemical screening in the zebrafish. Front Pharmacol 5:153
116. Tierney KB (2011) Behavioural assessments of neurotoxic effects and neurodegeneration in zebrafish. Biochim Biophys Acta 1812:381–389
117. Martineau PR, Mourrain P (2013) Tracking zebrafish larvae in group – status and perspectives. Methods (San Diego, Calif) 62:292–303
118. Ahrens MB, Li JM, Orger MB, Robson DN, Schier AF, Engert F, Portugues R (2012) Brain-wide neuronal dynamics during motor adaptation in zebrafish. Nature 485:471–477
119. Ahrens MB, Orger MB, Robson DN, Li JM, Keller PJ (2013) Whole-brain functional imaging at cellular resolution using light-sheet microscopy. Nat Methods 10:413–420
120. Del Bene F, Wyart C, Robles E, Tran A, Looger L, Scott EK, Isacoff EY, Baier H (2010) Filtering of visual information in the tectum by an identified neural circuit. Science 330:669–673
121. Kokel D, Bryan J, Laggner C, White R, Cheung CYJ, Mateus R, Healey D, Kim S, Werdich AA, Haggarty SJ, et al. (2010) Rapid behavior-based identification of neuroactive small molecules in the zebrafish. Nat Chem Biol 6:231–237
122. Rihel J, Prober DA, Arvanites A, Lam K, Zimmerman S, Jang S, Haggarty SJ, Kokel D, Rubin LL, Peterson RT, et al. (2010) Zebrafish behavioral profiling links drugs to biological targets and rest/wake regulation. Science 327:348–351

Hypocretins and Arousal

Shi-Bin Li, William J. Giardino, and Luis de Lecea

Abstract How the brain controls vigilance state transitions remains to be fully understood. The discovery of hypocretins, also known as orexins, and their link to narcolepsy has undoubtedly allowed us to advance our knowledge on key mechanisms controlling the boundaries and transitions between sleep and wakefulness. Lack of function of hypocretin neurons (a relatively simple and non-redundant neuronal system) results in inappropriate control of sleep states without affecting the total amount of sleep or homeostatic mechanisms. Anatomical and functional evidence shows that the hypothalamic neurons that produce hypocretins/orexins project widely throughout the entire brain and interact with major neuromodulator systems in order to regulate physiological processes underlying wakefulness, attention, and emotions. Here, we review the role of hypocretins/orexins in arousal state transitions, and discuss possible mechanisms by which such a relatively small population of neurons controls fundamental brain state dynamics.

Keywords Hypocretin • Narcolepsy • Outputs of hypocretin neurons • Probabilistic model of sleep and wake • Sleep-arousal transition

Contents

S.-B. Li, W.J. Giardino, and L. de Lecea (✉)
Department of Psychiatry and Behavioral Sciences, Stanford University, 1201 Welch Road, Stanford, CA 94305, USA
e-mail: llecea@stanford.edu

© Springer International Publishing AG 2016
Curr Topics Behav Neurosci (2017) 33: 93–104
DOI 10.1007/7854_2016_58
Published Online: 18 December 2016

1 Introduction

Almost 20 years after their discovery, the hypocretins (Hcrt, also known as orexins) have revealed an enormous amount of information about the regulation of brain states. In 1998, de Lecea and colleagues identified mRNA selectively expressed within the posterior hypothalamus that encoded the precursor of two peptides (Hcrt1, Hcrt2) that share marked amino acid identities with each other, and with a family of hormones known as incretins. These peptides were termed hypocretins, indicating that they are *hypo*thalamic members of the in*cretin* family [1]. Soon afterwards, Sakurai and colleagues used the term "orexins" to describe two neuropeptides expressed in the lateral hypothalamus (LH) that were identical to the hypocretins and activated two closely related orphan GPCRs. Because the term hypocretin predates orexin, and because the main function of these neuropeptides is not directly related to appetite, we will collectively refer to these peptides as hypocretins. Of the two GPCRs identified by Sakurai and colleagues, Hcrt receptor 1 (HcrtR1) has a higher affinity for Hcrt1 than Hcrt2, whereas Hcrt receptor 2 (HcrtR2) has a similar affinity for both peptides [2].

Anatomical analyses revealed that Hcrt neurons project broadly throughout the brain, including monoaminergic cell groups, cortex, septum, thalamus, amygdala, bed nucleus of the stria terminalis (BNST), and the reticular formation [3–5]. In situ hybridization histochemistry revealed that HcrtR1s are expressed in many brain regions including the prefrontal cortex, hippocampus, paraventricular thalamic nucleus, ventromedial hypothalamic nucleus, dorsal raphé nucleus (DRN), and locus coeruleus (LC) [6, 7]. Additional qRT-PCR analyses confirmed particularly robust HcrtR1 expression in many hypothalamic and thalamic nuclei, as well as LC. HcrtR2s have a general complementary distribution pattern of brain regions, such as cortex, septum, hippocampus, medial thalamus, raphé nuclei, and the tuberomammillary, dorsomedial, paraventricular, and ventral premammillary nuclei of the hypothalamus [6]. Selective functions for these receptors and intracellular signaling are discussed in other chapters in this issue.

In rodent models, the observation that optogenetic excitation of Hcrt neurons increased the probability of transitions from sleep to wake supplied a direct link between Hcrt and behavioral arousal [8]. Pharmacological manipulations with Hcrt receptor antagonists induced sleep in mice, rats, dogs, non-human primates and humans, indicating that the main function of Hcrt is excitatory [9]. One and a half years after the first report on the Hcrt system, Yanagisawa and colleagues reported that Hcrt peptide knockout mice showed behavioral arrests during the active phase that resembled cataplexy attacks [10] as seen in patients suffering from narcolepsy with cataplexy. Thus, Hcrt knockout mice are used as an animal model of the disorder narcolepsy/cataplexy, in which intrusion of sleep into wake accompanied by transient episodes of muscle weakness are often observed. In contrast to the early view that cataplexy is a REM-like state that can transition to REM sleep if the episode is of sufficient duration, Hcrt knockout mice display cataplexy episodes that begin with a brief wake-like electroencephalogram followed by high amplitude, irregular theta oscillations that do not directly precede REM sleep [11]. Overexpression of ataxin-3 in Hcrt neurons ablates these cells, therefore resulting in narcolepsy/cataplexy. However, this cataplexy state can be rescued by using recombinant adeno-associated virus (rAAV) to deliver the Hcrt gene to non-ablated neurons of the lateral hypothalamus or zona incerta [12, 13]. The role of each of the Hcrt receptors in the maintenance of vigilance states remains to be elucidated. Genetic deletion of both HcrtR1 and HcrtR2 exhibits a stronger effect on narcolepsy/cataplexy compared to the sole deletion of HcrtR2, where such fast transitions are very rare events. On the other hand, HcrtR1 gene knockout mice appear to have a similar sleep phenotype as wild type controls [14], with possibly some increased sleep fragmentation. There is a potential link between narcolepsy and the amygdala, as narcolepsy is often triggered by strong emotions. Indeed, cataplexy can be reduced by excitotoxic lesions of the amygdala in Hcrt knockout mice [15]. This potential link was further evidenced by the observation that Hcrt neurons discharge maximally during emotional and sensorimotor conditions that trigger cataplexy in narcoleptic animals [16].

How does hypocretin activity control normal sleep and wakefulness? Juxtacellular recordings of identified Hcrt neurons indicated that the firing of Hcrt cells is mostly phasic, lasting only a few seconds at frequencies 4–10 Hz during active wakefulness and during episodes of NREM or REM sleep preceding wakefulness in head-restricted rats. Otherwise, Hcrt neurons remain relatively quiescent during NREM and REM sleep [17]. In free-moving rats, micropipette-microwire recording with juxtacellular neurobiotin injection revealed that Hcrt cells have broad action potentials with elongated later positive deflections and discharge during active waking and phasic REM, but not in quiet waking and tonic periods of REM [16]. However, the extracellular waveforms of the Hcrt neuron spiking appear to be controversial as Takahashi and colleagues showed that Hcrt neurons fire biphasic broad spikes with early positive deflections at low frequency (<10 Hz) during wakefulness and cease firing during sleep by extracellular single unit recordings combined with neurobiotin juxtacellular labelling and Hcrt immunohistochemistry in head-fixed mice [18]. Thus, it is imperative to characterize the Hcrt neuron firing with a more reliable methodology, such as unit recording with optogenetic manipulation. Together, these studies indicated that Hcrt activity correlates with wakefulness, although a direct, causal relationship between Hcrt activity and wakefulness remained to be established. Adamantidis and colleagues were the first to use optogenetic methods in vivo to modulate the activity of Hcrt neurons in freely moving mice. Optogenetic stimulation of Hcrt neurons at a frequency of 5 Hz and higher dramatically increased the probability of sleep-to-wake transitions. This effect was mediated by Hcrt peptide release, since stimulations of Hcrt neurons in Hcrt-deficient animals did not show reductions in latencies to wakefulness [8]. From these studies, it became clear that Hcrt neurons do not maintain wakefulness by sending output signals tonically during wakefulness, but rather fire in short episodes to maintain the state only when it is appropriate. This raises the question of which signals are computed by Hcrt cells that drive their "appropriate" firing.

2 Hypocretin Neurons as Integrators

Hcrt neurons express receptors for neuropeptides (e.g., neuropeptide FF [NPFF], cholecystokinin [CCK], melanin-concentrating hormone [MCH], corticotropin-releasing [CRF]), and are sensitive to ambient stimuli including glucose [19], CO_2 and pH [20]. These signals are therefore considered primary inputs to Hcrt cells, and cover a broad range of physiological variables including metabolic state (glucose), stress (CRF), satiety (CCK), and circadian time. Direct evidence from a fasting experiment showed that Hcrt neurons monitor indicators (glucose, leptin, ghrelin) of energy balance and fasting augments arousal [21]. In addition to these primary Hcrt neuron on-site inputs, secondary signals are received by neurons that directly regulate Hcrt neurons through dense innervation from multiple brain regions. Retrograde tracing studies have demonstrated that Hcrt neurons receive afferents from the lateral septum, amygdala, and BNST among other structures [22, 23]. Other neuronal groups in the LH, including those expressing MCH also densely innervate Hcrt neurons. The Hcrt system has reciprocal connections with multiple neural systems and is involved in various physiological and pathophysiological modulations, thus the Hcrt system may function as a master integrator [24]. In particular, the metabolic, circadian, and limbic inputs might be integrated by the Hcrt neurons, which interact with a network of neuromodulators gating dynamics of sleep-to-wake transitions (Fig. 1) [25].

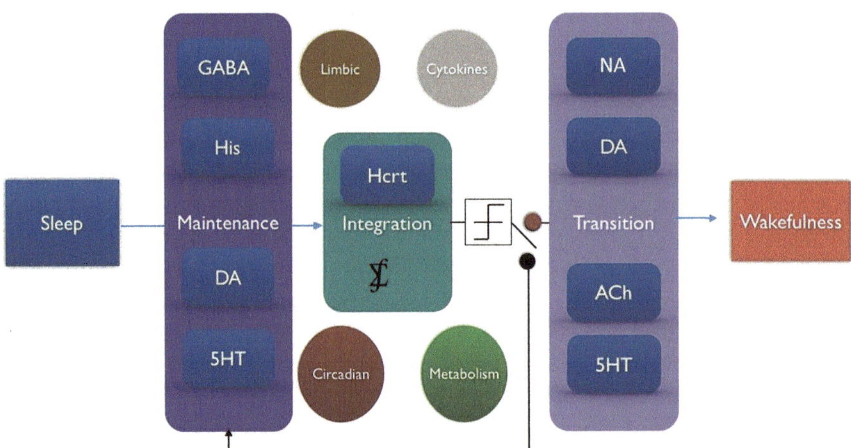

Fig. 1 Hcrt neurons integrate the inputs from the circadian, metabolic, limbic, and cytokine systems. The integration of the activities of these systems either leads to maintenance of sleep, in which the GABAergic, histaminergic, dopaminergic, and serotonergic systems are involved, or triggers the transition to arousal, in which the noradrenergic, dopaminergic, cholinergic, and serotonergic systems are implicated. Depending on the integrated signals arriving in Hcrt neurons, sleep will either be maintained or will transition into wakefulness

3 Arousal Circuits as Outputs of Hcrt Neurons

Hypothalamic Hcrt neurons project strongly to the noradrenergic LC [4], and sleep-to-wake transitions can be triggered by activating LC neurons in vivo with optogenetics [24]. In fact, sleep-to-wake transitions induced by Hcrt neuron stimulation can be reduced by simultaneous optogenetic inhibition of LC neurons [26, 27]. Furthermore, simultaneous optogenetic stimulations of LC neurons and Hcrt neurons synergistically reduce the latency to wakefulness (compared to Hcrt optogenetic stimulation alone) [26]. Overall, these findings emphasize the central role of Hcrt neurons in stabilization of arousal, and show that these effects are primarily mediated through Hcrt outputs to the LC.

Although earlier in vivo studies showed that the mean firing rate of midbrain dopamine neurons was not different across the sleep-wake states, later electrophysiological experiments showed that there were significant changes in temporal firing pattern across the sleep cycle [28]. Indeed, there are numerous pieces of evidence that point to dopamine neurons as critical modulators of arousal. For instance, it is well known that psychostimulants have a dramatic effect on wakefulness. In addition, dopaminergic neurons in the ventral periaqueductal gray matter (vPAG) have been shown to be active during natural wakefulness or environmental stimulation evoked wakefulness, and partial ablation of these dopaminergic neurons increased total daily sleep by ~20% [29]. It is thus possible that at least some of the wake-promoting effects of Hcrt are mediated by dopaminergic neurons, as dopamine transporter (DAT) inhibitors have been shown to be effective in promoting wakefulness [30, 31]. Furthermore, a model including adenosine and dopamine receptors in the nucleus accumbens of the basal ganglia has been proposed to integrate behavioral processes with wakefulness [32].

The histaminergic tuberomammillary nucleus (TMN) receives strong projections from Hcrt neurons [4, 33], and optogenetic stimulation of Hcrt neurons elicits the release of glutamate onto TMN neurons. Applying Hcrt-1 or Hcrt-2 depolarizes TMN neurons. Thus, the TMN appears to be a downstream effector of the Hcrt system driving the sleep-to-wake transitions [34, 35]. Indeed, output of HcrtR2-dependent histaminergic neurons increased linearly with optogenetic stimulation of Hcrt neurons through co-released Hcrt and glutamate by Hcrt neurons [36].

Hcrt neurons project to cholinergic brain nuclei, such as the basal forebrain and the laterodorsal and pedunculopontine tegmental nuclei [37–40]. A recent publication showed that optogenetic manipulation of basal forebrain cholinergic neurons is sufficient to elicit cortical activation and facilitate transitions from NREM sleep to wake [41]. Consistent with this observation, optogenetic activation of cholinergic, glutamatergic neurons rapidly induced wakefulness [42]. Thus it is possible that these neurons in the basal forebrain are orchestrated by Hcrt projections during promotion of wakefulness, as Hcrt neurons project strongly to the basal forebrain. In whole cell recordings from slice preparations, Hcrt depolarized neurons in the cholinergic laterodorsal tegmental nucleus exclusively through its action on HcrtR1 [43].

Opposite to the excitatory effect of the Hcrt neurons, the inhibitory GABAergic neurons reciprocally interact with the Hcrt system throughout the brain, including the hypothalamus [44], ventral tegmental area (VTA) [45], DRN [46], basal forebrain [47], and amygdala [48]. As the major inhibitory neurotransmitter, activation of $GABA_A$ receptors is clinically effective for the treatment of insomnia [49, 50]. The GABA receptors are important for stabilizing sleep. For example, conditionally deleting the $GABA_{B1}$ gene (thus deleting $GABA_B$ receptors) specifically in Hcrt neurons results in severe fragmentation of sleep and wake states [51]. In the perifornical-lateral hypothalamic area, the firing rate of the NREM-off neurons can be either decreased by $GABA_A$ receptor activation or increased by $GABA_A$ receptor blockade [44]. Using feedforward inhibition, Hcrt increases the firing rate of GABAergic neurons that target the dopaminergic neurons within the VTA, thus counteracting the direct excitatory effects of Hcrt on dopaminergic VTA neurons [45]. Likewise in the DRN, Hcrt projections directly target GABAergic inputs onto serotonergic neurons to modulate the activity of these neurons upon direct excitation of Hcrt neuron inputs [46].

In a local microcircuit of the LH, optogenetic activation of Hcrt neurons results in lower action potential firing of most MCH neurons through $GABA_A$ receptor signaling. These data suggest that intra-LH feedforward loops may mediate switching between sleep and wake signals [52]. Besides feedforward inhibition, there are also indirect feedback mechanisms underlying the modulation of Hcrt neuron activities. For example, Hcrt neurons receive inputs from GABAergic neurons that are activated by LC noradrenergic neurons. Thus, Hcrt neuron stimulation of the LC can lead to inhibition of Hcrt neurons themselves in this tri-synapse formation [53]. In the basal forebrain, blockade of Hcrt receptors increases GABA levels as determined by microdialysis in freely behaving rats [47]. Blockade of Hcrt-1 and Hcrt-2 receptors in vivo leads to NREM and REM sleep (essentially REM sleep in the clinic), whereas $GABA_A$ receptor blockade suppresses REM sleep [54]. Conversely, in another brain region innervated by Hcrt neurons, infusion of Hcrt-1 and Hcrt-2 into the central amygdala causes anxious behavior, which can be blocked by the co-infusion of flunitrazepam, a $GABA_A$ receptor α_2 subunit positive allosteric modulator [48].

4 Hypocretins and Narcolepsy

As Hcrt neurons fire rarely during sleep, alterations in the activities of the effectors innervated by the Hcrt neurons are plausible mechanisms underlying narcolepsy with cataplexy. Below, we summarize the studies investigating narcolepsy/cataplexy.

In a well-established canine model, disruption of HcrtR2 gene by positional cloning has been shown to cause canine narcolepsy [55]. Furthermore, in transgenic mouse models, Hcrt$^{-/-}$ and HcrtR2$^{-/-}$ mice resembled the abrupt arrests fulfilling criteria used for human cataplexy. In particular, HcrtR2s are critical for regulating wake/non-REM transitions, whereas the profound dysregulation of REM sleep control in narcolepsy-cataplexy syndrome attributes to both HcrtR2-dependent and HcrtR2-independent pathways [56].

As detailed above, noradrenergic neurons in the LC are one of the main targets of Hcrt cells. Even though LC neurons do not express HcrtR2 receptors, whose absence leads to narcolepsy in canine models [55], they likely play a role in the control of vigilance state boundaries. Optogenetic overstimulation of LC neurons leads to narcolepsy-like behavioral arrests [27] suggesting that HcrtR2-positive neurons inhibiting LC activity may prevent cataplexy attacks. In narcoleptic mice lacking Hcrt neurons, LC neurons fire more during wakefulness and during NREM sleep due to a functional decrease of GABAergic input; this is thought to significantly contribute to the insomnia and excessive daytime sleepiness exhibited by these mice [23, 57]. In a study with Hcrt receptor knockout mice investigating neurons downstream of Hcrt cells, selective restoration of HcrtR1s to noradrenergic neurons in the LC significantly improved the maintenance of wakefulness and NREM sleep [58]. Likewise, chemogenetic activation of noradrenergic neurons in LC of Hcrt neurons ablated mice stabilized the wakefulness episodes [58]. These observations indicate that the function of activating Hcrt receptors in LC neurons is related to the consolidation of fragmented wakefulness.

The serotonergic DRN receives projections from the Hcrt neurons [4], and Hcrt depolarizes DRN neurons in slice [59]. This depolarization is mediated by the combined effects of Hcrt-1 and Hcrt-2 receptors [43] generating a tetrodotoxin-insensitive nonselective cation current [46]. As aforementioned, Hcrt deficiency is associated with narcolepsy/cataplexy, which is characterized by the intrusion of sleep into wake and transient episodes of muscle weakness. Interestingly, in a Hcrt receptor knockout mouse model, cataplexy-like episodes were prevented by selectively expressing Hcrt receptors in serotonergic DRN neurons [58]. Using chemogenetic tools to selectively activate these neurons alleviated the narcoleptic phenotype normally seen in mice lacking Hcrt neurons [58]. Conversely, there might be reciprocal connections between Hcrt neurons and serotonergic neurons: in acute hypothalamic slice preparations, GFP-labeled Hcrt neurons were hyperpolarized by serotonin in a concentration-dependent manner. Single-channel recordings in these neurons revealed that serotonin-mediated hyperpolarization of Hcrt neurons was likely due to G-protein coupled inward rectifier potassium (GIRK) channels [60].

Narcolepsy and excessive daytime sleepiness are correlated with a decrease in cerebrospinal fluid (CSF) levels of histamine [61]. Furthermore, the increased need for sleep of severe head trauma patients may be due to the loss of histaminergic neurons. Therapies against daytime sleepiness by enhancing the histaminergic system have thus been proposed after such an injury [62]. However, patients with narcolepsy exhibit a marked increase in histaminergic TMN neurons compared to controls, particularly those patients with severe loss of Hcrt neurons. Thus, there may be a compensatory mechanism through which increased histaminergic neurons contribute to promotion of arousal in individuals with type I narcolepsy [63].

5   Engineering Behavioral State Transitions

Work by Saper and colleagues suggested that sleep-to-wake transitions are regulated by mutually inhibiting groups of neurons, the so-called flip-flop model of sleep regulation [64]. The model was based on anatomical evidence and on dynamic equations from systems that minimize intermediate states. Alternatively, we proposed a probabilistic model of sleep and wake based on information theory that establishes relationships (functions) between systems of neurons [25, 65]. A "codebook" of neurotransmitters and modulators would determine whether the conditions are appropriate for wakefulness [65]. This information is computed by an "integrator," presumably the Hcrt system, which is critical for appropriate sleep/wake cycle boundaries (Fig. 1). The integrator is clearly non-redundant, because its function cannot be compensated by other neurons in Hcrt-deficient narcoleptic/cataplectic patients. Ample evidence discussed above suggests that the output of the integrator (e.g., noradrenergic neurons in the LC) effectively elicits sleep-to-wake transitions [26, 27]. A question arises as to how long the integration interval must be to achieve stable and robust control of sleep/wake cycles. Short integration times would be very responsive to the environment, at the risk of fragmenting sleep too frequently, an effect that has deleterious effects on learning and memory [66], as well as the overall quality of sleep and its resting effects. By contrast, long integration times would favor long sleep episodes at the putative cost of not responding to environmental and metabolic challenges. In particular, a quantitative model of neural circuit including Hcrt neurons, noradrenergic neurons in the LC and GABAergic neurons proposed that there might be inhibitory neuropeptides involved in the modulation of LC neuron activities as $GABA_A$ receptor mediated inhibition is several orders of magnitude faster than the effect by Hcrt on LC neurons [67]. It may be proposed that sleep fragmentation may be beneficial in lower species that need to be able to efficiently switch from sleep to wake depending on environmental conditions or in response to predators, whereas in higher species, especially in humans, high sleep fragmentation equates with poor non-restorative sleep.

Another outstanding issue is whether sleep homeostatic mechanisms reside in some form in the Hcrt circuit. Whether synaptic downscaling, metaplastic phenomena affecting the entire brain drive sleep homeostasis is still a matter of debate [68], but it is clear that the consequences of such mechanisms specifically in master regulators of sleep architecture (i.e., Hcrt neurons) will have a significant impact in determining the length of sleep and the adaptation to sleep loss.

6   Conclusions

In this chapter, we discussed the Hcrt system and its interactions with other brain nuclei implicated in sleep/arousal regulation. We also discussed how neuromodulators and neurotransmitters in circadian, metabolic, and limbic systems participate in sleep maintenance and sleep-to-wake transitions. Based on our understanding of the role of Hcrt system in sleep modulation, we propose a probabilistic model of sleep and wakefulness that places Hcrt system in a central role as an integrator for appropriate thresholding of sleep/wake cycle boundaries.

References

1. de Lecea L et al (1998) The hypocretins: hypothalamus-specific peptides with neuroexcitatory activity. Proc Natl Acad Sci U S A 95:322–327. doi:10.1073/Pnas.95.1.322
2. Sakurai T et al (1998) Orexins and orexin receptors: a family of hypothalamic neuropeptides and G protein-coupled receptors that regulate feeding behavior. Cell 92(4):573–85
3. Nambu T et al (1999) Distribution of orexin neurons in the adult rat brain. Brain Res 827:243–260
4. Peyron C et al (1998) Neurons containing hypocretin (orexin) project to multiple neuronal systems. J Neurosci 18:9996–10015
5. Sakurai T et al (2005) Input of orexin/hypocretin neurons revealed by a genetically encoded tracer in mice. Neuron 46:297–308. doi:10.1016/j.neuron.2005.03.010
6. Marcus JN et al (2001) Differential expression of orexin receptors 1 and 2 in the rat brain. J Comp Neurol 435:6–25
7. Trivedi P, Yu H, MacNeil DJ, Van der Ploeg LH, Guan XM (1998) Distribution of orexin receptor mRNA in the rat brain. FEBS Lett 438:71–75
8. Adamantidis AR, Zhang F, Aravanis AM, Deisseroth K, de Lecea L (2007) Neural substrates of awakening probed with optogenetic control of hypocretin neurons. Nature 450:420–424. doi:10.1038/nature06310
9. Brisbare-Roch C et al (2007) Promotion of sleep by targeting the orexin system in rats, dogs and humans. Nat Med 13:150–155. doi:10.1038/nm1544
10. Chemelli RM et al (1999) Narcolepsy in orexin knockout mice: molecular genetics of sleep regulation. Cell 98:437–451
11. Vassalli A et al (2013) Electroencephalogram paroxysmal theta characterizes cataplexy in mice and children. Brain 136:1592–1608. doi:10.1093/brain/awt069
12. Hara J et al (2001) Genetic ablation of orexin neurons in mice results in narcolepsy, hypophagia, and obesity. Neuron 30:345–354. doi:10.1016/S0896-6273(01)00293-8
13. Liu M et al (2011) Orexin gene transfer into zona incerta neurons suppresses muscle paralysis in narcoleptic mice. J Neurosci 31:6028–6040. doi:10.1523/Jneurosci.6069-10.2011
14. Mieda M et al (2011) Differential roles of orexin receptor-1 and-2 in the regulation of non-REM and REM sleep. J Neurosci 31:6518–6526. doi:10.1523/Jneurosci.6506-10.2011
15. Burgess CR, Oishi Y, Mochizuki T, Peever JH, Scammell TE (2013) Amygdala lesions reduce cataplexy in orexin knock-out mice. J Neurosci 33:9734–9742. doi:10.1523/JNEUROSCI.5632-12.2013
16. Mileykovskiy BY, Kiyashchenko LI, Siegel JM (2005) Behavioral correlates of activity in identified hypocretin/orexin neurons. Neuron 46:787–798. doi:10.1016/j.neuron.2005.04.035
17. Lee MG, Hassani OK, Jones BE (2005) Discharge of identified orexin/hypocretin neurons across the sleep-waking cycle. J Neurosci 25:6716–6720. doi:10.1523/JNEUROSCI.1887-05.2005
18. Takahashi K, Lin JS, Sakai K (2008) Neuronal activity of orexin and non-orexin waking-active neurons during wake-sleep states in the mouse. Neuroscience 153:860–870. doi:10.1016/j.neuroscience.2008.02.058
19. Burdakov D, Karnani MM, Gonzalez A (2013) Lateral hypothalamus as a sensor-regulator in respiratory and metabolic control. Physiol Behav 121:117–124. doi:10.1016/j.physbeh.2013.03.023
20. Sakurai T (2007) The neural circuit of orexin (hypocretin): maintaining sleep and wakefulness. Nat Rev Neurosci 8:171–181. doi:10.1038/nrn2092

21. Yamanaka A et al (2003) Hypothalamic orexin neurons regulate arousal according to energy balance in mice. Neuron 38:701–713
22. Mochizuki T et al (2011) Orexin receptor 2 expression in the posterior hypothalamus rescues sleepiness in narcoleptic mice. Proc Natl Acad Sci U S A 108:4471–4476. doi:10.1073/pnas.1012456108
23. Mochizuki T et al (2004) Behavioral state instability in orexin knock-out mice. J Neurosci 24:6291–6300. doi:10.1523/JNEUROSCI.0586-04.2004
24. Li SB, Jones JR, de Lecea L (2016) Hypocretins, neural systems, physiology, and psychiatric disorders. Curr Psychiatry Rep 18:7. doi:10.1007/s11920-015-0639-0
25. de Lecea L, Huerta R (2014) Hypocretin (orexin) regulation of sleep-to-wake transitions. Front Pharmacol 5:16. doi:10.3389/fphar.2014.00016
26. Carter ME et al (2012) Mechanism for hypocretin-mediated sleep-to-wake transitions. Proc Natl Acad Sci U S A 109:E2635–E2644. doi:10.1073/Pnas.1202526109
27. Carter ME et al (2010) Tuning arousal with optogenetic modulation of locus coeruleus neurons. Nat Neurosci 13:1526–1533. doi:10.1038/Nn.2682
28. Dahan L et al (2007) Prominent burst firing of dopaminergic neurons in the ventral tegmental area during paradoxical sleep. Neuropsychopharmacology 32:1232–1241. doi:10.1038/sj.npp.1301251
29. Lu J, Jhou TC, Saper CB (2006) Identification of wake-active dopaminergic neurons in the ventral periaqueductal gray matter. J Neurosci 26:193–202. doi:10.1523/JNEUROSCI.2244-05.2006
30. Wisor J (2013) Modafinil as a catecholaminergic agent: empirical evidence and unanswered questions. Front Neurol 4. doi:10.3389/Fneur.2013.00139. Artn 139
31. Wisor JP et al (2001) Dopaminergic role in stimulant-induced wakefulness. J Neurosci 21:1787–1794
32. Lazarus M, Chen JF, Urade Y, Huang ZL (2013) Role of the basal ganglia in the control of sleep and wakefulness. Curr Opin Neurobiol 23:780–785. doi:10.1016/j.conb.2013.02.001
33. Torrealba F, Yanagisawa M, Saper CB (2003) Colocalization of orexin a and glutamate immunoreactivity in axon terminals in the tuberomammillary nucleus in rats. Neuroscience 119:1033–1044
34. Eriksson KS, Sergeeva O, Brown RE, Haas HL (2001) Orexin/hypocretin excites the histaminergic neurons of the tuberomammillary nucleus. J Neurosci 21:9273–9279
35. Schöne C et al (2012) Optogenetic probing of fast glutamatergic transmission from hypocretin/orexin to histamine neurons in situ. J Neurosci 32:12437–12443. doi:10.1523/JNEUROSCI.0706-12.2012
36. Schone C, Apergis-Schoute J, Sakurai T, Adamantidis A, Burdakov D (2014) Coreleased orexin and glutamate evoke nonredundant spike outputs and computations in histamine neurons. Cell Rep 7:697–704. doi:10.1016/j.celrep.2014.03.055
37. Eggermann E et al (2001) Orexins/hypocretins excite basal forebrain cholinergic neurones. Neuroscience 108:177–181
38. Fadel J, Burk JA (2010) Orexin/hypocretin modulation of the basal forebrain cholinergic system: role in attention. Brain Res 1314:112–123. doi:10.1016/j.brainres.2009.08.046
39. Fadel J, Deutch AY (2002) Anatomical substrates of orexin-dopamine interactions: lateral hypothalamic projections to the ventral tegmental area. Neuroscience 111:379–387
40. Ishibashi M et al (2015) Orexin receptor activation generates gamma band input to cholinergic and serotonergic arousal system neurons and drives an intrinsic Ca(2+)-dependent resonance in LDT and PPT cholinergic neurons. Front Neurol 6:120. doi:10.3389/fneur.2015.00120
41. Irmak SO, de Lecea L (2014) Basal forebrain cholinergic modulation of sleep transitions. Sleep 37:1941–1951. doi:10.5665/sleep.4246
42. Xu M et al (2015) Basal forebrain circuit for sleep-wake control. Nat Neurosci 18:1641–1647. doi:10.1038/nn.4143

43. Kohlmeier KA et al (2013) Differential actions of orexin receptors in brainstem cholinergic and monoaminergic neurons revealed by receptor knockouts: implications for orexinergic signaling in arousal and narcolepsy. Front Neurosci 7:246. doi:10.3389/fnins.2013.00246

44. Alam MN et al (2010) GABAergic regulation of the perifornical-lateral hypothalamic neurons during non-rapid eye movement sleep in rats. Neuroscience 167:920–928. doi:10.1016/j.neuroscience.2010.02.038

45. Balcita-Pedicino JJ, Sesack SR (2007) Orexin axons in the rat ventral tegmental area synapse infrequently onto dopamine and gamma-aminobutyric acid neurons. J Comp Neurol 503:668–684. doi:10.1002/cne.21420

46. Liu RJ, van den Pol AN, Aghajanian GK (2002) Hypocretins (orexins) regulate serotonin neurons in the dorsal raphe nucleus by excitatory direct and inhibitory indirect actions. J Neurosci 22:9453–9464

47. Vazquez-DeRose J et al (2014) Hypocretin/orexin antagonism enhances sleep-related adenosine and GABA neurotransmission in rat basal forebrain. Brain Struct Funct. doi:10.1007/s00429-014-0946-y

48. Avolio E, Alo R, Carelli A, Canonaco M (2011) Amygdalar orexinergic-GABAergic interactions regulate anxiety behaviors of the Syrian golden hamster. Behav Brain Res 218:288–295. doi:10.1016/J.Bbr.2010.11.014

49. Gottesmann C (2002) GABA mechanisms and sleep. Neuroscience 111:231–239

50. Harrison NL (2007) Mechanisms of sleep induction by GABA(A) receptor agonists. J Clin Psychiatry 68(Suppl 5):6–12

51. Matsuki T et al (2009) Selective loss of GABA(B) receptors in orexin-producing neurons results in disrupted sleep/wakefulness architecture. Proc Natl Acad Sci U S A 106:4459–4464. doi:10.1073/pnas.0811126106

52. Apergis-Schoute J et al (2015) Optogenetic evidence for inhibitory signaling from orexin to MCH neurons via local microcircuits. J Neurosci 35:5435–5441. doi:10.1523/JNEUROSCI.5269-14.2015

53. Li Y, van den Pol AN (2005) Direct and indirect inhibition by catecholamines of hypocretin/orexin neurons. J Neurosci 25:173–183. doi:10.1523/JNEUROSCI.4015-04.2005

54. Gotter AL et al (2014) Differential sleep-promoting effects of dual orexin receptor antagonists and GABAA receptor modulators. BMC Neurosci 15:109. doi:10.1186/1471-2202-15-109

55. Lin L et al (1999) The sleep disorder canine narcolepsy is caused by a mutation in the hypocretin (orexin) receptor 2 gene. Cell 98:365–376. doi:10.1016/S0092-8674(00)81965-0

56. Willie JT et al (2003) Distinct narcolepsy syndromes in orexin receptor-2 and orexin null mice: molecular genetic dissection of non-REM and REM sleep regulatory processes. Neuron 38:715–730

57. Tsujino N et al (2013) Chronic alterations in monoaminergic cells in the locus coeruleus in orexin neuron-ablated narcoleptic mice. PLoS One 8:e70012. doi:10.1371/journal.pone.0070012

58. Hasegawa E, Yanagisawa M, Sakurai T, Mieda M (2014) Orexin neurons suppress narcolepsy via 2 distinct efferent pathways. J Clin Invest 124:604–616. doi:10.1172/Jci71017

59. Brown RE, Sergeeva OA, Eriksson KS, Haas HL (2002) Convergent excitation of dorsal raphe serotonin neurons by multiple arousal systems (orexin/hypocretin, histamine and noradrenaline). J Neurosci 22:8850–8859

60. Muraki Y et al (2004) Serotonergic regulation of the orexin/hypocretin neurons through the 5-HT1A receptor. J Neurosci 24:7159–7166. doi:10.1523/Jneurosci.1027-04.2004

61. Bassetti CL et al (2010) Cerebrospinal fluid histamine levels are decreased in patients with narcolepsy and excessive daytime sleepiness of other origin. J Sleep Res 19:620–623. doi:10.1111/j.1365-2869.2010.00819.x

62. Valko PO et al (2015) Damage to histaminergic tuberomammillary neurons and other hypothalamic neurons with traumatic brain injury. Ann Neurol 77:177–182. doi:10.1002/ana.24298

63. Valko PO et al (2013) Increase of histaminergic tuberomammillary neurons in narcolepsy. Ann Neurol 74:794–804. doi:10.1002/ana.24019

                                                                S.-B. Li et al.

64. Saper CB, Fuller PM, Pedersen NP, Lu J, Scammell TE (2010) Sleep state switching. Neuron 68:1023–1042. doi:10.1016/j.neuron.2010.11.032
65. Sorooshyari S, Huerta R, de Lecea L (2015) A framework for quantitative modeling of neural circuits involved in sleep-to-wake transition. Front Neurol 6:32. doi:10.3389/fneur.2015.00032
66. Rolls A et al (2011) Optogenetic disruption of sleep continuity impairs memory consolidation. Proc Natl Acad Sci U S A 108:13305–13310. doi:10.1073/pnas.1015633108
67. Mosqueiro T, de Lecea L, Huerta R (2014) Control of sleep-to-wake transitions via fast aminoacid and slow neuropeptide transmission. New J Phys 16. doi:10.1088/1367-2630/16/11/115010
68. Tononi G, Cirelli C (2014) Sleep and the price of plasticity: from synaptic and cellular homeostasis to memory consolidation and integration. Neuron 81:12–34. doi:10.1016/j.neuron.2013.12.025

Orexin OX_2 Receptor Antagonists as Sleep Aids

Laura H. Jacobson, Sui Chen, Sanjida Mir, and Daniel Hoyer

Abstract The discovery of the orexin system represents the single major progress in the sleep field of the last three to four decades. The two orexin peptides and their two receptors play a major role in arousal and sleep/wake cycles. Defects in the orexin system lead to narcolepsy with cataplexy in humans and dogs and can be experimentally reproduced in rodents. At least six orexin receptor antagonists have reached Phase II or Phase III clinical trials in insomnia, five of which are dual orexin receptor antagonists (DORAs) that target both OX_1 and OX_2 receptors (OX_2Rs). All clinically tested DORAs induce and maintain sleep: suvorexant, recently registered in the USA and Japan for insomnia, represents the first hypnotic

L.H. Jacobson
The Florey Institute of Neuroscience and Mental Health, The University of Melbourne, 30 Royal Parade, Parkville, VIC 3052, Australia

Department of Pharmacology and Therapeutics, School of Biomedical Sciences, Faculty of Medicine, Dentistry and Health Sciences, The University of Melbourne, Parkville, VIC 3010, Australia

S. Chen and S. Mir
Department of Pharmacology and Therapeutics, School of Biomedical Sciences, Faculty of Medicine, Dentistry and Health Sciences, The University of Melbourne, Parkville, VIC 3010, Australia

D. Hoyer (✉)
Department of Pharmacology and Therapeutics, School of Biomedical Sciences, Faculty of Medicine, Dentistry and Health Sciences, The University of Melbourne, Parkville, VIC 3010, Australia

The Florey Institute of Neuroscience and Mental Health, The University of Melbourne, 30 Royal Parade, Parkville, VIC 3052, Australia

Department of Chemical Physiology, The Scripps Research Institute, 10550 N. Torrey Pines Road, La Jolla, CA 92037, USA
e-mail: d.hoyer@unimelb.edu.au

© Springer International Publishing AG 2016
Curr Topics Behav Neurosci (2017) 33: 105–136
DOI 10.1007/7854_2016_47
Published Online: 30 November 2016

principle that acts in a completely different manner from the current standard medications. It is clear, however, that in the clinic, all DORAs promote sleep primarily by increasing rapid eye movement (REM) and are almost devoid of effects on slow-wave (SWS) sleep. At present, there is no consensus on whether the sole promotion of REM sleep has a negative impact in patients suffering from insomnia. However, sleep onset REM (SOREM), which has been documented with DORAs, is clearly an undesirable effect, especially for narcoleptic patients and also in fragile populations (e.g. elderly patients) where REM-associated loss of muscle tone may promote an elevated risk of falls. Debate thus remains as to the ideal orexin agent to achieve a balanced increase in REM and non-rapid eye movement (NREM) sleep. Here, we review the evidence that an OX_2R antagonist should be at least equivalent, or perhaps superior, to a DORA for the treatment of insomnia. An OX_2R antagonist may produce more balanced sleep than a DORA. Rodent sleep experiments show that the OX_2R is the primary target of orexin receptor antagonists in sleep modulation. Furthermore, an OX_2R antagonist should, in theory, have a lower narcoleptic/cataplexic potential. In the clinic, the situation remains equivocal, since OX_2R antagonists are in early stages: MK-1064 has completed Phase I, and MIN202 is currently in clinical Phase II/III trials. However, data from insomnia patients have not yet been released. Promotional material suggests that balanced sleep is indeed induced by MIN-202, whereas in volunteers MK-1064 has been reported to act similarly to DORAs.

Keywords Insomnia • NREM sleep • Orexin receptor antagonist • OX_2R • REM sleep

Contents

Abbreviations

2-SORA	Selective orexin 2 receptor antagonist
BNST	Bed nucleus of the stria terminalis
CHMP	Committee for Medicinal Products for Human Use
CNS	Central nervous system
CSF	Cerebrospinal fluid
DMH	Dorsomedial hypothalamic nuclei
DORA	Dual orexin receptor antagonist
DOX	Doxycycline
DR	Dorsal raphe

DREADDs	Designer receptors exclusively activated by designer drugs
DREM	Direct transitions between wake and REM sleep
DSM-IV/ DSM-V	Diagnostic and Statistical Manual of Mental Disorders Fourth/ Fifth edition
DTA mice	Mice in which selective orexin neuron loss is engineered by inducible expression of diphtheria toxin A
EDS	Excessive daytime sleepiness
EMA	European Medicines Agency
FDA	US Food and Drug Administration
GABA$_A$R	Gamma-aminobutyric acid A receptor
HCRT	Hypocretin
ICV	Intracerebroventricular
KO	Knockout
LC	Locus coeruleus
LDT	Laterodorsal tegmental nucleus
LH	Lateral hypothalamus
LPT	Lateral pontine tegmentum
MnPO	Median preoptic nucleus
NREM	Non-rapid eye movement sleep
OX	Orexin
OXR	Orexin receptor
OX$_1$R	Orexin 1 receptor
OX$_2$R	Orexin 2 receptor
PB	Parabrachial nucleus
PC	Precoeruleus
PPT	Pedunculopontine tegmental area
PSG	Polysomnography
REM	Rapid eye movement sleep
SCN	Suprachiasmatic nucleus
SORA	Selective orexin receptor antagonist
SOREM	Sleep onset REM
SLD	Sublaterodorsal nucleus
SWS	Slow-wave sleep
TMN	Tuberomammillary nucleus
TST	Total sleep time
vlPAG	Ventrolateral periaqueductal grey
VLPO	Ventrolateral preoptic area
vPAG	Ventral periaqueductal grey
VTA	Ventral tegmental area
WASO	Wake after sleep onset
WT	Wild type
Z drugs	Non-benzodiazepine site GABA$_A$ receptor positive allosteric modulators; the current market-leading hypnotics (zopiclone, zolpidem, eszopiclone, zaleplon)
ZT	Zeitgeber time (ZT0 and ZT24 = lights on, ZT12 = lights off)

1   Introduction

Sleep disorders represent a major issue in modern society. It is estimated that 1/3rd of the population suffers from insomnia at one point or another in their life. Until recently, insomnia was clinically separated into primary and secondary insomnias; however, the latest edition of DSM-V (see [1]) no longer recommends this distinction. Most of the clinical trials carried out in the late twentieth and early twenty-first century were targeting DSM-IV-defined primary insomnia, including those with orexin receptor antagonists. This restriction may have complicated recruitment of patients, since insomnia in the absence of stress, anxiety or depression is a rare event. Insomnia patients are currently treated mainly with older hypnotics, essentially benzodiazepines or the newer Z drugs (zolpidem, zaleplon, zopiclone, eszopiclone) and/or various aspects of cognitive therapy. These drugs, in spite of their different labels, are all positive allosteric modulators at the same GABA$_A$ receptor, although the Z drugs are more subunit selective than the classical benzodiazepines [2]. This distinction may be of little relevance clinically, as side effects of these compounds are quite comparable [2, 3], although the dosing of Z drugs is generally "lighter". The issue with positive GABA$_A$ receptor modulators is that activation of the GABA$_A$ receptor leads to a general reduction of brain activity, with profound central and peripheral effects, including muscle relaxation. When overdosed, or when combined with alcohol or other hypnotics, these drugs can induce a state of complete lack of responsiveness to external stimuli. Other issues with drugs targeting the GABA$_A$ receptor relate to the development of tolerance (decreased efficacy), possible addiction, impairment of memory, next morning somnolence, increased risk of falls and other serious side effects (such as complex sleep behaviours) (see [2]).

Recently, the FDA has unilaterally recommended that the starting doses for Z drugs be reduced (FDA Drug Safety Communications 10 Jan. [4] and 14 May [5]); the labels for Z drugs specifically mention sleep driving, sleep eating, sleep sex and other abusive behaviours as possible side effects. In addition, although inducing and maintaining sleep, it is clear that Z drugs do not produce physiological sleep but instead promote slow-wave sleep (SWS) at the expense of REM sleep. It can be discussed at length whether this is desirable or not, or what negative effects may result from REM sleep deprivation, but overall, it is clear that Z drugs do not promote a balanced or physiological sleep pattern.

At the time when benzodiazepines and Z drugs were developed, clinical studies were limited to a few weeks, and thus the drugs were not indicated for chronic use. It is only more recently that long-term studies have been conducted in insomnia both with Z drugs and orexin receptor antagonists, since insomnia is frequently a long-lasting indication. Thus, investigator-driven studies have shown that zolpidem is active over 8 months of nightly use, is well tolerated and is free of insomnia rebound or withdrawal symptoms upon discontinuation for at least 1 year [6, 7]. The recent Phase III clinical trials performed with suvorexant (lasting 3, 6 or 12 months with a possible extension up to 14 months) represent the largest and longest company-sponsored clinical trials ever reported in insomnia [8–13].

Given the limitations of benzodiazepines, Z drugs and other hypnotics, there have been intense efforts to discover new and improved drug targets for insomnia, e.g. the histamine or melatonin systems. Histamine 1 receptor antagonists, however, have not shown substantial improvements over the GABA$_A$ modulators (there is still potential for sedation, drowsiness and cognitive impairment) for a number of reasons, predominantly relating to the paucity of selective H$_1$ receptor antagonists, the multi-target profile of the majority of sleep-promoting antihistaminergic agents and perhaps also to circadian rhythm-dependent efficacy of H$_1$ receptor antagonists [14]. On the other hand, melatonin receptor agonists such as ramelteon, tasimelteon or modified release forms of melatonin have modest effects on sleep induction and maintenance [15, 16]. Their main advantages are a good safety record and the absence of marked side effects. Given at the right time, a melatonin receptor agonist can reset the circadian clock, but this may not be sufficient to significantly improve sleep in an insomniac patient. Thus, the biggest recent contribution to sleep research and ultimately insomnia treatment was the discovery of orexin/hypocretin in 1998 independently by two groups [17–20].

From this discovery, it was rapidly established that orexin deficiency/impairment is a major cause for narcolepsy with cataplexy (see [21, 22]; see also [23, 24]). This was based on findings with narcoleptic dogs, where the orexin 2 receptor (OX$_2$R) is non-functional, on various orexin peptide or orexin receptor KO or orexin neuron ablation models in mice or rats, and on evidence from humans, showing (a) that the orexin peptide levels are low or undetectable in the cerebrospinal fluid (CSF) of narcolepsy patients and (b) that in post-mortem samples from patients, orexin-producing neurons in the lateral hypothalamus are almost completely absent [25–34].

These discoveries were made at rapid pace and thus attracted attention from drug developers in biotech and pharma. Only 9 years after the discovery of the orexin system and its two receptors, the results of the first Phase II study in insomnia using a dual OX$_{1/2}$ receptor antagonist ("DORA"; almorexant) were presented at an international meeting and subsequently published [35]. The compound tested, almorexant, was the first of a series of at least five DORAs that entered successful clinical Phase II/III trials in primary insomnia: the other compounds are SB-649868 [36], suvorexant [37, 38], filorexant [39, 40] and E2006/lemborexant [41]. Almorexant was halted from further development, most probably for safety reasons (see [42]). Suvorexant received approval in 2014 from the US FDA (as well as from the Japanese regulatory authorities: it has thus been available for the treatment of insomnia since early 2015, although the doses initially proposed by Merck had to be adapted ([43–45]; see [46] for a review). SB-649868 was also stopped from further development, although the reasons for this were never clearly stated by GlaxoSmithKline. Other drugs that were/are in the pipeline include the DORAs filorexant/MK-6096, lemborexant/E2006 and ACT-462206, as well as

OX$_2$R antagonists, sometimes referred to as "2-SORAs" by those in the field (MIN-202/JNJ-42847922, MK-1064, MK-3697, MK-8133 and others). Here, we review the evidence that supports the development of orexin 2 receptor antagonists to treat insomnia.

2 The Discovery of Orexin: Orexin Receptors as a Target for Sleep

The discovery of orexins A and B (hypocretins 1 and 2) and their receptors, orexins 1 and 2 (OX$_1$R and OX$_2$R), will be described in detail elsewhere in this book. Suffice to say that according to HUGO (Human Genome Organisation) the genes are called hypocretin (*HCRT* and *HCRTR*), whereas the peptides and their receptors are named orexin (OX and OXR) according to the IUPHAR (International Union of Basic and Clinical Pharmacology) (see [47, 48]). The two orexin peptides derive from a common precursor, prepro-orexin, which has similarities in amino acid sequence to some incretins, hence the name "hypocretin" [17, 49]. They are produced in a specific, localised area of the lateral hypothalamus (LH), by a few thousand cells in humans and other species [17–19, 50]. These LH neurons project throughout the central nervous system (CNS) and innervate a number of regions involved in motivation, feeding, arousal and sleep/wake functions such as prefrontal and limbic cortex, hippocampus, amygdala, LH, DMH, VTA, SCN, LDT, BNST, TMN, LC, VLPO, DR and others.

One of the first phenotypes observed in the orexin peptide knockout (KO) mouse was reduced food intake [28], while application of the peptide in the brain promoted arousal and feeding, thus the name "orexin" [18, 51]. Expectations that orexin receptor antagonists would be beneficial in eating disorders and obesity, however, have not been met, although there may still be work in progress, especially with more selective compounds [52, 53]. This may be due in part to the fact that increased feeding is difficult to separate from arousal [54, 55]. It is clear from the long-term clinical studies carried out with suvorexant (3, 6 and 12 months or up to 14 months in the extension studies) that this DORA had no marked effect on food intake or weight status. These were the most extensive studies performed in any sleep setting, and as is increasingly common these days, a good proportion of the probants were overweight or obese. It could be argued that the study was not designed for such purposes, since suvorexant was given at night, whereas a treatment for the reduction of food intake would normally be given during the day. However, it should also be recognised that at the higher doses, suvorexant tended to accumulate and was clearly present in the circulation during daytime. Thus, the issue of dual orexin receptor antagonism for the treatment of obesity is still an open question.

The other major feature noted upon the discovery of the orexin system was the effects on sleep/wake and arousal [17, 20, 28, 56]. Thus, it was found that orexin neurons are highly active during wake and stop firing during sleep ([57]; reviewed in [58, 59]). There is a very strong parallel between orexin neuronal activity and wake. In addition, intracerebroventricular (ICV) application of orexin causes arousal and active behaviours. The involvement of orexin in sleep/wake regulation became further evident when the first orexin peptide KO animals were carefully examined [27, 28]. Orexin KO mice showed an unusual sleep/wake pattern, with no structured sleep during either the normal active (night) or rest (day) phases. There were many sleep intrusions during the active phase, whereas sleep was highly fractionated during the resting phase. In addition, it was noticed that orexin KO mice showed frequent, rapid transitions from wake to what appeared to be REM sleep, with sudden loss of muscle tone. These features are reminiscent of what happens in narcoleptic dogs (golden retriever, Doberman) or human patients who suffer from narcolepsy with cataplexy. It was then soon established that narcoleptic dogs have a non-functional OX$_2$R [29]. In humans, the situation is different: there is no evidence that orexin receptor mutations are implicated in narcolepsy/cataplexy (reviewed in [60]). However, what rapidly became apparent is that orexin-producing neurons are absent (in the few autopsy cases that could be analysed) or that there is no or very low levels of orexin in the CSF of these patients [26, 28, 30–34]. The reasons for the loss of orexin neurons in narcolepsy/cataplexy remain unclear, although current hypotheses support an immune- or autoimmune-mediated pathophysiology (see [60, 61]). There is, however, a report suggesting that polymorphisms of the orexin 2 receptor may be associated with excessive daytime sleepiness (EDS; [62]).

Based on these and other compelling data collected by various groups both independently and in fruitful collaborations, a crucial role of the orexin system in sleep/wake regulation was hypothesised by Saper and colleagues [63, 64], better known as the flip/flop sleep switch model. The circuitry of sleep and wake has been a topic of research for more than 50 years, and the discovery of orexin prompted a rapid advance in the understanding of this circuitry, particularly with regard to the regulation and transition of vigilance states [65]. Wake-promoting networks include cholinergic neurons of the LDT and PPT, basal forebrain, monoaminergic nuclei LC (noradrenaline), DR (serotonin), TMN (histamine), vPAG (dopamine), the PB, forebrain regions, perhaps glutamatergic PC neurons and orexin-positive neurons of the LH. NREM sleep-associated regions include galanin-positive and GABAergic neurons of the VLPO and the MnPO. REM sleep-promoting areas ("REM-on" neurons) include the SLD and PC, which are modulated by the vlPAG and LPT (GABAergic) neurons ("REM-off" neurons) (reviewed in [64, 66]). Saper et al. [63, 64] have noted that many of these wake and sleep-promoting regions appeared to be reciprocally inhibitory, giving rise to their flip/flop switch model of vigilance state transitions. Orexin neurons play a central role in this model, sending dense projections across the brain, including to the wake-promoting monoaminergic nuclei (particularly the TMN and LC), as well as direct projections to the cortex and basal forebrain. Orexin neurons also project to the vlPAG and LPT REM-off regions. Orexin acts at these sites, perhaps to some degree, differentially through expression of its two receptors (e.g. LC, OX$_1$R; TMN, OX$_2$R; DR, OX$_{1/2}$R) to stabilise wake (reviewed in [64, 65]). This model is supported by the observations that loss of orexin neurons causes narcolepsy, which is characterised, amongst other things, by instability in vigilance states [64]. Thus, the issue with narcoleptic patients is that wake and sleep states are no longer stabilised by orexin, and they therefore keep switching between wake and sleep. These transitions can be very rapid, and switching from wake to REM sleep is not uncommon in these patients, contrary to normal subjects which may take up to 2 h to go through the normal phases from wake to deep sleep: active wake, passive wake, stage 1, followed by stages 2, 3 and 4 of NREM sleep and gradually back to stage 1, and only then transition into REM sleep.

Thus, within 5 years of its discovery, the lack of orexin in CSF has become a standard biomarker for the disease narcolepsy with cataplexy. In contrast, orexin levels are close to normal in other cases of narcolepsy (without cataplexy) and in a number of other neurological diseases, although the situation is more complex in Alzheimer's disease [67, 68]. The effects of orexin loss have been replicated many times using different approaches [69], such as the selective destruction of orexin-producing cells in the LH using an ataxin-3 transgenic approach or conditional, via inducible diphtheria toxin expression on orexin neurons [25, 70–72], or by using optogenetics to inactivate these neurons [73–76]. The results are consistent in rats and mice: silencing orexin neurons or inactivating the peptide results in a narcoleptic/cataplexic phenotype. It should be noted that in these rodent models, narcoleptic attacks are infrequent and may go unnoticed. Like in humans and dogs, these attacks can be induced by stimuli such as food or others with a strong emotional context, e.g. sex, joy, fear or even wheel running [77]. On the other hand, intracerebral or ICV application of orexin, ectopic overexpression of orexin, restoration of OX_2R expression in the posterior hypothalamus [70, 71, 78] or stimulation of the LH orexinergic neurons using optogenetics (see [58, 73]) will all produce a wake state, with elevated activity and feeding.

The functional features of orexin peptide KO in the mouse are almost equally reproduced by knocking out both OX_1R and OX_2R (see [59]). Double OXR KO results, again, in highly fractionated sleep, with rapid wake to sleep transitions and numerous and rapid transitions into REM sleep, which is highly unusual under normal conditions in higher vertebrates. By contrast, the OX_2R KO mouse does not show the same strength of narcoleptic phenotype observed with dual orexin receptor KO or the orexin peptide KO [79]. This is not to say that the OX_2R KO mouse has no sleep phenotype. These animals tend to sleep more often, but fast transitions from wake to REM sleep are absent in most mice and only very rarely observed in others (~0.1–0.2 direct transitions per 24 h), at an incidence 13-fold lower than that seen in orexin KO mice [79]. Globally, the proportion of REM sleep in the OX_2R KO mouse appears to be normal [79]. The situation is again different in the OX_1R KO mouse; on the surface, these animals appear to have no, or almost no, sleep phenotype, although a closer look suggests that there is increased sleep fractionation [55, 59].

In essence, data from KO mice indicate that the orexin peptides and the two receptors play a major role in controlling the sleep/wake cycle and arousal [80–82] and have a major impact on activities during wake, such as motivation and purposeful behaviours, including searching for and ingesting food [83]. On the other hand, it would appear that knocking out either the peptide or both of its receptors results in detrimental effects such as narcolepsy. Agreeably, the narcoleptic phenotype is prominent when the inactivation is constitutive, and the situation may be different when the inactivation is transient. Indeed, there is currently no evidence that pharmacological blockade of orexin receptors causes narcolepsy in normal animals or people ([35]; reviewed in [84]). This may appear surprising, given the known aetiology of narcolepsy. The lack of narcolepsy with pharmacological inhibition versus constitutive genetic ablation may perhaps be rationalised by the compensatory changes in neurotransmitter systems that could occur in a constitutive and complete loss of orexin signalling, but not with a transient or incomplete inhibition in an adult system, although experimental evidence for this hypothesis is currently not available. The situation, however, may also differ depending on the emotional status of the animal or patient and their innate predisposition to narcolepsy/cataplexy (see [85, 116]).

To our knowledge, there are no reports describing the effects of transient, conditional/inducible KO or knockdown of orexin, orexin receptors or their signalling pathways by using, for example, siRNA (see [86]) or DREADDs (see [87]). Such approaches are indicated to obviate any compensatory changes that may be present in constitutive knockouts. Interestingly, a new conditional animal model of orexin cell loss has recently been described, in which the Tet-Off system was used to control the expression of diphtheria toxin A in orexin neurons ("DTA" mice; [72]). Withholding doxycycline (DOX) in 3-month-old mice resulted in a rapid loss of orexin neurons, with an ~36% reduction of orexin immunoreactivity within 36 h. Progressive alterations in sleep/wake architecture appeared within 1 week of DOX withdrawal when orexin neuron loss was ~86% [72, 88]. Narcolepsy/cataplexy appeared 2 weeks after DOX withdrawal, when orexin neuron loss was 95% [72]. More work is required to determine whether or not a longer duration of continuously reduced orexin neuron activity is required to induce narcolepsy/cataplexy than to alter sleep architecture, as may be suggested from the early phenotyping of DTA mice.

In summary, based on receptor KO studies, it would appear that the OX$_1$R plays a minor role in sleep/wake regulation by itself, whereas the role of the OX$_2$R is more prominent. It remains to be seen why the simultaneous KO of both receptors has such a pronounced additive, or even synergistic, effect, resulting in an increased occurrence of narcolepsy and frequent intrusions into REM sleep during active wake. This additivity is suggestive of interactions between brain centres expressing OX$_1$Rs and OX$_2$Rs that are involved in REM and NREM sleep control (see [89–93]). The OX$_{1/2}$R double KO data are supported by the reduction in REM sleep latency and increase in the proportion of REM sleep seen when combining OX$_1$R and OX$_2$R antagonists, in comparison to the administration of either antagonist alone ([94, 95]; reviewed in [46, 84]).

3 Dual Orexin Receptor Antagonists as Clinical Candidates for Sleep Disorders

Based on (1) the marked sleep/wake phenotype observed in orexin peptide and/or dual orexin receptor KO mice; (2) the clear relationship between orexin and narcolepsy, a sleep disorder; and (3) the prominent effects that orexin has on arousal, the orexin system became a target of intense research in biotech and the pharma industry not only for the development of the next generation of hypnotics but also for the treatment of narcolepsy and possibly other sleep disorders. With regard to the latter indications, it could be expected that an orexin receptor agonist is the ideal tool to treat narcolepsy with cataplexy, a devastating disease that has no established disease-modifying treatment modality. Current treatments for narcolepsy with cataplexy include stimulants for EDS such as modafinil, methylphenidate or amphetamine derivatives, whereas cataplexy may be treated with antidepressants [61]. Sodium oxybate, the sodium salt of the weak $GABA_B$ receptor agonist, gamma-hydroxybutyrate, may be prescribed for both EDS and narcolepsy/cataplexy and is dosed at night [61]. Wake-promoting histamine H3 receptor inverse agonists are in development for narcolepsy (see, e.g. [96]): pitolisant (tiprolisant, Wakix) is currently pending an EC (European Commission) decision for marketing authorisation for the treatment of narcolepsy with or without cataplexy following a positive recommendation by the European Medicines Agency's (EMA) Committee for Medicinal Products for Human Use (CHMP) in November 2015. All of these treatments are symptomatic and do not address causative deficits in orexin. Yet it has to be kept in mind for an orexin replacement therapy that natural peptides have strong limitations with respect to general bioavailability, duration of action and brain penetration. The intranasal route has been investigated, but so far, there has been no breakthrough with natural orexin as a treatment for narcolepsy [97–99]. Recently, Yanagisawa's team described the synthesis of a non-peptide orexin receptor agonist [100]; the sleep research community awaits efficacy data with this compound, whether clinically or preclinically.

The real attraction of the orexin system as a target in sleep disorders is that it acts essentially on an arousal triggering system [54, 101], i.e. via a "wake on" modality, in contrast to the $GABA_A$ receptor which represents a "wake off" modality [102]. The electrophysiological signatures of OXR antagonists and $GABA_A$ receptor modulators are indeed quite different [35, 102, 103]. Overall, activation of the $GABA_A$ receptor causes a general dampening of the CNS, which results in sleep, but also muscle relaxation and eventually motor impairment by muscle inactivation. When the $GABA_A$ receptor is profoundly activated, e.g. with high doses of alcohol or hypnotics, the state reached is akin to anaesthesia, which is challenging to reverse immediately. By contrast, it is expected that an OXR antagonist simply blocks the wake state, leaving the CNS in a more "normal" state of activity; thus, sleep could be reversed instantly in case of need. This has proven to be the case, as repeatedly shown in mice, rats, dogs and eventually humans (e.g. [35, 104]). It is also expected that an OXR antagonist will lose its effects after a normal night of sleep, since plasma and brain levels of the drug decrease as the night progresses, whereas orexin levels increase significantly at the end of the normal night.

Furthermore, a major drawback of GABA$_A$ receptor positive allosteric modulators is their negative impact on memory/cognition, due to the general dampening of the CNS. In contrast, due to their anti-"wake on" targeting, OXR antagonists are not expected to show such effects and indeed have not demonstrated such to date, whether in preclinical models or in a clinical setting [35, 105–107]. Similarly, DORAs do not affect motor performance to the same extent as GABA$_A$ receptor modulators either alone or in combination with alcohol [35, 105, 108].

Thus, based on the narcolepsy discovery and the sleep phenotypes of orexin peptide and receptor KO animals, the biotech and pharma industry started antagonist programmes targeting both OX$_1$R and OX$_2$R, in other words dual antagonists or DORAs ([109, 110]; see Coleman and Winrow for a detailed account in this book). The publications on almorexant and subsequent DORAs and the registration of suvorexant (Belsomra) leave no doubt that dual antagonists are effective in inducing sleep in animals and humans (reviewed in [46, 84]). It could be argued that one of the reasons for promoting dual rather than selective receptor targeting was because the search for selective OXR antagonists was more arduous, although very early on, somewhat selective OX$_1$R and OX$_2$R antagonists had been reported. It is also clear that the fear that dual antagonists could induce narcolepsy/cataplexy (see [85]) have not been realized in "normal" subjects, whether mice, rats, dogs, monkeys, healthy volunteers or patients [8, 9, 11, 35, 111–115]. It must be kept in mind, however, that narcolepsy is generally triggered by strong positive or negative emotions. It was eventually revealed that suvorexant can produce narcoleptic attacks in susceptible dogs, and thus it is contraindicated in patients suffering from narcolepsy with cataplexy (FDA, Highlights of prescribing information). Indeed, it has been shown that both almorexant and suvorexant can produce narcolepsy in susceptible murine [116] and canine models (suvorexant flyer).

Given the initial reports on the positive effects of almorexant in animal models and in the clinic, our team at Novartis wondered about the main target(s) of almorexant in sleep [35, 114]. Thus, we studied the effects of almorexant in rats and mice using polysomnography. More specifically, we compared the effects of almorexant in normal and KO mice, either OX$_1$R KO, OX$_2$R KO or double OX$_1$R/OX$_2$R KO. The results were compelling [114]: almorexant, in a dose-dependent manner, induced and maintained apparently normal sleep in control wild-type (WT) animals and produced very similar effects in OX$_1$R KO mice. In contrast, almorexant had no effects on the sleep/wake cycle of OX$_2$R KO mice, nor did it affect sleep in dual orexin receptor KO mice [114]. The latter finding has also been demonstrated with DORA-12 and DORA-22 in dual orexin receptor KO mice [40, 48]. These studies indicate that almorexant and DORAs 12 and 22 require orexin receptors in order to be effective and, in doing so, put to rest once and for all any speculations that there could be an orexin receptor different from OX$_1$R or OX$_2$R behind the efficacy of almorexant. The other main finding from Mang et al. [114] was that the OX$_1$R appeared to play no role in the effects of almorexant and that the major contributor of almorexant in sleep was the OX$_2$R. To our knowledge, this is the only investigation on the effects of dual antagonists on the sleep profile of both of the individual OX_1R and OX_2R KO mouse strains. In the meantime, other teams have shown that selective OX_2R antagonists produce no sleep/wake effects in OX_2R KO mice: MK-3697 [117], MK-8133 [118] and JNJ-42847922 [119]. More recently, the OX_2R antagonist MK-1064 has joined this list, as it is devoid of sleep-inducing effects in OX_2R mice, whereas it produces sleep in WT mice, rats, dogs and human volunteers [120]. Thus, based on the data from Mang et al. [114] and other supporting data, Novartis decided to start a chemistry programme aiming at selective OX_2R antagonists. This resulted in the generation of IPSU, an OX_2R-preferring antagonist that was effective in inducing balanced sleep in the mouse [121, 122].

4 The Case for Selective Orexin 2 Receptor Antagonists: Preclinical Data

A number of OX_2R antagonists have now been described in the literature (Table 1). The first reported was Compound 29 ("TCS-OX2-29", [133]). A limited number of sleep studies have been conducted with this compound, likely due to poor pharmacokinetic properties [119]. When given ICV, however, opposing effects on sleep have been reported, with an increase in REM sleep with no effect on NREM sleep described in rats [135], while in mice, the proportion of NREM sleep was enhanced with no effects seen on REM [134]. Reasons for these differences are unclear. The second compound described in the literature was Compound 9 [125], subsequently known as JNJ-10397049. Using this compound, Dugovic and colleagues were the first to show that an OX_2R antagonist produced sleep in animals, whereas an OX_1R antagonist was largely inactive [94].

MIN-202/JNJ-42847922 [119], MK-1064 [120, 126], MK-3697 [117] and MK-8133 [118] are all subsequent, orally available, selective OX_2R antagonists which are efficacious in inducing sleep in a range of species, and which have been, or are being, explored in clinical studies, exemplified recently by MK-1064 [120]. In addition, other effective OX_2R antagonists have been described [128–131, 136] (Table 1).

JNJ-42847922/MIN-202 is the most advanced OX_2R antagonist and is currently in Phase II clinical development for insomnia. Bonaventure and colleagues [119, 123] described the preclinical features of JNJ-42847922 as producing balanced sleep in rodents, as may be expected from the efficacy of preceding OX_2R antagonists. The compound reduced NREM latency and active wake dose dependently in rats (3, 10, 30 mg/kg). It had no significant effect on REM latency or REM in general but increased NREM in a dose-dependent manner. The effects were constant when the compound was given for 7 days. Furthermore, JNJ-42847922 was inactive in OX_2R KO mice. Similarly to rats, JNJ-42847922 given to mice at 30 mg/kg reduced active wake and NREM latency, whereas NREM duration was

Table 1 Selective orexin 2 receptor antagonists inducing sleep

Compound	Institute/company	Binding or functional affinity (pK_D, pK_b, pIC$_{50}$)			Species tested	Sleep features	Clinical status	Reference
		Method	hOX$_1$R	hOX$_2$R				
MIN-202/ JNJ-42847922	Janssen/J&J/ Minerva	Radioligand binding, calcium	6.1 6.3	8.0 8.8	Mouse, rat, humans	Balanced in mice, main effects on NREM and NREM latency; no effect on REM latency. Inactive in OX$_2$R KO, balanced in humans	Phase I/II	[119, 123, 124]
JNJ-10397049	Janssen/J&J	Radioligand binding, calcium	5.5 NA	8.3 7.9	Rat	Balanced when applied alone, but REM latency decreased and REM duration increased in the presence of OX$_1$R antagonists GSK-1059865 or SB-408124, replicating effects of the DORA SB-649868	Research	[94, 95, 125]
MK-3697	Merck	Radioligand binding, calcium	5.44 5.7	8.96 7.8	Mouse, rat, dog	Balanced in mice and rats. Inactive in OX$_2$R KO. Affects only SWS in dogs	Clinical candidate	[117]
MK-1064	Merck	Radioligand binding, calcium	5.8 5.75	9.3 7.75	Mouse, rat, dog, rhesus monkey	Balanced: increases in NREM and REM in mice, limited effects on REM in dogs and monkeys	Clinical candidate	[126]
MK-8133	Merck	Calcium	5.31	7.55	Mouse, rat, dog	Balanced: In mice no effect on REM, increased SWS. Inactive in OX$_2$R KO. In rat, both SWS and REM increased. In dog, limited effect on REM, increased NREM	Clinical candidate	[118]
2-SORA-18	Merck	Radioligand binding, calcium	7.02 6.53	10.15 7.92	Rat	Balanced, main effects on delta sleep > REM	Research	[127]

(continued)

Table 1 (continued)

Compound	Institute/ company	Binding or functional affinity (pK_D, pK_b, pIC_{50})			Species tested	Sleep features	Clinical status	Reference
		Method	hOX$_1$R	hOX$_2$R				
2-SORA-19	Merck	Radioligand binding, calcium	5.74 >5	8.44 7.32	Mouse	Balanced, main effects on light and delta sleep > REM	Research	[128]
IPSU/cpd 26	Novartis	Calcium	6.29	7.85	Mouse	Balanced in mice when given during active phase, does not affect sleep in resting phase	Research	[121, 122]
c1m	Takeda	Calcium	5.52	7.57	Mouse	Balanced, drives NREM, limited effects on REM	Research	[129, 130]
Cpd 46	Actelion	Radioligand binding, calcium	6.58 6.13	8.52 8.40	Rat	Balanced, stimulates NREM and REM proportionally	Research	[131]
EMPA	Roche	Radioligand binding, calcium, IP1	6.05 >5 NA	8.96 9.1 8.6	Mouse, rat	EMPA was reported to be less efficacious than almorexant in rats, increasing NREM and not affecting NREM/REM balance	Research	[132]
Compound 29/TCS OX2 29	Banyu Pharmaceutical Co., Ltd	Calcium	>5	7.4	Rat, mouse (ICV delivery)	Rat: Increase in REM, no effect on NREM Mouse: increased NREM, no effect on REM	Research	[133–134]

increased; neither REM sleep duration nor REM latency were influenced. The compound, contrary to zolpidem, had no effect on alcohol-induced ataxia, neither did it induce ataxia on its own at 30 mg/kg.

Along the same lines, the Merck group has presented sleep data in rodents and dogs with OX$_2$R antagonists, including MK-1064. These compounds produce sleep in rodents that appears balanced with regard to REM and NREM sleep. For instance, MK-1064 [120, 126] was tested in mice, rats, dogs and rhesus monkeys. Across these different species, MK-1064 reduced active wake and increased light sleep (with the exception of the rat) and SWS and to some extent REM sleep (the latter more so in mice and rats than in dogs and rhesus monkeys). Gotter et al. [120] also demonstrated the influence of dosing time on responses to MK-1064, with rats dosed in the late active phase (ZT23) showing minimal sleep responses to MK-1064 in comparison to animals dosed in the mid-active phase (ZT17), in accordance with the theory that an advantage of OXR antagonists over other hypnotic mechanisms may be a reduced efficacy towards the end of the inactive period as orexin tone rises. Studies in the dog with MK-1064 demonstrated sleep efficacy, but not in a dose-dependent manner, with similar effects on active wake, NREM I and NREM II at 1 and 20 mg/kg (1 h after dosing), while the increase in REM sleep seen with 1 mg/kg was absent with the 20 mg/kg dose. When examining the sleep stage profile over the 10 h subsequent to dosing in these animals, however, a disconnect between NREM and REM responses to MK-1064 can be observed, whereby maximal NREM responses were seen within the 3 h after dosing, while REM sleep was enhanced (albeit to a lesser degree) only 4–9 h after dosing [120]. Reasons for the delayed REM response are not clear. The work in the dog in this study also demonstrated, once again, that animals are easily awakened from OXR antagonist-induced sleep, with cage cleaning and water presentation transiently increasing wakefulness in dogs. Furthermore, OX$_2$R antagonism by MK-1064, even at the very high dose of 20 mg/kg, did not show signs of an enhanced predisposition to cataplexy in the food-elicited cataplexy test [120]. The compound has now completed two Phase I studies focused on pharmacokinetics, safety and sleep in healthy volunteers (see next section).

MK-3697 [117] is another highly selective Merck OX$_2$R antagonist that was tested in mice, rats and dogs. MK-3697 was devoid of effects on sleep/wake in OX$_2$R KO mice. It reduced active wake markedly in all three nonmutant species, increased SWS and REM in mice and rats, whereas in dogs the increase in sleep was entirely due to SWS I and SWS II. A further OX$_2$R antagonist, MK-8133 [118], decreased active wake at 30 mg/kg, increased SWS and had no effect on REM sleep in mice. This compound was also ineffective in OX$_2$R KO mice. In rats, dose-dependent decreases in active wake were seen at 1 and 3 mg/kg of MK-8133 with significant increases in both SWS and REM sleep. In beagle dogs, 0.01 and 0.05 mg/kg, dose-dependently promoted sleep, with reductions in active wake and increases in NREM I and NREM II sleep, with only a trend towards increased REM sleep.

LSN2424100 is a recently described OX$_2$R antagonist from Eli Lilly, which showed efficacy in the delayed-reinforcement-of-low-rate (DRL) 72-s operant schedule for antidepressant-like activity [137], and has also been investigated for effects on ethanol consumption [138]. Its effects on sleep, however, have not yet been described.

Overall, these compounds illustrate that OX_2R antagonists are efficacious in inducing sleep in a range of species, while OX_1R antagonists alone appear near devoid of effects on sleep [53, 94, 95, 124, 139]. The importance of the OX_2R for OX_2R antagonists has been confirmed by both Janssen/J&J and the Merck groups, who showed that sleep-inducing OX_2R antagonists are inactive in OX_2R KO mice [117–120, 126, 128]. Together with our study using almorexant in OX_1R KO and OX_2R KO mice [114], which indicated that the OX_2R is the main, if not the sole, target for sleep induction by DORAs, these data cement that antagonism of the OX_2R is a necessary component of DORAs and that OX_2R antagonism alone is sufficient for inducing sleep.

One issue that remained after our experiments with almorexant in the OXR KO mice [114] was the nature of the sleep induced by the compound in the various mouse strains studied. The original data presented by Brisbare-Roch and colleagues suggested that almorexant induced sleep in rats that somewhat favoured REM sleep, in strong contrast to zolpidem, which markedly favours NREM sleep at the expense of REM sleep [35]. Our own data in wild-type mice suggested similarly that almorexant produced sleep that showed overproportional REM [114]. Interestingly, almorexant appeared to promote even more REM sleep in the OX_1R KO mice than in their respective controls, as if transient OX_2R antagonism on top of constitutive OX_1R receptor blockade had a specific impact on REM sleep [114].

In our opinion, a new and improved hypnotic should, ideally, induce sleep with proportionally equal increases in NREM and REM sleep (at least in healthy volunteers and normal animals), and at the appropriate time (i.e. no sleep onset REM (SOREM)), that is, it should induce "balanced" or "physiological" sleep. Within this definition, it should also be noted that absolute REM/NREM sleep contributions to total sleep vary between individuals and across zeitgeber time (ZT) and are influenced by aspects such as sleep debt, mental and physical activity and disease state.

In our subsequent investigations of the DORA suvorexant versus IPSU, an orexin receptor antagonist with some selectivity for the OX_2R (~six-fold selective for OX_2R versus OX_1R; Compound 26; [121]), dosed in mice in the dark (active) phase, suggested that OX_2R antagonism produced balanced sleep, whereas dual antagonists induce sleep with an overproportional contribution of REM sleep [121, 122]. In a parallel experiment, suvorexant and IPSU were dosed in the light (inactive) phase at doses that induced sleep when given in the active phase. In this situation, IPSU did not affect the sleep architecture of the mice, whereas suvorexant shifted the balance towards more REM sleep [122]. An overview of other OX_2R antagonists described in the literature to date suggests that the majority of them produce approximately balanced increases in REM and NREM (Table 1), which is not necessarily the case for DORAs (reviewed in [46, 84]).

Perhaps the most compelling case for the influence of dual versus OX_2R antagonism on REM sleep comes from the work of Dugovic, Bonaventure and colleagues at Janssen/J&J [94, 95, 124]. They showed that combined treatment with OX_1R and OX_2R antagonists increased REM sleep compared to that observed with either antagonist alone and increased sleep fractionation while reducing latency to REM [94, 95]. Interestingly, Compound 56, a potent and selective OX$_1$R antagonist that did not alter spontaneous sleep architecture in rats or wild-type mice, promoted REM sleep in OX$_2$R KO mice [124]. Furthermore, Piccoli et al. [53] compared the OX$_1$R antagonist GSK-1059865, the OX$_2$R antagonist JNJ-10397049 and the DORA SB-649868 in a rat PSG study. Data were summed over the 3 h after dosing in the mid-dark phase. GSK-1059865 had modest effects on vigilance states, inducing a significant reduction in wake and a small increase in NREM sleep at the highest dose of 25 mg/kg, but had no effects at 5 mg/kg. JNJ-10397049 reduced wake and increased NREM at 5 and 25 mg/kg but had no effect on the amount of REM sleep, and at 25 mg/kg reduced NREM and REM latency by ~30–40%. In comparison, SB-649868 significantly reduced wake and increased NREM and REM sleep, the latter by sevenfold at the highest dose investigated (10 mg/kg), and reduced REM latency from ~90 min (average over the three experiments) to 26.6 min, versus 52.3 min for JNJ-10397049 and 48.9 min and GSK-1059865. These data indicate that this DORA, at least, had a more profound effect on REM sleep than either of the OX$_1$R or OX$_2$R antagonists. More back-to-back studies of this nature would be informative. It should also be noted that induction of SOREM is a clinical feature of SB-649868 [113].

It is currently unclear whether or not induction of imbalanced REM sleep is clinically relevant in insomnia patients, and the matter is a topic of open discussion (see [8, 140]). What is clear, however, is that the data collected in rats or mice is relatively predictive, since all clinical data published so far shows that the DORAs that have been tested in insomnia in Phase II or Phase III produced an increase in REM sleep at the expense of NREM sleep, similar to that described in rodents, and reduce latency to REM ([8–13, 35, 111–113, 141–143]; see also [46, 84]).

It should be noted, however, that researchers at Merck have generally claimed that their DORAs produce balanced sleep in preclinical rodent models. Recent studies compared the REM and NREM sleep profiles of rats dosed with DORA-22 in the dark phase with that of rats dosed with vehicle in the light phase and drew the conclusion that the profiles were comparable [103]. The important profile comparison of rats dosed with DORA-22 in the light phase, however, was not provided. Yet our work dosing suvorexant in the light phase in mice clearly shows that a disproportionate increase in REM sleep remains present with the DORA suvorexant [122]. The fact of the matter is that increases in SWS have not been documented in the clinical setting for any of the DORAs to date, which is especially important considering that latency to REM is decreased by these compounds ([53, 111–113]; reviewed in [46, 84]). It remains unclear, however, as to what type of sleep profile induced in normal animals or humans would best translate to improved sleep in insomnia patients. There is certainly a paucity of valid animal models of insomnia, which should feature both a failure to sleep at the appropriate circadian time and the subsequent development of sleep debt [144]. Ideally, the question of NREM/REM balance in insomnia would be addressed in a well-controlled clinical study evaluating PSG of normal sleepers dosed with a DORA relative to insomniac patients on the same drug and dose and compared to respective placebos. Such a study has yet to be conducted but would be a valuable contribution to the field. More recent DORAs such as ACT-462206 [145], TASP0428980 [146] and others [128] show balanced increases in REM/NREM in animal models. ACT-462206 decreased wakefulness and increased both NREM and REM sleep while maintaining natural sleep architectures in rat and dog as measured by PSG [145]. Similar results were reported with TASP0428980 in rats [146]. Our studies and others regarding the kinetics of receptor binding of OX_1R and OX_2R suggest that this may be important in influencing the REM/NREM sleep balance induced by DORAs ([132, 147]; see later, this chapter), although more work is required in this area.

It is worth considering also that there remains some controversy about the respective contribution of individual OXR antagonism versus dual orexin receptor antagonism to sleep induction overall, as the Roche team has claimed that OX_2R antagonism alone was less efficacious than dual antagonism in increasing sleep [132]. In that study, the OX_2R antagonist EMPA [148, 149] increased cumulative NREM in rats (6 h) but only at the highest dose of 100 mg/kg, whereas the OX_1R antagonist SB-334867 showed some effects on cumulative REM and NREM sleep (3–30 mg/kg), and almorexant was efficacious in increasing both REM and NREM (30 and 100 mg/kg). None of the three compounds influenced REM/NREM ratios versus vehicle in the first 6 h after dosing. With regard to the hourly profiles, EMPA showed no significant effects on wake, REM or NREM sleep. The weak efficacy of EMPA may however have been due in part to its pharmacokinetic profile, since the oral availability of EMPA is low (1.1%, rat) and the $T_{1/2}$ short (0.37 h in the rat after oral administration at 20 mg/kg; [148]). The effects of SB-334867 on an hourly basis showed some effects on REM and NREM sleep, but not in a reliable time- or dose-dependent manner. Piccoli et al. [53] also described a weak NREM-enhancing effect by the OX_1R antagonist GSK-1059865 in rats (at 25 mg/kg, but not 5 mg/kg). A recent study [139], however, using similar doses of SB-334867 to Morairty et al. [132] was unable to reproduce the cumulative effects of this compound on NREM or REM sleep. Other studies with SB-334867 and other OX_1R antagonists have also failed to detect influences of OX_1R antagonism on sleep [92, 94, 95, 124, 150, 151]. The reason for this discrepancy is not clear, although Morairty et al. [132] pointed out that their experimental setup differed with regard to the time of administration, compared at least to Dugovic et al. [94]. Morairty et al. [132] administered compounds in the mid-dark period when animals had already accumulated a sleep debt by being awake, consequently allowing for greater sensitivity to sleep readouts.

A separate point of note about DORAs versus OX_2R antagonists for the treatment of insomnia is the risk of complex sleep disorders: narcolepsy/cataplexy, and related SOREM, direct transitions between wake and REM (DREM) and sleep paralysis. SB-649868 caused DREM in rats at sleep-inducing doses [95] and SOREM in clinical situations ([111–113]; reviewed in [84]). In clinical studies with suvorexant, rare occurrences of muscle weakness, sleep paralysis and REM sleep behaviour disorder (RBD) were reported (see [46]). DREM, SOREM, sudden muscle weakness, sleep paralysis and REM sleep without atonia are diagnostic features of narcolepsy [61, 152]. In this regard, it is worth reiterating that the phenotype of the OXR KO mice indicates that the dual orexin receptor KO shows a more pronounced narcolepsy/cataplexy phenotype than the OX$_2$R KO. Thus, it remains possible that OX$_2$R antagonists may have a lower risk of incurring narcolepsy/cataplexy-related complex sleep behaviours than DORAs. Ideally, this hypothesis would be tested by dosing OX$_2$R antagonists versus DORAs in narcolepsy/cataplexy-susceptible animal models, such as the ataxin-3 rat or mouse model, DTA mice or narcoleptic dogs.

In summary, there are several lines of preclinical evidence that support the selective targeting of OX$_2$R over DORAs for sleep disorders: (1) DORAs do not produce sleep in OX$_2$R KO mice although they are active in OX$_1$R KO mice, suggesting OX$_2$R are the primary target for sleep; (2) OX$_2$R antagonists are effective in producing sleep in preclinical models and humans; (3) OX$_2$R antagonists may produce a more physiologically balanced NREM/REM sleep profile than DORAs; (4) the theoretical risk of inducing narcolepsy/cataplexy-related complex sleep behaviours may be greater with DORAs than with OX$_2$R antagonists.

5 Clinical Data

MIN-202, MK-1064 and MK-3697 are OX$_2$R antagonists that have entered clinical development. To date, there are no refereed, published clinical data on any OX$_2$R antagonist in insomnia patients, although studies in healthy human volunteers have been conducted. The current development status of MK-3697 and MK-1064 is unclear: no clinical data have been published so far for MK-3697, and no OX$_2$R antagonists are listed in the pipeline published on Merck's website (6 May 2016). However, Phase I data for MK-1064 in healthy human volunteers has recently been published [120]. PK, tolerability and pharmacodynamic indicators of somnolence were investigated in a single ascending dose study (P001; 16 subjects), while the effects of MK-1064 on sleep were investigated in a separate PSG study (P003; 20 subjects). In the latter study, effects of MK-1064 on sleep parameters were moderate (like those of most orexin antagonists in volunteers, since it was not an insomnia trial), but statistically significant, representing the first demonstration of PSG-quantified sleep induced in humans by an OX$_2$R antagonist. MK-1064 at 50, 120 and 250 mg decreased latency to persistent sleep and wake after sleep onset, and increased total sleep time and sleep efficiency (although not in a dose-dependent manner); both NREM and REM sleep were increased moderately. The side effect profile was moderate, with reports of somnolence in P001, as is appropriate for a hypnotic, peaking 1–4 h after dosing in accordance with the $T_{\max}$ (mean $T_{\max}$: 1–2 h for the doses of 5–250 mg). Headache was reported in seven subjects in P001 (at 10, 400 and 750 mg doses) versus 1 in the placebo group, and in three subjects on active dose in the PSG study versus none in the placebo group. In addition, four subjects in the PSG active dose groups reported nightmares, which is also a reported adverse event with suvorexant, albeit also at a low level (reviewed in [46]). With regard to other sleep disorders, there was no mention of SOREM in the studies with MK-1064, although one subject in the PSG study reported sleep paralysis at the highest dose of that study of 250 mg. Overall, based on the aforementioned data, the authors concluded that MK-1064 did not offer any advantage over DORAs [120].

Minerva, on the other hand, is clearly proceeding with the development of MIN-202 in insomnia and major depressive disorders in collaboration with Janssen Pharmaceutical/J&J (see Minerva website and announcements). In the first Phase I study with MIN-202/JNJ-42847922, 10, 20, 40 or 80 mg was given to healthy volunteers and produced somnolence in a dose-dependent fashion. Adverse events included one incidence of sleep paralysis [119, 153]. Results of a Phase IIa study with MIN-202 have been announced in a press release [154]. The study was conducted in Europe and the USA as a randomized, placebo-controlled double-blind study. Twenty-eight insomnia patients (without psychiatric comorbidity) were given MIN-202 or placebo in a crossover design for 5 days, separated by a washout. Sleep efficiency was measured by PSG and was increased at the 40 mg dose of MIN-202 versus placebo, apparently "preserving key phases of sleep". Additional positive effects were seen for LPS, WASO and TST readouts. Adverse effects were listed as somnolence and abnormal dreams. In addition, Minerva has investigated MIN-202 in insomnia with major depressive disorder (MDD). In a double-blind, placebo-controlled, single-dose Phase Ib study conducted with patients suffering from major depressive disorder (MDD) and insomnia, MIN-202 reduced latency to persistent sleep and increased TST by approximately 45 min [153]. A further Phase Ib trial in MDD patients is ongoing, investigating MIN-202 coadministered with an antidepressant [154]. A recent press release [155] indicates that MIN-202 has progressed to a Phase I study in Japanese healthy volunteers, apparently demonstrating a similar PK to that observed in non-Asian populations and posting no serious clinically relevant safety concerns.

As stated above, all DORAs which have been tested in "primary insomnia" according to DSM-IV or simply "insomnia" according to DSM-5, induce and maintain sleep, whether in volunteers (where the effects are mild), in insomnia patients [8, 9, 12, 13, 113] or human models of insomnia (see, e.g. [112]). Thus, almorexant, SB-649868, suvorexant and filorexant all show positive outcomes in sleep studies. Suvorexant, the most advanced compound, received market approval in 2014 in the USA and Japan. The compound induces and maintains sleep; total sleep time (TST) is increased and wake after sleep onset (WASO) is reduced. It is noted that latency to REM is decreased, while REM sleep is increased; various components of NREM and especially slow-wave sleep (SWS) are little affected or even mildly reduced. Some patients show evidence of SOREM, i.e. entered REM sleep within 15 min after falling asleep. Side effects are considered acceptable although next day somnolence is dose dependently increased. The FDA recommended lower doses than proposed by Merck, 10 mg as a starting dose (or 5 mg in special patient populations), with a maximal dose of 15 or 20 mg.

The company had initially proposed 15–40 mg with differences between elderly adults (>65 years) and non-elderly adults, but the FDA makes no recommendations for different age groups. The prescribing recommendations are rather strict: suvorexant must be taken within 30 min after bedtime, i.e. once in bed and not before, and taken only once per night at least 7 h before the projected wakeup time. It is also recommended not to drive the next day. These limitations reflect a current trend at the FDA, which have asked that the starting dose of a number of hypnotics, including zolpidem and other Z drugs, be reduced compared to that which was common practice during the last 10–20 years. The FDA is concerned about safety, especially next morning residual effects of hypnotics and particularly on driving ([156]; and see [43]).

Suvorexant has a clinical profile that is very similar to that of all clinically tested DORAs. In essence, DORAs affect sleep by increasing REM, in contrast to the Z drugs which do so by promoting NREM sleep at the expense of REM sleep. Almorexant may be less extreme with respect to REM, but the Phase III results have not been published as extensively as those of suvorexant [143]. In addition, if receptor kinetics were to play a role, almorexant may constitute a special case, perhaps becoming a preferential OX$_2$R antagonist with increasing time after dosing: almorexant equilibrates very slowly at OX$_2$R and thus has a somewhat higher affinity at OX$_2$R compared to OX$_1$R at steady state (see [147]). What is clear, however, is that SB-649868, which is a very slow OX$_1$R binder [147], produced marked SOREM in insomnia patients in the Phase II trial [113]. It should be noted that SOREM and more generally elevated REM are characteristic of some patients with neurological or neuropsychiatric disorders. SOREM has even been proposed as an early biomarker for such disorders [157–159]. As discussed briefly above, it is unclear whether or not elevated REM sleep or SOREM in itself may have negative consequences in insomnia patients. SOREM, however, if associated with sudden muscle weakness, may present an elevated risk for falls in frail and/or the elderly, populations whose risks of falls are already elevated. The use of hypnotics and particularly rapidly acting hypnotics is already a noted risk for increased falls in the elderly and patients with dementia [160, 161]. Although none of the DORAs have produced evidence for cataplexy, suvorexant is contraindicated in patients suffering from narcolepsy with cataplexy.

The main issue for the FDA with suvorexant was next morning somnolence, which was dose dependent, prompting the recommendation to start with lower doses and not to exceed 20 mg/night. Somnolence with suvorexant is most likely PK related, given the relatively long half-life of the drug, but there may also be a kinetic-dependent effect as it is a very slow receptor binder that may reach equilibrium after only several hours [147]. Any OX$_2$R antagonists aspiring to registration in the future will thus require a specific PK profile, aimed at avoiding next day somnolence. On a positive note, there is no evidence that abrupt cessation of suvorexant medication will lead to rebound insomnia or withdrawal symptoms as often experienced with GABA$_A$ receptor modulators. It is unknown whether or not OX$_2$R antagonists will show a similar profile in this regard.

In conclusion, a number of OX$_2$R antagonists are now in development, although there is a paucity of insomnia patient efficacy data at the present time. Similarly, new DORAs have entered clinical studies, such as Lemborexant/E2006, currently in Phase II (presented at ACNP; see clinicaltrials.gov/show/NCT01995838 [162]), and ACT-462206 [163] in Phase I. Data on PSG parameters such as REM and NREM have not yet been released.

6 Future Considerations

MIN-202, MK-1064 and MK-3697 (OX_2R antagonists) have all entered clinical trials, and the community awaits the publication of the clinical data to compare their effects on sleep features in insomnia patients with those of DORAs. The development MK-1064 and MK-3697 is not listed in the Merck development pipeline as of mid-2016, suggesting they may have been stopped, although MK-1064 Phase I data have been published. No reasons were indicated for that interruption, as is common for big pharma. Minerva Neurosciences is developing MIN-202 in collaboration with Janssen Pharmaceutical, with promising claims that MIN-202 may produce a physiological sleep balance. At present, however, long-term studies are needed to make more definitive statements about the utility/potential advantages of OX_2R antagonists versus DORAs. Similarly, one will have to wait for Phase IV results to determine what impact, if any, overproportional REM sleep may have following long-term suvorexant administration. It is clear that SOREM is not a desired feature (a rare event in normal sleep, but which occurs frequently in patients with narcolepsy and cataplexy). SOREM induced by suvorexant is apparently infrequent and may even dissipate upon longer treatment duration (see [8, 11, 12]; see MSD Corporation [164] for the Suvorexant Advisory Committee Briefing Document). Further preclinical studies are indicated to probe the potential on-target liabilities of OX_2R antagonists in comparison with DORAs. Preclinical data for OX_2R antagonists, however, appears promising with regard to efficacy, in that OX_2R antagonism is sufficient and necessary for the generation of OXR antagonist-mediated sleep and may yet illustrate a propensity to produce more balanced/physiological sleep than other hypnotics.

References

1. APA (2013) Diagnostic and statistical manual of mental disorders, 5th edn. American Psychiatric Association, Arlington, VA
2. Greenblatt DJ, Roth T (2012) Zolpidem for insomnia. Expert Opin Pharmacother 13:879–893
3. Dündar Y, Dodd S, Strobl J, Boland A, Dickson R, Walley T (2004) Comparative efficacy of newer hypnotic drugs for the short-term management of insomnia: a systematic review and meta-analysis. Hum Psychopharmacol 19:305–322
4. FDA Drug Safety Communication (10 Jan 2013) Safety announcement: risk of next-morning impairment after use of insomnia drugs; FDA requires lower recommended doses for certain

drugs containing zolpidem (Ambien, Ambien CR, Edluar, and Zolpimist). U.S. food and drug administration. http://www.fda.gov/Drugs/DrugSafety/ucm334033.htm

5. FDA Drug Safety Communication (14 May 2013) FDA approves new label changes and dosing for zolpidem products and a recommendation to avoid driving the day after using Ambien CR. U.S. food and drug administration. http://www.fda.gov/Drugs/DrugSafety/ucm352085.htm

6. Randall S, Roehrs T, Harris E, Maan R, Roth T (2010) Chronic use of zolpidem is not associated with loss of efficacy. Sleep 33 Supplement: Abstract 0660, A221.

7. Roehrs TA, Randall S, Harris E, Maan R, Roth T (2012) Twelve months of nightly zolpidem does not lead to rebound insomnia or withdrawal symptoms: a prospective placebo-controlled study. J Psychopharmacol 26:1088–1095

8. Herring WJ, Connor KM, Ivgy-May N, Snyder E, Liu K, Snavely DB, Krystal AD, Walsh JK, Benca RM, Rosenberg R, Sangal RB, Budd K, Hutzelmann J, Leibensperger H, Froman S, Lines C, Roth T, Michelson D (2016) Suvorexant in patients with insomnia: results from two 3-month randomized controlled clinical trials. Biol Psychiatry 79:136–148

9. Herring WJ, Snyder E, Budd K, Hutzelmann J, Snavely D, Liu K, Lines C, Roth T, Michelson D (2012) Orexin receptor antagonism for treatment of insomnia: a randomized clinical trial of suvorexant. Neurology 79:2265–2274

10. Ma J, Svetnik V, Snyder E, Lines C, Roth T, Herring WJ (2014) Electroencephalographic power spectral density profile of the orexin receptor antagonist suvorexant in patients with primary insomnia and healthy subjects. Sleep 37:1609–1619

11. Michelson D, Snyder E, Paradis E, Chengan-Liu M, Snavely DB, Hutzelmann J, Walsh JK, Krystal AD, Benca RM, Cohn M, Lines C, Roth T, Herring WJ (2014) Safety and efficacy of suvorexant during 1-year treatment of insomnia with subsequent abrupt treatment discontinuation: a phase 3 randomised, double-blind, placebo-controlled trial. Lancet Neurol 13:461–471

12. Snyder E, Ma J, Svetnik V, Connor KM, Lines C, Michelson D, Herring WJ (2016) Effects of suvorexant on sleep architecture and power spectral profile in patients with insomnia: analysis of pooled phase-3 data. Sleep Med 19:93–100. doi:10.1016/j.sleep.2015.10.007

13. Sun H, Kennedy WP, Wilbraham D, Lewis N, Calder N, Li XD, Ma JS, Yee KL, Ermlich S, Mangin E, Lines C, Rosen L, Chodakewitz J, Murphy GM (2013) Effects of suvorexant, an orexin receptor antagonist, on sleep parameters as measured by polysomnography in healthy men. Sleep 36:259–267

14. Krystal AD, Richelson E, Roth T (2013) Review of the histamine system and the clinical effects of H1 antagonists: basis for a new model for understanding the effects of insomnia medications. Sleep Med Rev 17(4):263–272. doi:10.1016/j.smrv.2012.08.001

15. Ferracioli-Oda E, Qawasmi A, Bloch MH (2013) Meta-analysis: melatonin for the treatment of primary sleep disorders. PLoS One 8(5):e63773

16. Kuriyama A, Honda M, Hayashino Y (2014) Ramelteon for the treatment of insomnia in adults: a systematic review and meta-analysis. Sleep Med 15:385–392

17. De Lecea L, Kilduff TS, Peyron C, Gao XB, Foye PE, Danielson PE, Fukuhara C, Battenberg ELF, Gautvik VT, Bartlett FS, Frankel WN, van den Pol AN, Bloom FE, Gautvik KM, Sutcliffe JG (1998) The hypocretins: hypothalamus-specific peptides with neuroexcitatory activity. Proc Natl Acad Sci U S A 95:322–327

18. Sakurai T, Amemiya A, Ishii M, Matsuzaki I, Chemelli RM, Tanaka H, Williams SC, Richardson JA, Kozlowski GP, Wilson S, Arch JRS, Buckingham RE, Haynes AC, Carr SA, Annan RS, McNulty DE, Liu WS, Terrett JA, Elshourbagy NA, Bergsma DJ, Yanagisawa M (1998) Orexins and orexin receptors: A family of hypothalamic neuropeptides and G protein-coupled receptors that regulate feeding behavior. Cell 92:573–585

19. Sakurai T, Moriguchi T, Furuya K, Kajiwara N, Nakamura T, Yanagisawa M, Goto K (1999) Structure and function of human prepro-orexin gene. J Biol Chem 274:17771–17776

20. Sutcliffe JG, de Lecea L (2002) The hypocretins: setting the arousal threshold. Nat Rev Neurosci 3:339–349

21. Burgess CR, Scammell TE (2012) Narcolepsy: neural mechanisms of sleepiness and cataplexy. J Neurosci 32:12305–12311
22. Scammell TE, Willie JT, Guilleminault C, Siegel JM, Int Working Grp Rodent Models N (2009) A consensus definition of cataplexy in mouse models of narcolepsy. Sleep 32:111–116
23. Bassetti CL, Mathis J, Gugger M, Sturzenegger C, Mignot E, Radanov B (2001) Idiopathic hypersomnia and narcolepsy without cataplexy: a multimodal diagnostic approach in 20 patients including CSF orexin levels. Neurology 56:A8–A8
24. Baumann CR, Mignot E, Lammers GJ, Overeem S, Arnulf I, Rye D, Dauvilliers Y, Honda M, Owens JA, Plazzi G, Scammell TE (2014) Challenges in diagnosing narcolepsy without cataplexy: a consensus statement. Sleep 37:1035–1042
25. Beuckmann CT, Sinton CM, Williams SC, Richardson JA, Hammer RE, Sakurai T, Yanagisawa M (2004) Expression of a poly-glutamine-ataxin-3 transgene in orexin neurons induces narcolepsy-cataplexy in the rat. J Neurosci 24:4469–4477
26. Bourgin P, Zeitzer JM, Mignot E (2008) CSF hypocretin-1 assessment in sleep and neurological disorders. Lancet Neurol 7:649–662
27. Chemelli RM, Willie JT, Sinton CM, Elmquist JK, Scammell T, Lee C, Richardson JA, Williams SC, Xiong YM, Kisanuki Y, Fitch TE, Nakazato M, Hammer RE, Saper CB, Yanagisawa M (1999) Narcolepsy in orexin knockout mice: molecular genetics of sleep regulation. Cell 98:437–451
28. Hara J, Beuckmann CT, Nambu T, Willie JT, Chemelli RM, Sinton CM, Sugiyama F, Yagami K, Goto K, Yanagisawa M, Sakurai T (2001) Genetic ablation of orexin neurons in mice results in narcolepsy, hypophagia, and obesity. Neuron 30:345–354
29. Lin L, Faraco J, Li R, Kadotani H, Rogers W, Lin XY, Qiu XH, de Jong PJ, Nishino S, Mignot E (1999) The sleep disorder canine narcolepsy is caused by a mutation in the hypocretin (orexin) receptor 2 gene. Cell 98:365–376
30. Mignot E, Chen W, Black J (2003) On the value of measuring CSF hypocretin-1 in diagnosing narcolepsy. Sleep 26:646–649
31. Nishino S, Ripley B, Overeem S, Lammers GJ, Mignot E (2000) Hypocretin (orexin) deficiency in human narcolepsy. Lancet 355:39–40
32. Nishino S, Ripley B, Overeem S, Nevsimalova S, Lammers GJ, Vankova J, Okun M, Rogers W, Brooks S, Mignot E (2001) Low cerebrospinal fluid hypocretin (orexin) and altered energy homeostasis in human narcolepsy. Ann Neurol 50:381–388
33. Peyron C, Faraco J, Rogers W, Ripley B, Overeem S, Charnay Y, Nevsimalova S, Aldrich M, Reynolds D, Albin R, Li R, Hungs M, Pedrazzoli M, Padigaru M, Kucherlapati M, Fan J, Maki R, Lammers GJ, Bouras C, Kucherlapati R, Nishino S, Mignot E (2000) A mutation in a case of early onset narcolepsy and a generalized absence of hypocretin peptides in human narcoleptic brains. Nat Med 6:991–997
34. Thannickal TC, Moore RY, Nienhuis R, Ramanathan L, Gulyani S, Aldrich M, Cornford M, Siegel JM (2000) Reduced number of hypocretin neurons in human narcolepsy. Neuron 27:469–474
35. Brisbare-Roch C, Dingemanse J, Koberstein R, Hoever P, Aissaoui H, Flores S, Mueller C, Nayler O, van Gerven J, de Haas SL, Hess P, Qiu CB, Buchmann S, Scherz M, Weller T, Fischli W, Clozel M, Jenck F (2007) Promotion of sleep by targeting the orexin system in rats, dogs and humans. Nat Med 13:150–155
36. Di Fabio R, Pellacani A, Faedo S, Roth A, Piccoli L, Gerrard P, Porter RA, Johnson CN, Thewlis K, Donati D, Stasi L, Spada S, Stemp G, Nash D, Branch C, Kindon L, Massagrande M, Poffe A, Braggio S, Chiarparin E, Marchioro C, Ratti E, Corsi M (2011) Discovery process and pharmacological characterization of a novel dual orexin 1 and orexin 2 receptor antagonist useful for treatment of sleep disorders. Bioorg Med Chem Lett 21:5562–5567
37. Cox CD, Breslin MJ, Whitman DB, Schreier JD, McGaughey GB, Bogusky MJ, Roecker AJ, Mercer SP, Bednar RA, Lemaire W, Bruno JG, Reiss DR, Harrell CM, Murphy KL, Garson SL, Doran SM, Prueksaritanont T, Anderson WB, Tang CY, Roller S, Cabalu TD, Cui DH,

Hartman GD, Young SD, Koblan KS, Winrow CJ, Renger JJ, Coleman PJ (2010) Discovery of the dual orexin receptor antagonist (7R)-4-(5-Chloro-1,3-benzoxazol-2-yl)-7-methyl-1,4-diazepan-1-yl 5-met hyl-2-(2H-1,2,3-triazol-2-yl) phenyl methanone (MK-4305) for the treatment of insomnia. J Med Chem 53:5320–5332

38. Winrow CJ, Gotter AL, Cox CD, Doran SM, Tannenbaum PL, Breslin MJ, Garson SL, Fox SV, Harrell CM, Stevens J, Reiss DR, Cui DH, Coleman PJ, Renger JJ (2011) Promotion of sleep by suvorexant – a novel dual orexin receptor antagonist. J Neurogenet 25:52–61

39. Coleman PJ, Schreier JD, Cox CD, Breslin MJ, Whitman DB, Bogusky MJ, McGaughey GB, Bednar RA, Lemaire W, Doran SM, Fox SV, Garson SL, Gotter AL, Harrell CM, Reiss DR, Cabalu TD, Cui DH, Prueksaritanont T, Stevens J, Tannenbaum PL, Ball RG, Stellabott J, Young SD, Hartman GD, Winrow CJ, Renger JJ (2012) Discovery of (2R,5R)-5-{ (5-Fluoropyridin-2-yl)oxy methyl}-2-methylpiperidin-1-yl 5 -methyl-2-(pyrimidin-2-yl) phenyl methanone (MK-6096): a dual orexin receptor antagonist with potent sleep-promoting properties. ChemMedChem 7:415–424

40. Winrow CJ, Gotter AL, Cox CD, Tannenbaum PL, Garson SL, Doran SM, Breslin MJ, Schreier JD, Fox SV, Harrell CM, Stevens J, Reiss DR, Cui DH, Coleman PJ, Renger JJ (2012) Pharmacological characterization of MK 6096-A dual orexin receptor antagonist for insomnia. Neuropharmacology 62:978–987

41. Yoshida Y, Naoe Y, Terauchi T, Ozaki F, Doko T, Takemura A, Tanaka T, Sorimachi K, Beuckmann CT, Suzuki M, Ueno T, Ozaki S, Yonaga M (2015) Discovery of (1R,2S)-2-{ (2,4-Dimethylpyrimidin-5-yl)oxy methyl}-2-(3-fluorophenyl)-N -(5-fluoropyridin-2-yl) cyclopropanecarboxamide (E2006): a potent and efficacious oral orexin receptor antagonist. J Med Chem 58:4648–4664

42. Actelion/GlaxoSmithKline. Actelion and GSK discontinue clinical development of almorexant. http://www.actelion.com/en/investors/media-releases/index.page?newsId= 1483135

43. Farkas R (2013) Suvorexant safety and efficacy. U.S. food and drug administration. http://www.fda.gov/downloads/AdvisoryCommittees/CommitteesMeetingMaterials/Drugs/ PeripheralandCentralNervousSystemDrugsAdvisoryCommittee/UCM354215.pdf

44. FDA. Highlights of prescribing information: Belsomra® (suvorexant) tablets (2014) http:// www.accessdata.fda.gov/drugsatfda_docs/label/2014/204569s000lbledt.pdf

45. Suvorexant. Advisory Committee Briefing Document (2013) http://www.fda.gov/downloads/ AdvisoryCommittees/CommitteesMeetingMaterials/Drugs/PeripheralandCentralNervous SystemDrugsAdvisoryCommittee/UCM352970.pdf

46. Jacobson LH, Callander GE, Hoyer D (2014) Suvorexant for the treatment of insomnia. Expert Rev Clin Pharmacol 7:711–730

47. Alexander SPH, Catterall WA, Kelly E, Marrion N, Peters JA, Benson HE, Faccenda E, Pawson AJ, Sharman JL, Southan C, Davies JA, Collaborators CGTP (2015) The concise guide to pharmacology 2015/16: G protein coupled receptors. Br J Pharmacol 172 (24):5744–5869

48. Gotter AL, Webber AL, Coleman PJ, Renger JJ, Winrow CJ (2012) International union of basic and clinical pharmacology. LXXXVI. orexin receptor function, nomenclature and pharmacology. Pharmacol Rev 64:389–420

49. Alvarez CE, Sutcliffe JG (2002) Hypocretin is an early member of the incretin gene family. Neurosci Lett 324:169–172

50. Peyron C, Tighe DK, van den Pol AN, de Lecea L, Heller HC, Sutcliffe JG, Kilduff TS (1998) Neurons containing hypocretin (orexin) project to multiple neuronal systems. J Neurosci 18:9996–10015

51. Abe T, Mieda M, Yanagisawa M, Tanaka K, Sakurai T (2009) Both ox1r and ox2r orexin receptors mediate efferent signals of orexin neurons in eliciting food-anticipatory activity. J Physiol Sci 59:286–286

52. Lebold TP, Bonaventure P, Shireman BT (2013) Selective orexin receptor antagonists. Bioorg Med Chem Lett 23:4761–4769

53. Piccoli L, Micioni Di Bonaventura MV, Cifani C, Costantini VJ, Massagrande M, Montanari D, Martinelli P, Antolini M, Ciccocioppo R, Massi M, Merlo-Pich E, Di Fabio R, Corsi M (2012) Role of orexin-1 receptor mechanisms on compulsive food consumption in a model of binge eating in female rats. Neuropsychopharmacology 37:1999–2011
54. Bonnavion P, de Lecea L (2010) Hypocretins in the control of sleep and wakefulness. Curr Neurol Neurosci Rep 10:174–179
55. Willie JT, Chemelli RM, Sinton CM, Yanagisawa M (2001) To eat or to sleep? Orexin in the regulation of feeding and wakefulness. Annu Rev Neurosci 24:429–458
56. Yamanaka A, Sakurai T, Katsumoto T, Yanagisawa M, Goto K (1999) Chronic intracerebroventricular administration of orexin-A to rats increases food intake in daytime, but has no effect on body weight. Brain Res 849:248–252
57. Alam MN, Gong H, Alam T, Jaganath R, McGinty D, Szymusiak R (2002) Sleep-waking discharge patterns of neurons recorded in the rat perifornical lateral hypothalamic area. J Physiol 538:619–631
58. de Lecea L, Carter ME, Adamantidis A (2012) Shining light on wakefulness and arousal. Biol Psychiatry 71:1046–1052
59. Sakurai T (2007) The neural circuit of orexin (hypocretin): maintaining sleep and wakefulness. Nat Rev Neurosci 8:171–181
60. Liblau RS, Vassalli A, Seifinejad A, Tafti M (2015) Hypocretin (orexin) biology and the pathophysiology of narcolepsy with cataplexy. Lancet Neurol 14:318–328
61. Scammell TE (2015) Narcolepsy. N Engl J Med 373:2654–2662
62. Thompson MD, Comings DE, Abu-Ghazalah R, Jereseh Y, Lin L, Wade J, Sakurai T, Tokita S, Yoshida T, Tanaka H, Yanagisawa M, Burnham WM, Moldofsky H (2004) Variants of the orexin2/hcrt2 receptor gene identified in patients with excessive daytime sleepiness and patients with Tourette's syndrome comorbidity. Am J Med Genet B Neuropsychiatr Genet 129B:69–75
63. Saper CB, Chou TC, Scammell TE (2001) The sleep switch: hypothalamic control of sleep and wakefulness. Trends Neurosci 24:726–731
64. Saper CB, Fuller PM, Pedersen NP, Lu J, Scammell TE (2010) Sleep state switching. Neuron 68:1023–1042
65. de Lecea L, Heurta R (2014) Hypocretin (orexin) regulation of sleep-to-wake transitions. Front Pharmacol 5:16. doi:10.3389/fphar2014.00016
66. Luppi PH, Peyron C, Fort P (2016) Not a single but multiple populations of GABAergic neurons control sleep. Sleep Med Rev. doi:10.1016/j.smrv.2016.03.002
67. Roh JH, Jiang H, Finn MB, Stewart FR, Mahan TE, Cirrito JR, Heda A, Snider BJ, Li M, Yanagisawa M, de Lecea L, Holtzman DM (2014) Potential role of orexin and sleep modulation in the pathogenesis of Alzheimer's disease. J Exp Med 211:2487–2496
68. Liguori C (2016) Orexin and Alzheimer's disease. Curr Top Behav Neurosci. doi:10.1007/7854_2016_50
69. Mochizuki T, Crocker A, McCormack S, Yanagisawa M, Sakurai T, Scammell TE (2004) Behavioral state instability in orexin knock-out mice. J Neurosci 24:6291–6300
70. Mieda M, Willie JT, Hara J, Sinton CM, Sakurai T, Yanagisawa M (2004) Orexin peptides prevent cataplexy and improve wakefulness in an orexin neuron-ablated model of narcolepsy in mice. Proc Natl Acad Sci U S A 101:4649–4654
71. Mieda M, Willie JT, Sakurai T (2005) Rescue of narcoleptic orexin neuron-ablated mice by ectopic overexpression of orexin peptides. In: Nishino S, Sakurai T (eds) Orexin/Hypocretin system: physiology and pathophysiology, Contemporary clinical neuroscience. Humana Press, Totowa, pp 359–366
72. Tabuchi S, Tsunematsu T, Black SW, Tominaga M, Maruyama M, Takagi K, Minokoshi Y, Sakurai T, Kilduff TS, Yamanaka A (2014) Conditional ablation of orexin/hypocretin neurons: a new mouse model for the study of narcolepsy and orexin system function. J Neurosci 34:6495–6509

73. Adamantidis A, Carter ME, de Lecea L, Shaw P, Tafti M, Thorpy M (2013) Optogenetic control of arousal neurons. In: Shaw P, Tafti M, Thorpy M (eds) Genetic basis of sleep and sleep disorders. Cambridge University Press, Cambridge, pp 66–72
74. Adamantidis AR, Zhang F, Aravanis AM, Deisseroth K, De Lecea L (2007) Neural substrates of awakening probed with optogenetic control of hypocretin neurons. Nature 450:420–U429
75. Carter ME, Yizhar O, Chikahisa S, Nguyen H, Adamantidis A, Nishino S, Deisseroth K, de Lecea L (2010) Tuning arousal with optogenetic modulation of locus coeruleus neurons. Nat Neurosci 13:1526–U1117
76. Tsunematsu T, Kilduff TS, Boyden ES, Takahashi S, Tominaga M, Yamanaka A (2011) Acute optogenetic silencing of orexin/hypocretin neurons induces slow-wave sleep in mice. J Neurosci 31:10529–10539
77. Espana RA, McCormack SL, Mochizuki T, Scammell TE (2007) Running promotes wakefulness and increases cataplexy in orexin knockout mice. Sleep 30:1417–1425
78. Mochizuki T, Arrigoni E, Marcus JN, Clark EL, Yamamoto M, Honer M, Borroni E, Lowell BB, Elmquist JK, Scammell TE (2011) Orexin receptor 2 expression in the posterior hypothalamus rescues sleepiness in narcoleptic mice. Proc Natl Acad Sci U S A 108:4471 4476
79. Willie JT, Chemelli RM, Sinston CM, Tokita H, Williams SC, Kisanuki YY, Marcus JN, Lee C, Elmquist JK, Kohlmeier KA, Leonard CS, Richardson JA, Hammer RE, Yanagisawa M (2003) Distinct narcolepsy syndromes in orexin receptor-2 and Orexin null mice: molecular genetic dissection of non-REM and REM sleep regulatory processes. Neuron 38:715–730
80. Adamantidis A, de Lecea L (2008) Physiological arousal: a role for hypothalamic systems. Cell Mol Life Sci 65:1475–1488
81. Adamantidis A, de Lecea L (2008) Sleep and metabolism: shared circuits, new connections. Trends Endocrinol Metab 19:362–370
82. Adamantidis A, de Lecea L (2009) The hypocretins as sensors for metabolism and arousal. J Physiol 587:33–40
83. Sakurai T (2014) The role of orexin in motivated behaviours. Nat Rev Neurosci 15:719–731
84. Hoyer D, Jacobson LH (2013) Orexin in sleep, addiction and more: is the perfect insomnia drug at hand? Neuropeptides 47:477–488
85. Tafti M (2007) Reply to 'promotion of sleep by targeting the orexin system in rats, dogs and humans'. Nat Med 13:525–526
86. Hoyer D (2007) RNA interference for studying the molecular basis of neuropsychiatric disorders. Curr Opin Drug Discov Devel 10:122–129
87. Urban DJ, Roth BL (2015) DREADDs (designer receptors exclusively activated by designer drugs): chemogenetic tools with therapeutic utility. Annu Rev Pharmacol Toxicol 55:399–417
88. Branch AF, Navidi W, Tabuchi S, Terao A, Yamanaka A, Scammell TE, Diniz Behn C (2016) Progressive loss of the orexin neurons reveals dual effects on wakefulness. Sleep 39:369–377
89. Bourgin P, Huitron-Resendiz S, Spier AD, Fabre V, Morte B, Criado JR, Sutcliffe JG, Henriksen SJ, de Lecea L (2000) Hypocretin-1 modulates rapid eye movement sleep through activation of locus coeruleus neurons. J Neurosci 20:7760–7765
90. Chen LC, McKenna JT, Bolortuya Y, Brown RE, McCarley RW (2013) Knockdown of orexin type 2 receptor in the lateral pontomesencephalic tegmentum of rats increases REM sleep. Eur J Neurosci 37:957–963
91. Chen LC, McKenna JT, Bolortuya Y, Winston S, Thakkar MM, Basheer R, Brown RE, McCarley RW (2010) Knockdown of orexin type 1 receptor in rat locus coeruleus increases REM sleep during the dark period. Eur J Neurosci 32:1528–1536
92. Gozzi A, Turrini G, Piccoli L, Massagrande M, Amantini D, Antolini M, Martinelli P, Cesari N, Montanari D, Tessari M, Corsi M, Bifone A (2011) Functional magnetic resonance

imaging reveals different neural substrates for the effects of orexin-1 and orexin-2 receptor antagonists. PLoS One 6(1):e16406. doi:10.1371/journal.pone.0016406

93. Mieda M, Hasegawa E, Kisanuki YY, Sinton CM, Yanagisawa M, Sakurai T (2011) Differential roles of orexin receptor-1 and-2 in the regulation of non-REM and REM sleep. J Neurosci 31:6518–6526

94. Dugovic C, Shelton JE, Aluisio LE, Fraser IC, Jiang X, Sutton SW, Bonaventure P, Yun S, Li X, Lord B, Dvorak CA, Carruthers NI, Lovenberg TW (2009) Blockade of orexin-1 receptors attenuates orexin-2 receptor antagonism-induced sleep promotion in the rat. J Pharmacol Exp Ther 330:142–151

95. Dugovic C, Shelton JE, Yun S, Bonaventure P, Shireman BT, Lovenberg TW (2014) Orexin-1 receptor blockade dysregulates REM sleep in the presence of orexin-2 receptor antagonism. Front Neurosci 8:28. doi:10.3389/fnins.2014.00028

96. Lin JS, Dauvilliers Y, Arnulf I, Bastuji H, Anaclet C, Parmentier R, Kocher L, Yanagisawa M, Lehert P, Ligneau X, Perrin D, Robert P, Roux M, Lecomte JM, Schwartz JC (2008) An inverse agonist of the histamine H-3 receptor improves wakefulness in narcolepsy: studies in orexin(-/-) mice and patients. Neurobiol Dis 30:74–83

97. Baier PC, Hallschmid M, Seeck-Hirschner M, Weinhold SL, Burkert S, Diessner N, Goder R, Aldenhoff JB, Hinze-Selch D (2011) Effects of intranasal hypocretin-1 (orexin A) on sleep in narcolepsy with cataplexy. Sleep Med 12:941–946

98. Baier PC, Weinhold SL, Huth V, Gottwald B, Ferstl R, Hinze-Selch D (2008) Olfactory dysfunction in patients with narcolepsy with cataplexy is restored by intranasal orexin A (hypocretin-1). Brain 131:2734–2741

99. Weinhold SL, Seeck-Hirschner M, Nowak A, Hallschmid M, Goder R, Baier PC (2014) The effect of intranasal orexin-A (hypocretin-1) on sleep, wakefulness and attention in narcolepsy with cataplexy. Behav Brain Res 262:8–13

100. Nagahara T, Saitoh T, Kutsumura N, Irukayama-Tomobe Y, Ogawa Y, Kuroda D, Gouda H, Kumagai H, Fujii H, Yanagisawa M, Nagase H (2015) Design and synthesis of non-peptide, selective orexin receptor 2 agonists. J Med Chem 58:7931–7937

101. Carter ME, de Lecea L, Adamantidis A (2013) Functional wiring of hypocretin and LC-NE neurons: implications for arousal. Front Behav Neurosci 7:43. doi:10.3389/fnbeh.2013.00043

102. Fox SV, Gotter AL, Tye SJ, Garson SL, Savitz AT, Uslaner JM, Brunner JI, Tannenbaum PL, McDonald TP, Hodgson R, Yao L, Bowlby MR, Kuduk SD, Coleman PJ, Hargreaves R, Winrow CJ, Renger JJ (2013) Quantitative electroencephalography within sleep/wake states differentiates GABA(A) modulators eszopiclone and zolpidem from dual orexin receptor antagonists in rats. Neuropsychopharmacology 38:2401–2408

103. Gotter AL, Garson SL, Stevens J, Munden RL, Fox SV, Tannenbaum PL, Yao L, Kuduk SD, McDonald T, Uslaner JM, Tye SJ, Coleman PJ, Winrow CJ, Renger JJ (2014) Differential sleep-promoting effects of dual orexin receptor antagonists and GABA(A) receptor modulators. BMC Neurosci 15:109. doi:10.1186/1471-2202-15-109

104. Tannenbaum PL, Stevens J, Binns J, Savitz AT, Garson SL, Fox SV, Coleman P, Kuduk SD, Gotter AL, Marino M, Tye SJ, Uslaner JM, Winrow CJ, Renger JJ (2014) Orexin receptor antagonist-induced sleep does not impair the ability to wake in response to emotionally salient acoustic stimuli in dogs. Front Behav Neurosci 8:182

105. Ramirez AD, Gotter AL, Fox SV, Tannenbaum PL, Yao L, Tye SJ, McDonald T, Brunner J, Garson SL, Reiss DR, Kuduk SD, Coleman PJ, Uslaner JM, Hodgson R, Browne SE, Renger JJ, Winrow CJ (2013) Dual orexin receptor antagonists show distinct effects on locomotor performance, ethanol interaction and sleep architecture relative to gamma-aminobutyric acid-A receptor modulators. Front Neurosci 7:254. doi:10.3389/fnins.2013.00254

106. Tye SJ, Uslaner JM, Eddins DM, Savitz AT, Fox SV, Cannon CE, Binns J, Garson SL, Stevens J, Yao L, Hodgson R, Brunner J, McDonald TP, Bowlby MR, Tannenbaum PL, Kuduk SD, Gotter AL, Winrow CJ, Renger JJ (2012) Sleep-promoting doses of diazepam, eszopiclone and zolpidem disrupt Rhesus macaque memory and attention as compared to DORA-22, a dual orexin receptor antagonist. Society for Neuroscience Annual Meeting, New Orleans, USA, Abstract 921.20/EEE48

107. Uslaner JM, Tye SJ, Eddins DM, Wang XH, Fox SV, Savitz AT, Binns J, Cannon CE, Garson SL, Yao LH, Hodgson R, Stevens J, Bowlby MR, Tannenbaum PL, Brunner J, McDonald TP, Gotter AL, Kuduk SD, Coleman PJ, Winrow CJ, Renger JJ (2013) Orexin receptor antagonists differ from standard sleep drugs by promoting sleep at doses that do not disrupt cognition. Sci Transl Med 5(179):179ra44. doi:10.1126/scitranslmed.3005213

108. Steiner MA, Lecourt H, Strasser DS, Brisbare-Roch C, Jenck F (2011) Differential effects of the dual orexin receptor antagonist almorexant and the GABA(A)-alpha 1 receptor modulator zolpidem, alone or combined with ethanol, on motor performance in the rat. Neuropsychopharmacology 36:848–856

109. Coleman PJ, Cox CD, Roecker AJ (2011) Discovery of dual orexin receptor antagonists (DORAs) for the treatment of insomnia. Curr Top Med Chem 11:696–725

110. Scammell TE, Winrow CJ (2011) Orexin receptors: pharmacology and therapeutic opportunities. Annu Rev Pharmacol Toxicol 51:243–266

111. Bettica P, Nucci G, Pyke C, Squassante L, Zamuner S, Ratti E, Gomeni R, Alexander R (2012) Phase I studies on the safety, tolerability, pharmacokinetics and pharmacodynamics of SB-649868, a novel dual orexin receptor antagonist. J Psychopharmacol 26:1058–1070

112. Bettica P, Squassante L, Groeger JA, Gennery B, Winsky-Sommerer R, Dijk DJ (2012) Differential effects of a dual orexin receptor antagonist (SB-649868) and zolpidem on sleep initiation and consolidation, SWS, REM sleep, and EEG power spectra in a model of situational insomnia. Neuropsychopharmacology 37:1224–1233

113. Bettica P, Squassante L, Zamuner S, Nucci G, Danker-Hopfe H, Ratti E (2012) The orexin antagonist SB-649868 promotes and maintains sleep in men with primary insomnia. Sleep 35:1097–1104

114. Mang GM, Durst T, Burki H, Imobersteg S, Abramowski D, Schuepbach E, Hoyer D, Fendt M, Gee CE (2012) The dual orexin receptor antagonist almorexant induces sleep and decreases orexin-induced locomotion by blocking orexin 2 receptors. Sleep 35:1625–1635

115. Sun H, Yee KL, Gill S, Liu W, Li X, Panebianco D, Mangin E, Morrison D, McCrea J, Wagner JA, Troyer MD (2015) Psychomotor effects, pharmacokinetics and safety of the orexin receptor antagonist suvorexant administered in combination with alcohol in healthy subjects. J Psychopharmacol 29:1159–1169

116. Black SW, Morairty SR, Fisher SP, Chen TM, Warrier DR, Kilduff TS (2013) Almorexant promotes sleep and exacerbates cataplexy in a murine model of narcolepsy. Sleep 36:325–336

117. Roecker AJ, Reger TS, Mattern MC, Mercer SP, Bergman JM, Schreier JD, Cube RV, Cox CD, Li D, Lemaire W, Bruno JG, Harrell CM, Garson SL, Gotter AL, Fox SV, Stevens J, Tannenbaum PL, Prueksaritanont T, Cabalu TD, Cui D, Stellabott J, Hartman GD, Young SD, Winrow CJ, Renger JJ, Coleman PJ (2014) Discovery of MK-3697: a selective orexin 2 receptor antagonist (2-SORA) for the treatment of insomnia. Bioorg Med Chem Lett 24:4884–4890

118. Kuduk SD, Skudlarek JW, DiMarco CN, Bruno JG, Pausch MH, O'Brien JA, Cabalu TD, Stevens J, Brunner J, Tannenbaum PL, Garson SL, Savitz AT, Harrell CM, Gotter AL, Winrow CJ, Renger JJ, Coleman PJ (2015) Identification of MK-8133: an orexin-2 selective receptor antagonist with favorable development properties. Bioorg Med Chem Lett 25:2488–2492

119. Bonaventure P, Shelton J, Yun S, Nepomuceno D, Sutton S, Aluisio L, Fraser I, Lord B, Shoblock J, Welty N, Chaplan SR, Aguilar Z, Halter R, Ndifor A, Koudriakova T, Rizzolio M, Letavic M, Carruthers NI, Lovenberg T, Dugovic C (2015) Characterization of JNJ-42847922, a selective orexin-2 receptor antagonist, as a clinical candidate for the treatment of insomnia. J Pharmacol Exp Ther 354:471–482

120. Gotter AL, Forman MS, Harrell CM, Stevens J, Svetnik V, Yee KL, Li X, Roecker AJ, Fox SV, Tannenbaum PL, Garson SL, Lepeleire ID, Calder N, Rosen L, Struyk A, Coleman PJ, Herring WJ, Renger JJ, Winrow CJ (2016) Orexin 2 receptor antagonism is sufficient to

promote NREM and REM sleep from mouse to man. Sci Rep 6:27147. doi:10.1038/srep27147

121. Betschart C, Hintermann S, Behnke D, Cotesta S, Fendt M, Gee CE, Jacobson LH, Laue G, Ofner S, Chaudhari V, Badiger S, Pandit C, Wagner J, Hoyer D (2013) Identification of a novel series of orexin receptor antagonists with a distinct effect on sleep architecture for the treatment of insomnia. J Med Chem 56:7590–7607

122. Hoyer D, Durst T, Fendt M, Jacobson LH, Betschart C, Hintermann S, Behnke D, Cotesta S, Laue G, Ofner S, Legangneux E, Gee CE (2013) Distinct effects of IPSU and suvorexant on mouse sleep architecture. Front Neurosci 7:235. doi:10.3389/fnins.2013.00235

123. Letavic MA, Bonaventure P, Carruthers NI, Dugovic C, Koudriakova T, Lord B, Lovenberg TW, Ly KS, Mani NS, Nepomuceno D, Pippel DJ, Rizzolio M, Shelton JE, Shah CR, Shireman BT, Young LK, Yun S (2015) Novel octahydropyrrolo 3,4-c pyrroles are selective orexin-2 antagonists: SAR leading to a clinical candidate. J Med Chem 58:5620–5636

124. Bonaventure P, Yun S, Johnson PL, Shekhar A, Fitz SD, Shireman BT, Lebold TP, Nepomuceno D, Lord B, Wennerholm M, Shelton J, Carruthers N, Lovenberg T, Dugovic C (2015) A selective orexin-1 receptor antagonist attenuates stress-induced hyperarousal without hypnotic effects. J Pharmacol Exp Ther 352(3):590–601

125. McAtee LC, Sutton SW, Rudolph DA, Li X, Aluisio LE, Phuong VK, Dvorak CA, Lovenberg TW, Carruthers NI, Jones TK (2004) Novel substituted 4-phenyl-[1,3]dioxanes: potent and selective orexin receptor 2 (OX(2)R) antagonists. Bioorg Med Chem Lett 14:4225–4229

126. Roecker AJ, Mercer SP, Schreier JD, Cox CD, Fraley ME, Steen JT, Lemaire W, Bruno JG, Harrell CM, Garson SL, Gotter AL, Fox SV, Stevens J, Tannenbaum PL, Prueksaritanont T, Cabalu TD, Cui DH, Stellabott J, Hartman GD, Young SD, Winrow CJ, Renger JJ, Coleman PJ (2014) Discovery of 5 "-Chloro-N-(5,6-dimethoxypyridin-2-yl)methyl-2,2′:5′,3 ″-terpyridine-3′-carboxamide (MK-1064): a selective orexin 2 receptor antagonist (2-SORA) for the treatment of insomnia. ChemMedChem 9:311–322

127. Mercer SP, Roecker AJ, Garson S, Reiss DR, Meacham Harrell C, Murphy KL, Bruno JG, Bednar RA, Lemaire W, Cui D, Cabalu TD, Tang C, Prueksaritanont T, Hartman GD, Young SD, Winrow CJ, Renger JJ, Coleman PJ (2013) Discovery of 2,5-diarylnicotinamides as selective orexin-2 receptor antagonists (2-SORAs). Bioorg Med Chem Lett 23:6620–6624

128. Roecker AJ, Mercer SP, Bergman JM, Gilbert KF, Kuduk SD, Harrell CM, Garson SL, Fox SV, Gotter AL, Tannenbaum PL, Prueksaritanont T, Cabalu TD, Cui D, Lemaire W, Winrow CJ, Renger JJ, Coleman PJ (2015) Discovery of diazepane amide DORAs and 2-SORAs enabled by exploration of isosteric quinazoline replacements. Bioorg Med Chem Lett 25:4992–4999

129. Fujimoto T, Kunitomo J, Tomata Y, Nishiyama K, Nakashima M, Hirozane M, Yoshikubo S, Hirai K, Marui S (2011) Discovery of potent, selective, orally active benzoxazepine-based orexin-2 receptor antagonists. Bioorg Med Chem Lett 21:6414–6416

130. Etori K, Saito YC, Tsujino N, Sakurai T (2014) Effects of a newly developed potent orexin-2 receptor-selective antagonist, compound 1 m, on sleep/wakefulness states in mice. Front Neurosci 8:8. doi:10.3389/fnins.2014.00008

131. Sifferlen T, Boller A, Chardonneau A, Cottreel E, Gatfield J, Treiber A, Roch C, Jenck F, Aissaoui H, Williams JT, Brotschi C, Heidmann B, Siegrist R, Boss C (2015) Substituted pyrrolidin-2-ones: centrally acting orexin receptor antagonists promoting sleep. Part 2. Bioorg Med Chem Lett 25:1884–1891

132. Morairty SR, Revel FG, Malherbe P, Moreau JL, Valladao D, Wettstein JG, Kilduff TS, Borroni E (2012) Dual hypocretin receptor antagonism is more effective for sleep promotion than antagonism of either receptor alone. PLoS One 7(7):e39131. doi:10.1371/journal.pone.0039131

133. Hirose M, Egashira S, Goto Y, Hashihayata T, Ohtake N, Iwaasa H, Hata M, Fukami T, Kanatani A, Yamada K (2003) N-acyl 6,7-dimethoxy-1,2,3,4-tetrahydroisoquinoline: the first orexin-2 receptor selective non-peptidic antagonist. Bioorg Med Chem Lett 13:4497–4499

134. Moore MW, Akladious A, Hu Y, Azzam S, Feng P, Strohl KP (2014) Effects of orexin 2 receptor activation on apnea in the C57BL/6J mouse. Respir Physiol Neurobiol 200:118–125

135. Kummangal BA, Kumar D, Mallick HN (2013) Intracerebroventricular injection of orexin-2 receptor antagonist promotes REM sleep. Behav Brain Res 237:59–62

136. Raheem IT, Breslin MJ, Bruno J, Cabalu TD, Cooke A, Cox CD, Cui D, Garson S, Gotter AL, Fox SV, Harrell CM, Kuduk SD, Lemaire W, Prueksaritanont T, Renger JJ, Stump C, Tannenbaum PL, Williams PD, Winrow CJ, Coleman PJ (2015) Discovery of piperidine ethers as selective orexin receptor antagonists (SORAs) inspired by filorexant. Bioorg Med Chem Lett 25:444–450

137. Fitch TE, Benvenga MJ, Jesudason CD, Zink C, Vandergriff AB, Menezes MM, Schober DA, Rorick-Kehn LM (2014) LSN2424100: a novel, potent orexin-2 receptor antagonist with selectivity over orexin-1 receptors and activity in an animal model predictive of antidepressant-like efficacy. Front Neurosci 8:5. doi:10.3389/fnins.2014.00005

138. Anderson RI, Becker HC, Adams BL, Jesudason CD, Rorick-Kehn LM (2014) Orexin-1 and orexin-2 receptor antagonists reduce ethanol self-administration in high-drinking rodent models. Front Neurosci 8:33. doi:10.3389/fnins.2014.00033

139. Brodnik ZD, Bernstein DL, Prince CD, España RA (2015) Hypocretin receptor 1 blockade preferentially reduces high effort responding for cocaine without promoting sleep. Behav Brain Res 291:377–384

140. Boss C, Roch C (2015) Recent trends in orexin research-2010 to 2015. Bioorg Med Chem Lett 25:2875–2887

141. Hoever P, de Haas S, Dorffner G, Chiossi E, van Gerven J, Dingemanse J (2012) Orexin receptor antagonism: an ascending multiple-dose study with almorexant. J Psychopharmacol 26:1071–1080

142. Hoever P, de Haas S, Winkler J, Schoemaker RC, Chiossi E, van Gerven J, Dingemanse J (2010) Orexin receptor antagonism, a new sleep-promoting paradigm: an ascending single-dose study with almorexant. Clin Pharmacol Ther 87:593–600

143. Hoever P, Dorffner G, Benes H, Penzel T, Danker-Hopfe H, Barbanoj MJ, Pillar G, Saletu B, Polo O, Kunz D, Zeitlhofer J, Berg S, Partinen M, Bassetti CL, Hogl B, Ebrahim IO, Holsboer-Trachsler E, Bengtsson H, Peker Y, Hemmeter UM, Chiossi E, Hajak G, Dingemanse J (2012) Orexin receptor antagonism, a new sleep-enabling paradigm: a proof-of-concept clinical trial. Clin Pharmacol Ther 91:975–985

144. Toth LA, Bhargava P (2013) Animal models of sleep disorders. Comp Med 63:91–10

145. Boss C, Brisbare-Roch C, Steiner MA, Treiber A, Dietrich H, Jenck F, von Raumer M, Sifferlen T, Brotschi C, Heidmann B, Williams JT, Aissaoui H, Siegrist R, Gatfield J (2014) Structure-activity relationship, biological, and pharmacological characterization of the pro-line sulfonamide ACT-462206: a potent, brain-penetrant dual orexin 1/orexin 2 receptor antagonist. ChemMedChem 9:2486–2496

146. Suzuki R, Nozawa D, Futamura A, Nishikawa-Shimono R, Abe M, Hattori N, Ohta H, Araki Y, Kambe D, Ohmichi M, Tokura S, Aoki T, Ohtake N, Kawamoto H (2015) Discovery and in vitro and in vivo profiles of N-ethyl-N- 2- 3-(5-fluoro-2-pyridinyl)-1H-pyrazol-1-yl ethyl -2-(2H-1,2, 3-triazol-2-yl)-benzamide as a novel class of dual orexin receptor antagonist. Bioorganic & Medicinal Chemistry 23:1260–1275

147. Callander GE, Olorunda M, Monna D, Schuepbach E, Langenegger D, Betschart C, Hintermann S, Behnke D, Cotesta S, Fendt M, Laue G, Ofner S, Briard E, Gee CE, Jacobson LH, Hoyer D (2013) Kinetic properties of "dual" orexin receptor antagonists at OX1R and OX2R orexin receptors. Front Neurosci 7:230. doi:10.3389/fnins.2013.00230

148. Malherbe P, Borroni E, Gobbi L, Knust H, Nettekoven M, Pinard E, Roche O, Rogers-Evans M, Wettstein JG, Moreau JL (2009) Biochemical and behavioural characterization of EMPA, a novel high-affinity, selective antagonist for the OX(2) receptor. Br J Pharmacol 156:1326–1341

149. Malherbe P, Borroni E, Pinard E, Wettstein JG, Knoflach F (2009) Biochemical and electrophysiological characterization of almorexant, a dual orexin 1 receptor (OX(1))/orexin 2 receptor (OX(2)) antagonist: comparison with selective OX(1) and OX(2) antagonists. Mol Pharmacol 76:618–631
150. Smith MI, Piper DC, Duxon MS, Upton N (2003) Evidence implicating a role for orexin-1 receptor modulation of paradoxical sleep in the rat. Neurosci Lett 341:256–258
151. Steiner MA, Gatfield J, Brisbare-Roch C, Dietrich H, Treiber A, Jenck F, Boss C (2013) Discovery and characterization of ACT-335827, an orally available, brain penetrant orexin receptor type 1 selective antagonist. ChemMedChem 8:898–903
152. Dauvilliers Y, Jennum P, Plazzi G (2013) Rapid eye movement sleep behavior disorder and rapid eye movement sleep without atonia in narcolepsy. Sleep Med 14:775–781
153. Minerva Neurosciences Inc (2015) Press release 21 Jan 2015: 'Minerva neuroscience reports positive phase 1 data with MIN-202, selective orexin-2 antagonist for treatment of sleep disorders including primary and comorbid insomnia'. http://ir.minervaneurosciences.com/releases.cfm?ReleaseID=892355
154. Minerva Neurosciences Inc (2016) Press release 11 Jan 2016: 'Minerva neurosciences announces favorable top line results from MIN-202 phase 2A clinical trial in insomnia disorder'. http://ir.minervaneurosciences.com/releasedetail.cfm?ReleaseID=949714
155. Minerva Neurosciences Inc (2016) Press release 1 Feb 2016: 'Minerva neurosciences announces top line data from MIN-202 phase I clinical study in Japanese patients'. http://ir.minervaneurosciences.com/releasedetail.cfm?ReleaseID=952576
156. Vermeeren A, Sun H, Vuurman EF, Jongen S, Van Leeuwen CJ, Van Oers ACM, Palcza J, Li X, Laethem T, Heirman I, Bautmans A, Troyer MD, Wrishko R, McCrea J (2015) On-the-road driving performance the morning after bedtime use of suvorexant 20 and 40 mg: a study in non-elderly healthy volunteers. Sleep 38:1803–1813
157. Riemann D, Spiegelhalder K, Nissen C, Hirscher V, Baglioni C, Feige B (2012) REM sleep instability – a new pathway for insomnia? Pharmacopsychiatry 45:167–176
158. Staner L, Van Veeren C, Stefos G, Hubain PP, Linkowski P, Mendlewicz J (1998) Neuroendocrine and clinical characteristics of major depressed patients exhibiting sleep-onset REM. Biol Psychiatry 43:817–821
159. Stefos G, Staner L, Kerkhofs M, Hubain P, Mendlewicz J, Linkowski P (1998) Shortened REM latency as a psychobiological marker for psychotic depression? An age-, gender-, and polarity-controlled study. Biol Psychiatry 44:1314–1320
160. Park H, Satoh H, Miki A, Urushihara H, Sawada Y (2015) Medications associated with falls in older people: systematic review of publications from a recent 5-year period. Eur J Clin Pharmacol 71:1429–1440
161. Tamiya H, Yasunaga H, Matusi H, Fushimi K, Ogawa S, Akishita M (2015) Hypnotics and the occurrence of bone fractures in hospitalized dementia patients: a matched case-control study using a national inpatient database. PLoS One 10(6):e0129366. doi:10.1371/journal.pone.0129366
162. https://clinicaltrials.gov/show/NCT01995838: A multicenter, randomized, double-blind, placebo-controlled, parallel-group, Bayesian adaptive randomization design, dose response study of the efficacy of E2006 in adults and elderly subjects with chronic insomnia
163. Hoch M, van Gorsel H, van Gerven J, Dingemanse J (2014) Entry-into-humans study with ACT-462206, a novel dual orexin receptor antagonist, comparing its pharmacodynamics with almorexant. J Clin Pharmacol 54:979–986
164. MSD Corporation (2013) Suvorexant advisory committee briefing document. http://www.fda.gov/downloads/AdvisoryCommittees/CommitteesMeetingMaterials/Drugs/PeripheralandCentralNervousSystemDrugsAdvisoryCommittee/UCM352970.pdf

Roles for Orexin/Hypocretin in the Control of Energy Balance and Metabolism

Paulette B. Goforth and Martin G. Myers

Abstract The neuropeptide hypocretin is also commonly referred to as orexin, since its orexigenic action was recognized early. Orexin/hypocretin (OX) neurons project widely throughout the brain and the physiologic and behavioral functions of OX are much more complex than initially conceived based upon the stimulation of feeding. OX most notably controls functions relevant to attention, alertness, and motivation. OX also plays multiple crucial roles in the control of food intake, metabolism, and overall energy balance in mammals. OX signaling not only promotes food-seeking behavior upon short-term fasting to increase food intake and defend body weight, but, conversely, OX signaling also supports energy expenditure to protect against obesity. Furthermore, OX modulates the autonomic nervous system to control glucose metabolism, including during the response to hypoglycemia. Consistently, a variety of nutritional cues (including the hormones leptin and ghrelin) and metabolites (e.g., glucose, amino acids) control OX neurons. In this chapter, we review the control of OX neurons by nutritional/metabolic cues, along with our current understanding of the mechanisms by which OX and OX neurons contribute to the control of energy balance and metabolism.

Keywords Energy expenditure · Food intake · Glucose homeostasis · Orexin/ hypocretin

P.B. Goforth
Department of Pharmacology, University of Michigan, 1000 Wall St, 5131 Brehm Tower, Ann Arbor, MI 48105, USA

M.G. Myers (✉)
Departments of Internal Medicine, and Molecular and Integrative Physiology, University of Michigan, 1000 Wall St, 6317 Brehm Tower, Ann Arbor, MI 48105, USA
e-mail: mgmyers@umich.edu

© Springer International Publishing AG 2016
Curr Topics Behav Neurosci (2017) 33: 137–156
DOI 10.1007/7854_2016_51
Published Online: 30 November 2016

Contents

1   Introduction

Two distinct neuropeptides, orexin-A and -B (a.k.a. hypocretin-1 and -2), derive from a common precursor peptide encoded by a single *Hcrt* gene [1]. The orexin/hypocretin (OX) peptides act via two receptors (OX1R/HCRT1R and OX2R/HCRT2R), which are encoded by distinct genes. There is evidence for some selectivity of each isoform of OX and each receptor for a subset of actions ascribed to OX [2, 3]. While we note some instances of peptide or receptor specificity in the control of energy balance by OX, this chapter will generally consider the regulation and function of the two OX peptides and receptors in aggregate.

A relatively small and circumscribed set of neurons in the brain contain OX; these lie in the lateral hypothalamic area (LHA), which conveys information integrated from hypothalamic homeostatic systems to limbic, behavioral, and autonomic centers elsewhere in the brain [1, 4–7]. Early lesioning experiments suggested a role for the LHA in the control of arousal and motivation, since ablation of the LHA produces a profound psychomotor retardation and lack of motivation (including the motivation to feed) [8]. Conversely, the application of stimulating current to the LHA is rewarding/reinforcing and promotes food intake [9–11].

Although the LHA has long been known to contain a variety of glutamatergic and GABAergic neurons, more discretely defined LHA-specific neuronal subtypes were not identified until the discovery of melanin-concentrating hormone (MCH)-containing cells in the dorsal LHA and zona incerta and the identification of OX cells surrounding the fornix; both MCH and OX neurons are glutamatergic [12–16]. Additional subsets of (primarily GABAergic) LHA neurons have been defined by the expression of other neuropeptides, including neurotensin (NT), galanin

(GAL), and corticotrophin releasing factor (CRF), among others; unlike MCH and OX, the expression of these other peptides is not restricted to the LHA [17–19].

Importantly, while others and we often consider the regulation and function of OX and OX neurons in aggregate, OX neurons express other neuropeptides, including dynorphin (an opioid peptide) and the calcitonin receptor ligand, amylin, and also contain the machinery for glutamatergic transmission [20]. Thus, the function of OX is not always identical to the function of OX neurons, which can mediate downstream effects via multiple distinct neurotransmitters.

2   Role(s) for OX in Food Intake

The "orexin" moniker often applied to OX peptides derives from the early seminal work in which Sakurai et al. revealed the OX peptides to be the endogenous ligands for the previously orphan G-protein coupled receptors that we now know as the OX receptors [1]. In this paper, they also demonstrated that the central administration of OX peptides acutely and profoundly increased food consumption in rats and mice [21]. Similarly, suppression of OX signaling via knockdown or the administration of an OX1 receptor antagonist blunts food consumption in food-restricted mice [22–24].

Interestingly, however, OX infusion acutely stimulates feeding when delivered during the light, but not dark phase, and produces no significant increase in total food intake over 24 h [25]. Some initial insight into the potential basis for the diurnal nature of the orexigenic function of OX came from the recognition (simultaneous with the description of "orexin" and its function) of the causative role for OX receptor mutations in canine narcolepsy [26]. Subsequent work has demonstrated that narcolepsy in humans and other species generally results from interference with the function of the OX neuron [27, 28]. Narcolepsy is characterized by disrupted sleep–wake cycles, daytime somnolence, and cataplexy (the acute loss of muscle tone, usually occurring with strong emotions) [29]. Thus, OX signaling promotes wakefulness and alertness and modulates sleep/wake cycles [30]. Indeed, diurnal inputs represent major non-nutritional/metabolic regulators of OX neuron activity; OX neurons become active at the time of awakening and increased activity (at the onset of the dark cycle in rodents) [30]. Injection of OX into the brain increases wakefulness and activity, while interference with OX receptor signaling promotes somnolence and inactivity [30].

An early conclusion based upon the orexigenic action of OX and the role for OX in the control of alertness and diurnal rhythms was that OX-induced feeding in the light cycle is simply due to inducing an aroused state when rodents would normally be asleep. Waking animals during the light cycle without OX administration does not similarly increase food intake, however, suggesting that OX may prolong the wakeful period to increase food consumption, or increase feeding by other mechanisms [31]. Indeed, OX-induced increases in food intake and locomotor activity (e.g., wheel running), an indicator of enhanced arousal, are not always coincident

[31]. Thus, equating the promotion of feeding by OX with the promotion of wakefulness underestimates the potential importance of OX in the control of feeding and energy balance. In fact, OX-induced feeding does not simply follow arousal state, but is augmented by signals of low energy state and hunger cues (Fig. 1) [32, 33]. For example, administration of an OX1R antagonist blunts the increased food intake associated with acute insulin-induced hypoglycemia (IIH, which increases *Hcrt* mRNA expression and activates OX neurons), consistent with the notion that OX contributes to the increased arousal and feeding that is provoked by hypoglycemia [34–38].

3 The Regulation of OX Neurons by Nutritional Status: The Role of Hormones

OX neuronal activity is modulated by numerous nutrients, neuropeptides, and peripheral factors, such as leptin and ghrelin, which report energy status independent of arousal state. Fasting (e.g., overnight food deprivation in rodents) induces the immunohistochemically detectable expression of c-Fos (a surrogate for increased neuronal activity) in OX neurons [17, 39]. Consistent with this increased activity, fasting remodels synaptic input onto OX neurons, enhancing excitatory and reducing inhibitory input [40]. Hormones that signal nutritional status, such as leptin and ghrelin, play important roles in the regulation of OX neurons, including in the response of OX neurons to fasting.

4 Leptin Indirectly Suppresses OX Activity

Leptin, which is produced by adipocytes in approximate proportion to their triglyceride content, signals the repletion of energy (fat) stores to the CNS [41–44]. While there is relatively little variation in leptin levels under normal circumstances, the depletion of fat stores (e.g., during prolonged caloric restriction) suppresses leptin production and thus decreases circulating leptin concentrations [41–44]. Decreased leptin defends against the depletion of energy stores by initiating a host of physiologic and behavioral responses, including stimulating food intake (increased motivation to feed, decreased satiety) and reducing energy expenditure via the suppression of the sympathetic nervous system (SNS) and the neuroendocrine thyroid, growth, and reproductive axes [41–44].

Leptin administration during fasting blunts the induction of c-Fos and the remodeling of synaptic inputs of OX neurons [17, 39]. Furthermore, the application of leptin to slice preparations decreases excitatory input and opens a K_{ATP} channel in OX neurons, leading to their hyperpolarization and inhibition (Fig. 1) [45]. While OX neurons do not express the leptin receptor (LepRb), neighboring LHA LepRb

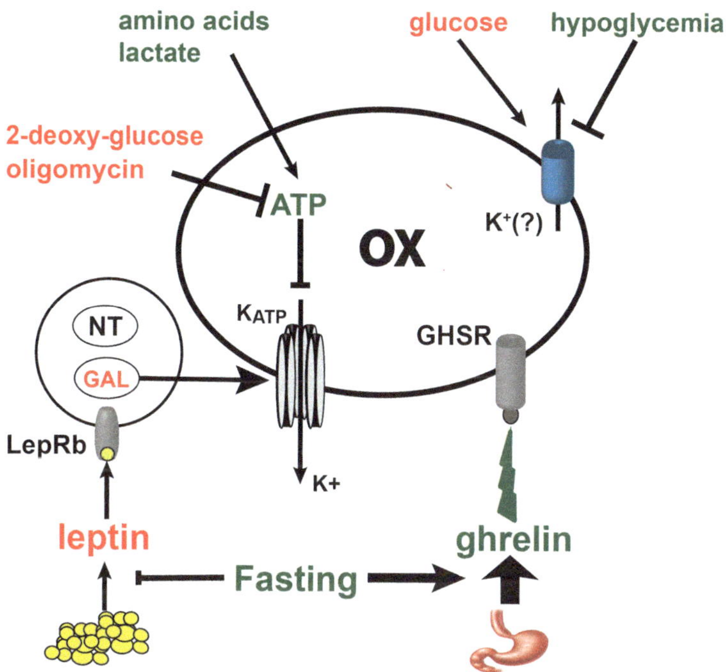

Fig. 1 Modulation of OX neuronal activity by hormones and nutrients. Signals of nutritional status modulate the activity of OX neurons. These signals include the hormones leptin and ghrelin, which are decreased and increased, respectively, during short-term fasting, thereby disinhibiting (low leptin) and activating (high ghrelin) OX neurons. Leptin controls OX neurons indirectly, by promoting the release of GAL from neighboring LepRb neurons, which activates OX neuron K_{ATP} channels. Ghrelin may act directly via GHSRs on OX neurons. Low glucose activates and high glucose inhibits OX neurons. The inhibition of OX neurons by high glucose depends upon other metabolites; however, their inhibition by glucose (mediated by an as yet unidentified K+ channel) is augmented when other metabolic fuels are low (as during treatment with 2-DG or oligomycin). Conversely, their inhibition by glucose is blunted in the presence of activating metabolites, such as amino acids and lactate. Like leptin, amino acids and lactate modulate OX neurons via K_{ATP} channels. *Abbreviations*: *LepRb* long form of the leptin receptor, *GAL* galanin, *NT* neurotensin, K_{ATP} ATP-sensitive potassium channel, *GHSR* growth hormone secretagogue receptor

neurons lie in synaptic contact with OX cells, consistent with a potential role for LHA LepRb neurons in the control of OX cells by leptin [17, 46]. Several other leptin-regulated cell types, including arcuate nucleus (ARC) proopiomelanocortin (POMC)- or agouti-related peptide/neuropeptide Y (AgRP/NPY)-expressing cells, have been suggested to contribute to the control of OX neurons, as well [47, 48]. Disruption of LepRb in an NT-expressing subset of LHA LepRb cells interferes with the ability of leptin to control the activity of OX neurons in vivo and in slice preparations, however, suggesting that LHA LepRb neurons play a crucial role in the regulation of OX neuron electrical activity by leptin [17, 45].

While LHA LepRb neurons, including those that contain NT, display markers of GABAergic cells, GABA transmission is not required for the electrical inhibition of OX neurons by leptin in slice preparations [17, 45]. Many NT-expressing LHA LepRb neurons also contain GAL [18]; while NT does not attenuate the activity of OX neurons, GAL hyperpolarizes and inhibits OX neurons, and inhibitors of GAL signaling block the suppression of OX neuron firing by leptin, suggesting that leptin inhibits the electrical activity of OX neurons by promoting the release of GAL from LHA LepRb neurons [45, 49]. Indeed, GAL activates K_{ATP} channels in other cell types, and the ablation of LepRb from GAL-containing neurons decreases LHA *Gal* expression and increases the activity of OX neurons [49–52].

5 Ghrelin Stimulates OX Activity

The stomach-derived hormone, ghrelin, acts oppositely to leptin in many ways: while leptin is regulated chronically and varies little over the diurnal cycle, circulating ghrelin rises from very low levels to peak just prior to the predicted onset of feeding and falls rapidly to baseline following food intake [53]. Fasting increases ghrelin over the short term, but active (acylated) ghrelin is decreased following prolonged fasting, presumably because recently ingested lipids are needed to mediate its acylation. Ghrelin acutely increases hunger and food intake, as well as locomotor activity [53].

Ghrelin administration induces c-Fos in OX neurons [54]. OX neurons express the receptor for ghrelin (GHSR) and ghrelin promotes membrane depolarization and increased action potential firing in dissociated OX neurons, suggesting that ghrelin directly activates OX neurons (Fig. 1) [55].

6 Hormonal Regulation of OX Expression

Fasting not only increases the activity of OX neurons, it also increases *Hcrt* mRNA expression [21, 56]. Interestingly, unlike fasting, chronic caloric restriction (reduction, but not absolute lack, of food intake) does not increase *Hcrt* expression [57]. Thus, some physiologic or metabolic parameter that changes with fasting must be required for increased *Hcrt* expression; the activation of the OX neurons per se does not suffice. Interestingly, while leptin inhibits the electrical activity of OX neurons, it promotes *increased Hcrt* expression: not only is *Hcrt* expression decreased in leptin-deficient *ob/ob* mice, but also leptin administration increases *Hcrt* expression in wild-type and *ob/ob* mice [17, 46, 55]. Since leptin decreases with fasting, the fasting-induced signal that increases *Hcrt* expression must be distinct from leptin; indeed, fasting increases *Hcrt* expression in *ob/ob* mice [58]. As for the electrical control of OX neurons by leptin, LHA LepRb neurons mediate the control of *Hcrt* expression by leptin [17, 46], yet the nature of the neurotransmitter(s) by which LHA LepRb neurons control *Hcrt* expression remains unclear. While leptin-independent increases in *Hcrt* expression, and concomitant excitation of OX neuronal activity, during fasting would increase OX tone and prompt feeding, the leptin-mediated rise in *Hcrt* expression may contribute to OX-dependent stimulation of energy expenditure over the long term.

7 OX Regulates Feeding Circuits

The projection of OX neurons onto ARC NPY and POMC cells, and their modulation by OX, provides one potential structural and functional basis for the control of feeding by OX [23, 59–61]. Consistent with this model, injection of OX into the ARC boosts food intake, and exogenous OX excites orexigenic NPY-containing neurons, and depresses the activity of anorexigenic POMC-containing neurons, via signaling at the OX1R [62, 63].

OX signaling also mediates food anticipatory behavior and cue-dependent food consumption and drives the hedonic intake of palatable food, such as sucrose and chocolate (in an OX1R dependent manner), suggesting a role for the OX system in hedonic (rather than homeostatic) feeding [64–74]. OX neurons innervate the ventral tegmental area (VTA), the source of dopaminergic (DA) projections to the nucleus accumbens (NAc), which are implicated in motivation and reward [5, 75]. OX depolarizes VTA DA neurons, while OX receptor antagonists decrease the basal firing rate of these cells, suggesting that VTA OX signaling promotes DA signaling and could modulate motivation and reward [76–79]. Thus, OX is well positioned to modulate motivation and reward, including the motivation to feed. Indeed, reward cues increase the activity of OX neurons and OX injection into the VTA reinstates drug seeking (e.g., for cocaine), while an OX1R antagonist blocks the reinstatement of drug seeking [80, 81].

A variety of data suggest that this pathway plays a role in the motivation to feed, as well: the application of OX receptor antagonists (including into the VTA) blocks conditioned place preference and operant responding for palatable food in response to ghrelin injection, and blocks NAc-driven intake of palatable food [82]. Furthermore, this pathway is important for the response to leptin, since ablation of LepRb in LHA GAL neurons, which project specifically within the LHA (including to OX neurons), but not to the VTA, not only increases the activity of OX neurons, but also increases hedonic feeding as measured by sucrose preference [49]. Thus, the control of OX neurons by leptin and ghrelin modulates the motivation to seek and consume rewarding foods, consistent with the notion of a role for OX in the control of food intake beyond simply controlling wakefulness (Fig. 2). Importantly, fasting also increases the activity of the hypothalamic–pituitary–adrenal stress axis; the disinhibition of LHA LepRb neurons and the activation of OX neurons play a crucial role in this response to fasting, as well [83].

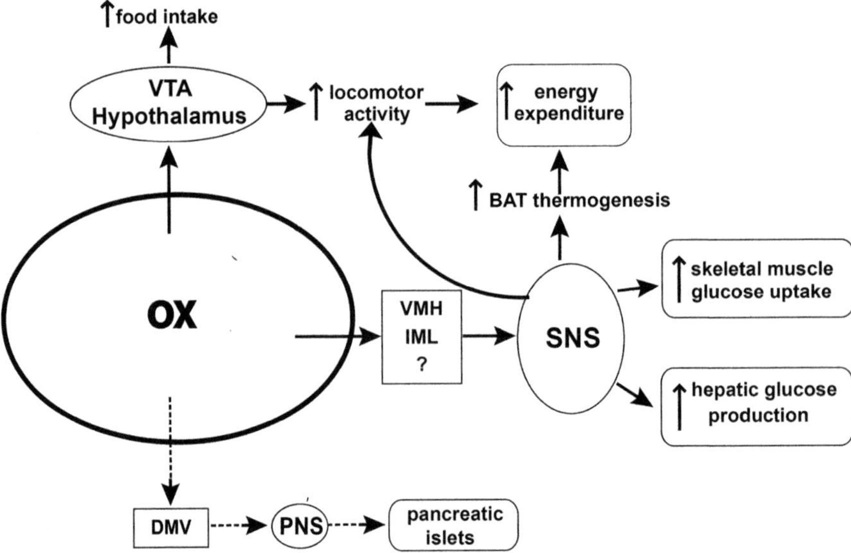

Fig. 2 Pathways of OX-dependent regulation of energy balance. OX signaling increases feeding at least in part by acting in the VTA and at several hypothalamic sites, including the LH and ARC, to increase food-seeking. OX also promotes energy expenditure in part via the VTA, NAc, and hypothalamus and the control of locomotor activity, as well as by acting via sites (potentially including the VMH and IML, among others) to increase SNS activity, which increases activity and basal metabolic rate. OX-mediated activation of the SNS also increases hepatic glucose production and skeletal muscle glucose uptake. Modulation of the PNS by OX (potentially via the DMV) may contribute to the control of pancreatic hormone secretion. *Abbreviations*: *VTA* ventral tegmental area, *VMH* ventromedial hypothalamus, *LH* lateral hypothalamus, *NAc* nucleus accumbens, *ARC* arcuate nucleus, *IML* intermediolateral nucleus, *SNS* sympathetic nervous system, *PNS* parasympathetic nervous system, *BAT* brown adipose tissue

8 Role(s) for OX in Energy Expenditure

While the role for OX in stimulating food intake suggests that mice null for *Hcrt* or the OX receptors should be hypophagic and lean, these animals (like other narcoleptic models and narcoleptic humans) display increased body weight and adiposity compared to controls – even though they do consume less food [84–86]. This obesity results from decreased energy expenditure, which is substantial enough to overcome the decrease in food intake in animals null for OX action [84–86]. Similarly, mice that overexpress *Hcrt* throughout the body display increased energy expenditure and are resistant to high fat diet-induced obesity; this effect is mediated by OX2R [86]. Thus, while OX stimulates feeding, acutely, the promotion of energy expenditure by OX dominates for the control of energy balance over the long term, and deficits in OX signaling ultimately result in increased susceptibility to weight gain.

9 OX-Stimulated Spontaneous Physical Activity (SPA) Provides Obesity Resistance

OX signaling promotes leanness and confers obesity resistance, in part by regulating SPA. Brain injection of OX-A acutely stimulates arousal and SPA when given during the light or dark cycle, and site-specific injections into the LHA, PVN, or NAc are sufficient to drive SPA, implicating these regions as important sites for OX-stimulated energy expenditure [31, 66, 87, 88]. Multiple studies demonstrate an ability for OX to protect against high fat diet induced obesity (DIO) via enhanced energy expenditure [86]. Indeed, dysfunctional OX signaling blunts energy expenditure, leading to obesity [84, 85, 89, 90]. Similarly, loss of OX signaling via OX neuron ablation results in decreased SPA and increased body weight despite hypophagia [84]. Thus, obesity is linked to diminished SPA and OX-mediated SPA is required for body weight maintenance.

10 OX Regulation of Autonomic Function

Not only does OX increase locomotor activity [31, 88, 91–93], but OX signaling may impact resting energy expenditure by supporting sympathetic tone, since centrally administered OX increases blood pressure, heart rate, and sympathetic nerve activity [94–96]. Furthermore, OX injection increases SNS tone, BAT thermogenesis, and body temperature, while *Hcrt*-null animals display impaired BAT thermogenesis [96–104]. In addition, human patients with narcolepsy/cataplexy and low CSF OX exhibit lower muscle sympathetic nerve activity [105]. Indeed, retrograde tracing from liver, white adipose tissue, and brown adipose tissue (BAT) identifies OX neurons as part of the neural circuitry that projects polysynaptically to these tissues [57, 106]. OX neurons innervate hypothalamic and extra-hypothalamic areas (including the VMH, DMH, PVH, NTS, RVLM, and DMV) that regulate autonomic outflow to peripheral tissues [5]. Although the brain regions and mechanisms that govern OX regulation of ANS function are mostly unknown, several studies propose the direct action of OX action in the medulla oblongata, hypothalamus, and on sympathetic preganglionic neurons in the intermediolateral (IML) column of the spinal cord [97, 100, 107–109].

OX signaling works over the long term to promote energy expenditure by increasing SNS tone and increasing locomotor activity (Fig. 2), and OX-dependent energy expenditure is essential to maintain normal energy balance and body weight, especially under conditions of high fat diet challenge. Thus, the elucidation of the mechanisms underlying OX-induced SPA versus autonomic regulation of thermogenesis (including OXR subtype involvement), as well as their relative importance for energy expenditure will be crucial for increasing our understanding of that processes that contribute to, or protect against, obesity.

11 The Role for OX in the Response to Nutritional Alterations

Given the opposing OX-mediated actions on food intake and energy expenditure on overall energy balance, how does the OX system coordinate energy availability with nutritional status? Under normal, ad libitum, feeding conditions, where ghrelin is low and leptin is normal, baseline *Hcrt* expression is supported, but there is no specific metabolic cue to increase the activity of the OX neuron. Thus, in the absence of strong metabolic/nutritional signals to the OX neurons, the control of activity, attention, motivation, and autonomic function by OX neurons depends primarily on other cues, such as circadian inputs [110].

In response to a short-term fast, ghrelin increases, while leptin falls rapidly [42, 53]. This increases the activity of the OX neuron; since *Hcrt* mRNA expression is also increased in this circumstance (by an unknown mechanism), overall OX action in the brain will be increased, promoting alertness, activity, and motivation (including food-seeking and consumption), as is observed in fasted animals [42, 53]. This represents an appropriate behavioral response to increase foraging and thus attempt to restore feeding to normal levels.

In contrast, during extreme, longer-term caloric restriction that exhausts adipose and other energy stores, leptin will have been low for an extended time and acylated ghrelin be low [42, 53]. Thus, the activity of the OX neuron and (because of low leptin) the expression of *Hcrt* may be suppressed, further decreasing OX tone. This would decrease alertness, activity, motivation, and overall energy expenditure, as is observed in animals that have exhausted their nutritional reserves. This response presumably serves to decrease the expenditure of energy on food-seeking in an environment where food is not available, preserving any remaining energy stores until the food supply is restored.

The *ob/ob* mouse represents another instructive example, somewhat similar to long-term caloric restriction. These animals are hyperphagic, display reduced energy expenditure, and are obese, but have low ghrelin (due to continual feeding) and completely lack leptin [42, 43, 53]. As with animals that have exhausted their nutritional reserves, the OX neurons in these animals do not display increased activity (presumably because the low ghrelin counteracts the effects of low leptin) [55]. Also, because leptin is absent, *Hcrt* mRNA expression is decreased in *ob/ob* animals [17]. Note also that *ob/ob* mice have elevated blood glucose, which inhibits OX neurons (see below) [55]. Thus overall OX tone in *ob/ob* mice is low, consistent with their decreased activity, disrupted circadian rhythm, decreased SNS output, and decreased motivation (including decreased willingness to work for food).

12 OX Neurons as Metabolic Sensors

In addition to leptin and ghrelin, which relay information about long-term energy stores and/or recent caloric ingestion, respectively, to OX neurons, OX neurons also respond to nutrients, including glucose, amino acids, and lactate (Fig. 1) [111–115].

Acute IIH elicits c-Fos activation in OX neurons [38], suggesting that they are glucose-inhibited (thus, activated when glucose is low). Indeed, OX neurons directly sense changes in ambient glucose concentrations in slice preparations, exhibiting decreased activity with rising glucose and increased activity in low glucose concentrations [113, 115]. There is some cell-to-cell variation in the response of OX neurons to glucose; however, high glucose only transiently suppresses the activity of most OX cells, but produces sustained inhibition in a smaller group of OX neurons. Interestingly, the effect of high glucose is mimicked by the non-metabolizable glucose analog, 2-deoxyglucose (2DG), indicating that the inhibition of OX neurons by elevated glucose occurs independently of its cellular metabolism [116], as does their activation by low glucose. While tandem pore potassium (K_{2P}) channels were originally posited to mediate glucose inhibition in OX neurons, glucose sensing in OX is not altered by ablation of the K_{2P} subunits, TASK1 and TASK3, leaving the molecular mechanisms that mediate OX glucose sensing unresolved [115, 117, 118].

OX neurons are not activated by low glucose under all circumstances, however, but are "conditional glucosensors": their inhibition by glucose is augmented when metabolism is suppressed by 2DG or the mitochondrial inhibitor, oligomycin; conversely, increasing lactate and pyruvate, or intracellular ATP, prevents the inhibition of OX neurons by glucose [119–121]. Likewise, amino acids, which signal nutrient availability, excite OX neurons and blunt glucose sensing in OX neurons in slice preparations [114]. Interestingly, leptin and ghrelin also modulate the ability of OX neurons to respond to changes in glucose, as leptin blunts and ghrelin enhances the activation of OX neurons by low glucose [121]. Similarly, acute IIH increases *Hcrt* mRNA expression in fasted (6 h), but not freely fed, rats [34].

OX neurons are also stimulated by lactate, which is released by astrocytes in proportion to neuronal activity in the surrounding area; hence, lactate concentrations mirror local energy demands [111], suggesting that OX neurons monitor and respond to signals of local, as well as systemic, metabolism. Amino acids and lactate decrease the activity of K_{ATP} channels in OX neurons; since leptin increases K_{ATP} channel activity in OX neurons, these K_{ATP} channels integrate signals from multiple nutritional and metabolic cues to control OX neuron activity and glucose sensitivity [111, 114, 122].

13 OX and the Control of Glucose Homeostasis

Given the evident role for OX in the coordination of energy balance with arousal state and the glucose sensitivity of OX neurons, it is not surprising that there is substantial evidence implicating the OX system in the regulation of glucose homeostasis, independent of feeding behavior [123]. The impact of OX signaling on glucose homeostasis appears complex, however, since OX can exert both glucose-elevating and glucose-lowering effects [via the promotion of hepatic glucose production (HGP) and enhanced glucose uptake, respectively] [123]. Subcutaneous or intracerebroventricular (icv) administration of OX increases blood glucose in ad libitum-fed rats [124, 125]; this effect is not mediated by decreased insulin, since circulating insulin rises appropriately for increased glucose concentrations. A continuous 5 h icv infusion of OX induces a similar rise in blood glucose in fasted rats [126], revealing that OX can increase blood glucose in both fed and fasted states.

Consistently, retrodialysis of the $GABA_A$ receptor antagonist, bicuculline into the LHA to block inhibitory neurotransmission and activate OX neurons enhances blood glucose levels by augmenting HGP, and co-infusion of the OX1R antagonist, SB-408124, blocks this effect [125]. Also, while the existing OX-cre mouse lines available display imperfect specificity for OX neurons, activation of OX-cre neurons by the use of Designer Receptors Activated by Designer Drugs (DREADDs) sustainably increases blood glucose [127]. Conversely, the cre-dependent ablation of most OX neurons lowers fasting blood glucose [127].

While most observations suggest that OX signaling elevates blood glucose, Tsuneki et al. observed a *reduction* in blood glucose in response to tail vein injection or icv OX treatment in fasted (but not fed), normal and streptozotocin (STZ)-induced diabetic mice [128]. In a similar manner, *Hcrt*-null mice display an exaggerated rise in blood glucose following conditioned glucose ingestion [129]. Hence, OX can decrease blood glucose under some circumstances.

14 OX Influences Blood Glucose via Autonomic Regulation

The apparently opposite actions of OX on blood glucose under different physiologic conditions presumably reflect the differential activation or importance of distinct autonomic outputs under these conditions (Fig. 2). OX-induced elevations in blood glucose result from enhanced HGP following OX-dependent stimulation of the SNS system [125, 130], since hepatic sympathetic (but not parasympathetic) denervation or treatment with adrenergic antagonists prevents OX-induced HGP and hyperglycemia [125, 130]. Alternatively, OX injection into the ventromedial hypothalamic nucleus (VMH) stimulates glucose uptake into skeletal muscle, also in an adrenergic receptor-dependent manner [129, 131]. Thus, the abilities of OX to increase and decrease blood glucose are both SNS-dependent, and the exact physiologic condition may dictate the exact response. For instance, HGP is maximal and glucose uptake into muscle is very low in insulin-deficient mice, and glucose uptake is predicted to be low secondary to decreased SNS tone in *Hcrt*-null mice, so the effect of OX-mediated activation of skeletal muscle glucose uptake may dominate under both of these models, whereas the induction of HGP would dominate during normal conditions, or during hypoglycemia.

In addition, OX signaling impacts the parasympathetic nervous system to influence the response to hypoglycemia or glucoprivation by increasing vagal pancreatic efferent nerve activity (Fig. 2) [132]. Microinjection of an OX1R antagonist into the dorsal motor nucleus of the vagus (DMV) blunts IIH-induced increases in pancreatic efferent nerve firing, and intra-DMV injection of OX-A enhances pancreatic nerve firing [132]. Interestingly, recurrent IIH decreases *Hcrt* mRNA expression, along with and *Hcrt1r* expression in the DMV, suggesting a potential role for the dysfunction of this circuit in the decreased strength of the counter-regulatory response following recurrent hypoglycemic events [37].

This system could also underlie the feeding-independent daily rhythms in blood glucose concentrations, which are blunted following treatment with an OX1R antagonist and in *Hcrt*-null animals [133]. Diurnal variations in *Hcrt* expression and OX neuronal activity may differentially engage the sympathetic and parasympathetic branches of the autonomic system across the diurnal cycle to raise and lower blood glucose via the modulation of HGP relative to insulin (and glucagon) secretion [133].

During hypoglycemia, the response of the OX neuron will differ by metabolic state [119, 120]. The most common cause of hypoglycemia is an overdose of therapeutic insulin in diabetic individuals [134]. Under normal, ad libitum-fed conditions (taking too much insulin with a meal), leptin is normal and ghrelin is low. While low glucose will tend to increase the activity of the OX neuron in this circumstance, the response will not be as robust as in the case of an overdose of insulin during fasting (for instance, not eating soon enough after the pre-meal insulin injection). In the latter case, metabolic fuel should be normal (at least in the short term) but ghrelin will be high, augmenting the activation of the OX neuron by hypoglycemia, even with relatively normal leptin levels. This increased OX tone will increase activity, alertness, motivation (including the motivation to feed, reversing hypoglycemia) and also augment glucose production to combat hypoglycemia.

15 Summary and Future Directions

OX neurons respond to central and peripheral signals to coordinate energy status with arousal state, playing crucial roles in the control of food intake, energy expenditure, and glucose homeostasis. Acute reductions in energy availability (such as fasting) stimulate OX signaling to promote food seeking behavior and intake. In contrast, long-term depletion of energy stores decreases OX tone to blunt arousal, motivation and thus energy expenditure – conserving any remaining energy until food availability is restored. OX neurons also control multiple aspects of glucose homeostasis, including by increasing arousal and SNS output to respond to hypoglycemia. Also, baseline OX tone modulates ANS balance to promote glucose uptake and control the secretion of islet hormones, contributing to physiologic glucose homeostasis over the longer-term. Importantly, the short-term effects of OX on feeding behavior and glucose homeostasis appear to be mostly OX1R-dependent, while long-term signaling via OX2R predominates in the control of energy expenditure. It should be noted, however, that most studies of OXR involvement rely upon pharmacological interrogation using a well-studied OX1R selective antagonist; the lack of a corresponding OX2R selective antagonist until recent times has prohibited a similarly clear analysis of potential OX2R involvement.

Clearly, many questions remain regarding the neural mechanisms by which OX mediates its effects on feeding, energy balance, and metabolism. Within the OX neuron itself, it will be important to determine the fasting-induced signal that increases *Hcrt* mRNA expression, as well as to define the molecular details of the mechanisms by which OX neurons detect glucose.

While it is tempting to speculate that the motivation/food-seeking aspects of OX action are mediated via the VTA and the control of DA circuits, a great deal of additional genetic work will be required to prove this, along with any potential contributions of other projections from OX neurons. Additionally, the circuits by which OX controls ANS function (both for the control of energy expenditure and glucose metabolism) will be crucial to determine. The contributions of brainstem/ spinal cord versus hypothalamic (especially in the VMN) OX action will be particularly important to parse – especially if the effects of OX on glucose uptake can be dissociated from its stimulation of the counter-regulatory response to hypoglycemia.

References

1. Sakurai T et al (1998) Orexins and orexin receptors: a family of hypothalamic neuropeptides and G protein-coupled receptors that regulate feeding behavior. Cell 92(5):1 page following 696
2. Marcus JN et al (2001) Differential expression of orexin receptors 1 and 2 in the rat brain. J Comp Neurol 435(1):6–25
3. Tsujino N, Sakurai T (2013) Role of orexin in modulating arousal, feeding, and motivation. Front Behav Neurosci 7:28
4. de Lecea L, Sutcliffe JG (1999) The hypocretins/orexins: novel hypothalamic neuropeptides involved in different physiological systems. Cell Mol Life Sci 56(5–6):473–480
5. Peyron C et al (1998) Neurons containing hypocretin (orexin) project to multiple neuronal systems. J Neurosci 18(23):9996–10015
6. Date Y et al (1999) Orexins, orexigenic hypothalamic peptides, interact with autonomic, neuroendocrine and neuroregulatory systems. Proc Natl Acad Sci U S A 96(2):748–753

7. Nambu T et al (1999) Distribution of orexin neurons in the adult rat brain. Brain Res 827 (1–2):243–260
8. Anand BK, Brobeck JR (1951) Localization of a "feeding center" in the hypothalamus of the rat. Proc Soc Exp Biol Med 77(2):323–324
9. Delgado JM, Anand BK (1953) Increase of food intake induced by electrical stimulation of the lateral hypothalamus. Am J Physiol 172(1):162–168
10. Mogenson GJ, Stevenson JA (1967) Drinking induced by electrical stimulation of the lateral hypothalamus. Exp Neurol 17(2):119–127
11. Fulton S, Woodside B, Shizgal P (2000) Modulation of brain reward circuitry by leptin. Science 287(5450):125–128
12. Bittencourt JC et al (1992) The melanin-concentrating hormone system of the rat brain: an immuno- and hybridization histochemical characterization. J Comp Neurol 319(2):218–245
13. Rosin DL et al (2003) Hypothalamic orexin (hypocretin) neurons express vesicular glutamate transporters VGLUT1 or VGLUT2. J Comp Neurol 465(4):593–603
14. Torrealba F, Yanagisawa M, Saper CB (2003) Colocalization of orexin a and glutamate immunoreactivity in axon terminals in the tuberomammillary nucleus in rats. Neuroscience 119(4):1033–1044
15. Meister B (2007) Neurotransmitters in key neurons of the hypothalamus that regulate feeding behavior and body weight. Physiol Behav 92(1–2):263–271
16. Qu D et al (1996) A role for melanin-concentrating hormone in the central regulation of feeding behaviour. Nature 380(6571):243–247
17. Leinninger GM et al (2011) Leptin action via neurotensin neurons controls orexin, the mesolimbic dopamine system and energy balance. Cell Metab 14(3):313–323
18. Laque A et al (2013) Leptin receptor neurons in the mouse hypothalamus are colocalized with the neuropeptide galanin and mediate anorexigenic leptin action. Am J Physiol Endocrinol Metab 304(9):E999–E1011
19. Allison MB et al (2015) TRAP-seq defines markers for novel populations of hypothalamic and brainstem LepRb neurons. Mol Metab 4(4):299–309
20. Li Z et al (2015) Hypothalamic amylin acts in concert with leptin to regulate food intake. Cell Metab 22(6):1059–1067
21. Sakurai T et al (1998) Orexins and orexin receptors: a family of hypothalamic neuropeptides and G protein-coupled receptors that regulate feeding behavior. Cell 92(4):573–585
22. Haynes AC et al (2000) A selective orexin-1 receptor antagonist reduces food consumption in male and female rats. Regul Pept 96(1–2):45–51
23. Yamada H et al (2000) Inhibition of food intake by central injection of anti-orexin antibody in fasted rats. Biochem Biophys Res Commun 267(2):527–531
24. Sharf R et al (2010) Orexin signaling via the orexin 1 receptor mediates operant responding for food reinforcement. Biol Psychiatry 67(8):753–760
25. Yamanaka A et al (1999) Chronic intracerebroventricular administration of orexin-A to rats increases food intake in daytime, but has no effect on body weight. Brain Res 849 (1–2):248–252
26. Lin L et al (1999) The sleep disorder canine narcolepsy is caused by a mutation in the hypocretin (orexin) receptor 2 gene. Cell 98(3):365–376
27. Nishino S et al (2000) Hypocretin (orexin) deficiency in human narcolepsy. Lancet 355 (9197):39–40
28. Peyron C et al (2000) A mutation in a case of early onset narcolepsy and a generalized absence of hypocretin peptides in human narcoleptic brains. Nat Med 6(9):991–997
29. Mignot E (2004) Sleep, sleep disorders and hypocretin (orexin). Sleep Med 5(Suppl 1):S2–S8
30. de Lecea L (2012) Hypocretins and the neurobiology of sleep-wake mechanisms. Prog Brain Res 198:15–24
31. Kotz CM et al (2002) Feeding and activity induced by orexin A in the lateral hypothalamus in rats. Regul Pept 104(1–3):27–32

32. Thorpe AJ et al (2003) Peptides that regulate food intake: regional, metabolic, and circadian specificity of lateral hypothalamic orexin A feeding stimulation. Am J Physiol Regul Integr Comp Physiol 284(6):R1409–R1417
33. Thorpe AJ, Teske JA, Kotz CM (2005) Orexin A-induced feeding is augmented by caloric challenge. Am J Physiol Regul Integr Comp Physiol 289(2):R367–R372
34. Cai XJ et al (1999) Hypothalamic orexin expression: modulation by blood glucose and feeding. Diabetes 48(11):2132–2137
35. Griffond B et al (1999) Insulin-induced hypoglycemia increases preprohypocretin (orexin) mRNA in the rat lateral hypothalamic area. Neurosci Lett 262(2):77–80
36. Moriguchi T et al (1999) Neurons containing orexin in the lateral hypothalamic area of the adult rat brain are activated by insulin-induced acute hypoglycemia. Neurosci Lett 264 (1–3):101–104
37. Paranjape SA et al (2006) Habituation of insulin-induced hypoglycemic transcription activation of lateral hypothalamic orexin-A-containing neurons to recurring exposure. Regul Pept 135(1–2):1–6
38. Otlivanchik O, Le Foll C, Levin BE (2015) Perifornical hypothalamic orexin and serotonin modulate the counterregulatory response to hypoglycemic and glucoprivic stimuli. Diabetes 64(1):226–235
39. Diano S et al (2003) Fasting activates the nonhuman primate hypocretin (orexin) system and its postsynaptic targets. Endocrinology 144(9):3774–3778
40. Horvath TL, Gao XB (2005) Input organization and plasticity of hypocretin neurons: possible clues to obesity's association with insomnia. Cell Metab 1(4):279–286
41. Myers MG Jr et al (2009) The geometry of leptin action in the brain: more complicated than a simple ARC. Cell Metab 9(2):117–23
42. Flak JN, Myers MG Jr (2016) Minireview: CNS mechanisms of leptin action. Mol Endocrinol 30(1):3–12
43. Friedman JM (2009) Obesity: causes and control of excess body fat. Nature 459(7245):340–342
44. Williams KW et al (2011) The acute effects of leptin require PI3K signaling in the hypothalamic ventral premammillary nucleus. J Neurosci 31(37):13147–13156
45. Goforth PB et al (2014) Leptin acts via lateral hypothalamic area neurotensin neurons to inhibit orexin neurons by multiple GABA-independent mechanisms. J Neurosci 34 (34):11405–11415
46. Louis GW et al (2010) Direct innervation and modulation of orexin neurons by lateral hypothalamic LepRb neurons. J Neurosci 30(34):11278–11287
47. Cui H et al (2012) Neuroanatomy of melanocortin-4 receptor pathway in the lateral hypothalamic area. J Comp Neurol 520(18):4168–4183
48. Elias CF et al (1998) Chemically defined projections linking the mediobasal hypothalamus and the lateral hypothalamic area. J Comp Neurol 402(4):442–459
49. Laque A et al (2015) Leptin modulates nutrient reward via inhibitory galanin action on orexin neurons. Mol Metab 4(10):706–717
50. de Weille J et al (1988) ATP-sensitive K+ channels that are blocked by hypoglycemia-inducing sulfonylureas in insulin-secreting cells are activated by galanin, a hyperglycemia-inducing hormone. Proc Natl Acad Sci U S A 85(4):1312–1316
51. Dunne MJ et al (1989) Galanin activates nucleotide-dependent K+ channels in insulin-secreting cells via a pertussis toxin-sensitive G-protein. EMBO J 8(2):413–420
52. Zini S et al (1993) Galanin reduces release of endogenous excitatory amino acids in the rat hippocampus. Eur J Pharmacol 245(1):1–7
53. Muller TD et al (2015) Ghrelin. Mol Metab 4(6):437–460
54. Toshinai K et al (2003) Ghrelin-induced food intake is mediated via the orexin pathway. Endocrinology 144(4):1506–1512
55. Yamanaka A et al (2003) Hypothalamic orexin neurons regulate arousal according to energy balance in mice. Neuron 38(5):701–713

56. Lopez M et al (2000) Leptin regulation of prepro-orexin and orexin receptor mRNA levels in the hypothalamus. Biochem Biophys Res Commun 269(1):41–45

57. Stanley S et al (2010) Identification of neuronal subpopulations that project from hypothalamus to both liver and adipose tissue polysynaptically. Proc Natl Acad Sci U S A 107 (15):7024–7029

58. Yamamoto Y et al (2000) Effects of food restriction on the hypothalamic prepro-orexin gene expression in genetically obese mice. Brain Res Bull 51(6):515–521

59. Ma X et al (2007) Electrical inhibition of identified anorexigenic POMC neurons by orexin/ hypocretin. J Neurosci 27(7):1529–1533

60. Kohno D et al (2003) Ghrelin directly interacts with neuropeptide-Y-containing neurons in the rat arcuate nucleus: Ca2+ signaling via protein kinase A and N-type channel-dependent mechanisms and cross-talk with leptin and orexin. Diabetes 52(4):948–956

61. Muroya S et al (2004) Orexins (hypocretins) directly interact with neuropeptide Y, POMC and glucose-responsive neurons to regulate Ca 2+ signaling in a reciprocal manner to leptin: orexigenic neuronal pathways in the mediobasal hypothalamus. Eur J Neurosci 19(6):1524–1534

62. Jain MR et al (2000) Evidence that NPY Y1 receptors are involved in stimulation of feeding by orexins (hypocretins) in sated rats. Regul Pept 87(1–3):19–24

63. Morello G et al (2016) Orexin-A represses satiety-inducing POMC neurons and contributes to obesity via stimulation of endocannabinoid signaling. Proc Natl Acad Sci U S A 113 (17):4759–4764

64. Mieda M et al (2004) Orexin neurons function in an efferent pathway of a food-entrainable circadian oscillator in eliciting food-anticipatory activity and wakefulness. J Neurosci 24 (46):10493–10501

65. Akiyama M et al (2004) Reduced food anticipatory activity in genetically orexin (hypocretin) neuron-ablated mice. Eur J Neurosci 20(11):3054–3062

66. Thorpe AJ et al (2005) Centrally administered orexin A increases motivation for sweet pellets in rats. Psychopharmacology (Berl) 182(1):75–83

67. Harris GC, Wimmer M, Aston-Jones G (2005) A role for lateral hypothalamic orexin neurons in reward seeking. Nature 437(7058):556–559

68. Zheng H, Patterson LM, Berthoud HR (2007) Orexin signaling in the ventral tegmental area is required for high-fat appetite induced by opioid stimulation of the nucleus accumbens. J Neurosci 27(41):11075–11082

69. Nair SG, Golden SA, Shaham Y (2008) Differential effects of the hypocretin 1 receptor antagonist SB 334867 on high-fat food self-administration and reinstatement of food seeking in rats. Br J Pharmacol 154(2):406–416

70. Richards JK et al (2008) Inhibition of orexin-1/hypocretin-1 receptors inhibits yohimbine-induced reinstatement of ethanol and sucrose seeking in Long-Evans rats. Psychopharmacology (Berl) 199(1):109–117

71. Borgland SL et al (2009) Orexin A/hypocretin-1 selectively promotes motivation for positive reinforcers. J Neurosci 29(36):11215–11225

72. Petrovich GD, Hobin MP, Reppucci CJ (2012) Selective Fos induction in hypothalamic orexin/hypocretin, but not melanin-concentrating hormone neurons, by a learned food-cue that stimulates feeding in sated rats. Neuroscience 224:70–80

73. Piccoli L et al (2012) Role of orexin-1 receptor mechanisms on compulsive food consumption in a model of binge eating in female rats. Neuropsychopharmacology 37(9):1999–2011

74. Kay K et al (2014) Hindbrain orexin 1 receptors influence palatable food intake, operant responding for food, and food-conditioned place preference in rats. Psychopharmacology (Berl) 231(2):419–427

75. DiLeone RJ, Georgescu D, Nestler EJ (2003) Lateral hypothalamic neuropeptides in reward and drug addiction. Life Sci 73(6):759–768

76. Vittoz NM, Berridge CW (2006) Hypocretin/orexin selectively increases dopamine efflux within the prefrontal cortex: involvement of the ventral tegmental area. Neuropsychopharmacology 31 (2):384–395

77. Vittoz NM, Schmeichel B, Berridge CW (2008) Hypocretin/orexin preferentially activates caudomedial ventral tegmental area dopamine neurons. Eur J Neurosci 28(8):1629–1640

78. Korotkova TM et al (2003) Excitation of ventral tegmental area dopaminergic and nondopaminergic neurons by orexins/hypocretins. J Neurosci 23(1):7–11

79. Srinivasan S et al (2012) The dual orexin/hypocretin receptor antagonist, almorexant, in the ventral tegmental area attenuates ethanol self-administration. PLoS One 7(9):e44726

80. Borgland SL et al (2006) Orexin A in the VTA is critical for the induction of synaptic plasticity and behavioral sensitization to cocaine. Neuron 49(4):589–601

81. Wang B, You ZB, Wise RA (2009) Reinstatement of cocaine seeking by hypocretin (orexin) in the ventral tegmental area: independence from the local corticotropin-releasing factor network. Biol Psychiatry 65(10):857–862

82. Perello M et al (2010) Ghrelin increases the rewarding value of high-fat diet in an orexin-dependent manner. Biol Psychiatry 67(9):880–886

83. Bonnavion P et al (2015) Antagonistic interplay between hypocretin and leptin in the lateral hypothalamus regulates stress responses. Nat Commun 6:6266

84. Hara J et al (2001) Genetic ablation of orexin neurons in mice results in narcolepsy, hypophagia, and obesity. Neuron 30(2):345–354

85. Hara J, Yanagisawa M, Sakurai T (2005) Difference in obesity phenotype between orexin-knockout mice and orexin neuron-deficient mice with same genetic background and environmental conditions. Neurosci Lett 380(3):239–242

86. Funato H et al (2009) Enhanced orexin receptor-2 signaling prevents diet-induced obesity and improves leptin sensitivity. Cell Metab 9(1):64–76

87. Hagan JJ et al (1999) Orexin A activates locus coeruleus cell firing and increases arousal in the rat. Proc Natl Acad Sci U S A 96(19):10911–10916

88. Kiwaki K et al (2004) Orexin A (hypocretin 1) injected into hypothalamic paraventricular nucleus and spontaneous physical activity in rats. Am J Physiol Endocrinol Metab 286(4): E551–E559

89. Novak CM, Kotz CM, Levine JA (2006) Central orexin sensitivity, physical activity, and obesity in diet-induced obese and diet-resistant rats. Am J Physiol Endocrinol Metab 290(2): E396–E403

90. Teske JA et al (2006) Elevated hypothalamic orexin signaling, sensitivity to orexin A, and spontaneous physical activity in obesity-resistant rats. Am J Physiol Regul Integr Comp Physiol 291(4):R889–R899

91. Lubkin M, Stricker-Krongrad A (1998) Independent feeding and metabolic actions of orexins in mice. Biochem Biophys Res Commun 253(2):241–245

92. Jones DN et al (2001) Effects of centrally administered orexin-B and orexin-A: a role for orexin-1 receptors in orexin-B-induced hyperactivity. Psychopharmacology (Berl) 153 (2):210–218

93. Perez-Leighton CE et al. (2012) Behavioral responses to orexin, orexin receptor gene expression, and spontaneous physical activity contribute to individual sensitivity to obesity. Am J Physiol Endocrinol Metab 303(7):E865–E874

94. Samson WK et al (1999) Cardiovascular regulatory actions of the hypocretins in brain. Brain Res 831(1–2):248–253

95. Matsumura K, Tsuchihashi T, Abe I (2001) Central orexin-A augments sympathoadrenal outflow in conscious rabbits. Hypertension 37(6):1382–1387

96. Shirasaka T et al (1999) Sympathetic and cardiovascular actions of orexins in conscious rats. Am J Physiol 277(6 Pt 2):R1780–R1785

97. Dun NJ et al (2000) Orexins: a role in medullary sympathetic outflow. Regul Pept 96 (1–2):65–70

98. Zheng H, Patterson LM, Berthoud HR (2005) Orexin-A projections to the caudal medulla and orexin-induced c-Fos expression, food intake, and autonomic function. J Comp Neurol 485 (2):127–142

99. Yasuda T et al (2005) Dual regulatory effects of orexins on sympathetic nerve activity innervating brown adipose tissue in rats. Endocrinology 146(6):2744–2748
100. Monda M et al (2007) Sympathetic and hyperthermic reactions by orexin A: role of cerebral catecholaminergic neurons. Regul Pept 139(1–3):39–44
101. Verty AN, Allen AM, Oldfield BJ (2010) The endogenous actions of hypothalamic peptides on brown adipose tissue thermogenesis in the rat. Endocrinology 151(9):4236–4246
102. Madden CJ, Tupone D, Morrison SF (2012) Orexin modulates brown adipose tissue thermogenesis. Biomol Concepts 3(4):381–386
103. Morrison SF, Madden CJ, Tupone D (2012) An orexinergic projection from perifornical hypothalamus to raphe pallidus increases rat brown adipose tissue thermogenesis. Adipocytes 1(2):116–120
104. Sellayah D, Sikder D (2012) Orexin receptor-1 mediates brown fat developmental differentiation. Adipocytes 1(1):58–63
105. Donadio V et al (2014) Lower wake resting sympathetic and cardiovascular activities in narcolepsy with cataplexy. Neurology 83(12):1080–1086
106. Oldfield BJ et al (2002) The neurochemical characterisation of hypothalamic pathways projecting polysynaptically to brown adipose tissue in the rat. Neuroscience 110(3):515–526
107. Korim WS et al (2014) Orexinergic activation of medullary premotor neurons modulates the adrenal sympathoexcitation to hypothalamic glucoprivation. Diabetes 63(6):1895–1906
108. Machado BH et al (2002) Pressor response to microinjection of orexin/hypocretin into rostral ventrolateral medulla of awake rats. Regul Pept 104(1–3):75–81
109. Antunes VR et al (2001) Orexins/hypocretins excite rat sympathetic preganglionic neurons in vivo and in vitro. Am J Physiol Regul Integr Comp Physiol 281(6):R1801–R1807
110. de Lecea L, Huerta R (2014) Hypocretin (orexin) regulation of sleep-to-wake transitions. Front Pharmacol 5:16
111. Parsons MP, Hirasawa M (2010) ATP-sensitive potassium channel-mediated lactate effect on orexin neurons: implications for brain energetics during arousal. J Neurosci 30(24):8061–8070
112. Venner A et al (2011) Orexin neurons as conditional glucosensors: paradoxical regulation of sugar sensing by intracellular fuels. J Physiol 589(Pt 23):5701–5708
113. Burdakov D, Gerasimenko O, Verkhratsky A (2005) Physiological changes in glucose differentially modulate the excitability of hypothalamic melanin-concentrating hormone and orexin neurons in situ. J Neurosci 25(9):2429–2433
114. Karnani MM et al (2011) Activation of central orexin/hypocretin neurons by dietary amino acids. Neuron 72(4):616–629
115. Burdakov D et al (2006) Tandem-pore K+ channels mediate inhibition of orexin neurons by glucose. Neuron 50(5):711–722
116. Gonzalez JA et al (2008) Metabolism-independent sugar sensing in central orexin neurons. Diabetes 57(10):2569–2576
117. Gonzalez JA et al (2009) Deletion of TASK1 and TASK3 channels disrupts intrinsic excitability but does not abolish glucose or pH responses of orexin/hypocretin neurons. Eur J Neurosci 30(1):57–64
118. Guyon A et al (2009) Glucose inhibition persists in hypothalamic neurons lacking tandem-pore K+ channels. J Neurosci 29(8):2528–2533
119. Liu ZW et al (2011) Intracellular energy status regulates activity in hypocretin/orexin neurones: a link between energy and behavioural states. J Physiol 589(Pt 17):4157–4166
120. Venner A et al (2011) Orexin neurons as conditional glucosensors: paradoxical regulation of sugar sensing by intracellular fuels. J Physiol 589(Pt 23):5701–5708
121. Sheng Z et al (2014) Metabolic regulation of lateral hypothalamic glucose-inhibited orexin neurons may influence midbrain reward neurocircuitry. Mol Cell Neurosci 62:30–41
122. Goforth PB et al (2011) Excitatory synaptic transmission and network activity are depressed following mechanical injury in cortical neurons. J Neurophysiol 105(5):2350–2363
123. Tsuneki H, Wada T, Sasaoka T (2012) Role of orexin in the central regulation of glucose and energy homeostasis. Endocr J 59(5):365–374

124. Yoshimichi G et al (2001) Orexin-A regulates body temperature in coordination with arousal status. Exp Biol Med (Maywood) 226(5):468–476
125. Yi CX et al (2009) A major role for perifornical orexin neurons in the control of glucose metabolism in rats. Diabetes 58(9):1998–2005
126. Nowak KW et al (2000) Acute orexin effects on insulin secretion in the rat: in vivo and in vitro studies. Life Sci 66(5):449–454
127. Inutsuka A et al (2014) Concurrent and robust regulation of feeding behaviors and metabolism by orexin neurons. Neuropharmacology 85:451–460
128. Tsuneki H et al (2002) Reduction of blood glucose level by orexins in fasting normal and streptozotocin-diabetic mice. Eur J Pharmacol 448(2–3):245–252
129. Shiuchi T et al (2009) Hypothalamic orexin stimulates feeding-associated glucose utilization in skeletal muscle via sympathetic nervous system. Cell Metab 10(6):466–480
130. Tsuneki H et al (2015) Hypothalamic orexin prevents hepatic insulin resistance via daily bidirectional regulation of autonomic nervous system in mice. Diabetes 64(2):459–470
131. Ramadori G et al (2011) SIRT1 deacetylase in SF1 neurons protects against metabolic imbalance. Cell Metab 14(3):301–312
132. Wu X et al (2004) Hypothalamus-brain stem circuitry responsible for vagal efferent signaling to the pancreas evoked by hypoglycemia in rat. J Neurophysiol 91(4):1734–1747
133. Tsuneki H et al (2013) Hypothalamic orexin prevents hepatic insulin resistance induced by social defeat stress in mice. Neuropeptides 47(3):213–219
134. McCrimmon RJ, Sherwin RS (2010) Hypoglycemia in type 1 diabetes. Diabetes 59 (10):2333–2339

Orexin and Central Modulation of Cardiovascular and Respiratory Function

Pascal Carrive and Tomoyuki Kuwaki

Abstract Orexin makes an important contribution to the regulation of cardiorespiratory function. When injected centrally under anesthesia, orexin increases blood pressure, heart rate, sympathetic nerve activity, and the amplitude and frequency of respiration. This is consistent with the location of orexin neurons in the hypothalamus and the distribution of orexin terminals at all levels of the central autonomic and respiratory network. These cardiorespiratory responses are components of arousal and are necessary to allow the expression of motivated behaviors. Thus, orexin contributes to the cardiorespiratory response to acute stressors, especially those of a psychogenic nature. Consequently, upregulation of orexin signaling, whether it is spontaneous or environmentally induced, can increase blood pressure and lead to hypertension, as is the case for the spontaneously hypertensive rat and the hypertensive BPH/2J Schlager mouse. Blockade of orexin receptors will reduce blood pressure in these animals, which could be a new pharmacological approach for the treatment of some forms of hypertension. Orexin can also magnify the respiratory reflex to hypercapnia in order to maintain respiratory homeostasis, and this may be in part why it is upregulated during obstructive sleep apnea. In this pathological condition, blockade of orexin receptors would make the apnea worse. To summarize, orexin is an important modulator of cardiorespiratory function. Acting on orexin signaling may help in the treatment of some cardiovascular and respiratory disorders.

P. Carrive (✉)
School of Medical Sciences, University of New South Wales, Sydney, NSW 2052, Australia
e-mail: p.carrive@unsw.du.au

T. Kuwaki
Department of Physiology, Graduate School of Medical & Dental Sciences, Kagoshima University, Kagoshima, Japan

© Springer International Publishing AG 2016
Curr Topics Behav Neurosci (2017) 33: 157–196
DOI 10.1007/7854_2016_46
Published Online: 30 November 2016

Keywords Blood pressure • Chemoreflex • Heart rate • Hypercapnia • Hypocretin • Obstructive sleep apnea • Ox1R • Ox2R • Psychological stress • Respiration • Rostral ventrolateral medulla • Schlager mouse • SHR • Sympathetic

Contents

1   Introduction

Orexin plays a key role in the control of arousal and the maintenance of wakefulness [1, 2]. It is also important in the control of hyperarousal when the organism engages in motivated behaviors [3, 4]. Arousal/hyperarousal have many components, and it appears that the role of orexin is to integrate these components into a coordinated response. One of these components is the adjustment of cardiovascular and respiratory function, because without such adjustments, behavioral activity or preparation for behavioral activity would not be possible.

In this chapter, we review previous work showing that orexin and its receptors are directly involved in cardiovascular and respiratory function. Orexin terminals and orexin receptors are found in the central networks regulating cardiovascular and respiratory function, and their activation can produce cardiorespiratory responses independently of behavioral changes, i.e., under anesthesia. Emerging evidence shows that orexin likely contributes to stress-induced cardiovascular responses, the development of hypertension, and the protection of breathing during sleep apnea.

2 Anatomy of the Orexin System in Relation to Central Cardiorespiratory Control

2.1 Orexin Neurons and Their Connections

There are two orexin isoforms, orexin-A (OxA, 33 amino acids) and orexin-B (OxB, 28 amino acids) that are derived from the prepro-orexin precursor. OxA and OxB are made by a restricted group of neurons located in the dorsal part of the tuberal hypothalamus. The group is centered on the perifornical area (PeF) and extends medially into the dorsomedial hypothalamic nucleus and laterally into the lateral hypothalamic area [5, 6]. This area corresponds relatively well to the classic hypothalamic defense area, a region identified more than 50 years ago and from which powerful behavioral and cardiovascular responses can be evoked ([7–10]; for a review see [11]). At least 60% of orexin neurons are glutamatergic, and none are GABAergic [12]. However, practically all orexin neurons also contain the inhibitory neuropeptide dynorphin which is found in the same vesicles as orexin [13, 14]. The functional significance of a co-release of these two peptides is not known but certainly intriguing, because orexin neurons could be either excitatory or inhibitory depending on the balance of the receptors for these two peptides on the postsynaptic side [14].

Orexin can be detected in the rat embryo as early as embryonic day 18, but starting from the third postnatal week, a marked increase is observed both in orexin expression and in the number of orexin neurons [15, 16]. Interestingly, the third week corresponds to the weaning period, when the young rat starts interacting with its environment.

Studies using a genetically encoded retrograde tracer [17] or mapping of appositions made by anterogradely labeled terminals on orexin neurons [18] have shown that inputs to orexin neurons originate mostly from forebrain areas, either limbic (lateral septum, bed nucleus of the stria terminalis [BNST], amygdala, infralimbic cortex) or hypothalamic (preoptic area, posterior hypothalamus). These regions are well known for their role in emotional and autonomic control [19, 20]. Inputs from the brainstem differ between studies. Yoshida et al. [18] reported projections from the periaqueductal gray (PAG) and parabrachial nucleus, while Sakurai et al. [17] reported dense projections from the median raphé, as well as from the raphé pallidus and the C1 group. The input from C1 neurons of the rostral ventrolateral medulla was later confirmed [21]. It may also be mentioned that orexin neurons can be directly inhibited by glucose and excited by mixtures of dietary amino acids [22, 23].

On the output side, orexin terminals are found not only in the limbic structures described above where they make reciprocal connections but also in all the central autonomic and respiratory regions of the hypothalamus and brainstem, including the paraventricular nucleus of the hypothalamus (Pa), periaqueductal gray (PAG), parabrachial/Kölliker-Fuse nucleus, nucleus of the solitary tract (Sol), the premotor sympathetic centers of the Pa, rostral ventrolateral medulla (RVLM), rostral ventromedial medulla (RVMM) and medullary raphé, the premotor respiratory centers of the retrotrapezoid, and pre-Bötzinger complex [5, 6, 24–38]. Orexin neurons are also themselves i) premotor sympathetic neurons since they directly innervate sympathetic preganglionic neurons (SPNs) [39–41] ii) premotor somatic neurons since they project directly to the hypoglossal and phrenic nuclei [36, 42, 43]. Projections are also found to the dorsal motor nucleus of the vagus, although the projection may be weak [6, 44]. Thus, orexin neurons can act at all levels of the central autonomic and respiratory network, from limbic structures to premotor autonomic and respiratory centers and finally to SPNs and phrenic motoneurons themselves. These anatomical features support the idea that orexin contributes to the link between the regulatory systems of consciousness (sleep-wake cycle or emotional stress) and unconscious homeostatic reflexes ([45, 46].

2.2 Orexin Receptors and Their Distribution

There are two orexin receptors, Ox1R and Ox2R [47–49]. OxA can act on both Ox1R and Ox2R, while OxB acts primarily on Ox2R. The two receptors have a different, partly overlapping distribution in the brain, including within the central autonomic and respiratory network.

Four main in situ hybridization studies have compared the distribution of the two receptors in the brain and hypothalamus [50–53]. They show that the BNST expresses both receptors, while the amygdala primarily expresses Ox1R and the septum primarily Ox2R. In the hypothalamus, most areas express both receptors, except Pa, which appears to express Ox2R exclusively. In the brainstem, most of the autonomic and respiratory areas described above also express both receptors, except for the A5 catecholaminergic cell group, which like the A6 group of the locus coeruleus expresses Ox1R exclusively and the Kölliker-Fuse and hypoglossal and trapezoid nuclei (both are respiratory centers) which express mostly Ox2R [51]. Single-cell RT-PCR in SPNs shows that Ox1R is easier to detect than Ox2R suggesting that it is present there in greater amount [54].

Immunohistochemical studies of Ox1R and Ox2R distribution [55–57] have confirmed the overall distribution of the two receptors, although they show more overlap than suggested by the in situ hybridization studies. For instance, the amygdala also contains Ox2R and the septum Ox1R. More importantly, Pa contains a significant amount of Ox1R in both its magno- and parvocellular areas, which colocalizes with vasopressin/oxytocin and CRF, respectively [57, 58]. Interestingly, orexin neurons themselves have both receptors as autoreceptors [58, 59]. A triple-labeling study by Beig et al. [60] estimated that approximately 71% of orexin neurons possess both receptors. Other reports confirm Ox1R labeling in the RVMM and nearby areas of the retrotrapezoid and pre-Bötzinger nuclei [26, 27, 36] and of both Ox1R and Ox2R in 77–85% of C1 neurons of the RVLM [33, 61] as well as in trapezoid and trigeminal nuclei [56]. In contrast, only Ox1R could be detected in SPNs, although Ox2R could be found colocalized with Ox1R in nearby somatic motoneurons [60]. Phrenic motoneurons also have Ox1R, but it is not known if they also have Ox2R [36].

Immunohistochemical staining of orexin receptors may however be difficult to interpret due to the poor sensitivity of their antibodies (see discussion in [62]). A different approach was recently used with transgenic reporter mice in which GFP (green fluorescent protein)-coupled Ox1R [63, 64]. GFP expression in these mice reveals a distribution that is consistent with immunohistochemical studies of Ox1R distribution at least in the forebrain. Thus, regions such as the septum, arcuate, and Pa which were Ox1R immunopositive but showed no in situ hybridization for Ox1R were GFP positive [63]. In the brainstem, strong GFP expression was confirmed in the dorsal raphé, locus coeruleus, raphé pallidus, and RVMM; however, no GFP-labeled cells could be found in RVLM contrary to Shahid et al. [33] report [64]. However, a large proportion of the C1 neurons located medially in the RVMM were Ox1R-GFP positive.

2.3 Tools for the Study of Orexin Cardiorespiratory Function

Both OxA and OxB have been injected centrally, sometimes in combination with Ox1R and Ox2R antagonists. OxA binds to Ox1R and Ox2R with equal affinity, while OxB preferentially binds to Ox2R. Those antagonists that are selective for Ox1R or Ox2R are referred to as SORA (SORA1 and SORA2, respectively, for selective orexin receptor antagonist), while those that antagonize both receptors are known as DORA (dual orexin receptor antagonist). The SORA1 that have been used in the study of orexin cardiorespiratory effects are, in chronological order, SB334867, the first commercially available SORA, SB408124 [47, 65, 66], ACT335827 [67], and compound 56 [68]. The stability and specificity of SB334867 and SB408124 have been questioned, and results obtained with these drugs should be interpreted with caution [47, 65, 69]. SORA2 are TCS-OX2 29 [70] and EMPA [71]. A number of reports have also used the DORA almorexant [72, 73]. Finally, [Ala11, D-Leu15]-OxB is a selective Ox2R agonist [74].

Orexin knockout and/or transgenic mice have also played a major role in our understanding of orexin cardiovascular and respiratory function [9, 75]. Orexin-ataxin3 transgenic mice in which orexin neurons degenerate after birth have been used as well [75, 76]; however, these studies do not necessarily reveal orexin function since orexin neurons also contain other neurotransmitters and modulators.

3 Orexin and Central Modulation of Cardiovascular Function

3.1 Cardiovascular Responses to Centrally Injected Orexin

3.1.1 Cardiovascular Effects of Orexin Microinjected into the Ventricles or Subarachnoid Space

Lateral Ventricle The cardiovascular effects of orexin were first demonstrated in 1999 by Shirasaka et al. [77] and Samson et al. [78] (Fig. 1A). These studies showed that injection of OxA and OxB (1–5 nmol) in the lateral ventricle of freely moving rats could evoke marked and sustained increases in blood pressure. Both reported a stronger effect with OxA than OxB. Most importantly, Shirasaka et al. [77] showed (1) that this pressor effect was associated with an increase in heart rate and renal sympathetic nerve activity and (2) that the same effects could still be evoked under anesthesia. The latter two findings established without doubt that the cardiovascular effect of orexin was due to a direct central sympathoexcitatory action. Hirota et al. [81] later demonstrated that prior blockade of Ox1R with intracerebroventricular (icv) injections of the SORA1 SB334867 (50 nmol) could almost completely block the cardiovascular response of icv OxA (50 nmol). This was confirmed in the awake rat using smaller doses of icv OxA (3 nmol) and another SORA1, SB408124 (3 nmol) [82]. Regrettably, no SORA2 was tested in these studies.

Cisterna Magna Similar pressor and tachycardic changes were reported after intracisternal injections of OxA (0.03–7 nmol) and OxB (0.03–0.3 nmol), with again a stronger effect from OxA compared to OxB [79]. Later work by Huang et al. [83] showed that the SORA2 TCS-OX2 29 (3 nmol) was more potent than SB334867 (up to 10 nmol) at antagonizing this effect (OxA 3 nmol), suggesting a predominantly Ox2R-mediated response in contrast to what had been proposed after lateral ventricle injections. Still, SB408124 (10 nmol) practically abolished the pressor and tachycardic effect of 3 nmol OxA [61].

Intrathecal Upper thoracic intrathecal injections of OxA (0.3–10 nmol) and OxB (3 nmol) also evoke strong pressor and tachycardic effects [80, 84] (Fig. 1C). Shahid et al. [80] also showed that these effects could be abolished with intrathecal preinjection of SB334867 (200 nmol); however, no SORA2 was tested.

It is interesting to compare the magnitude of OxA effects at these three different levels of injection. In the anesthetized mouse, intracisternal injections do not produce weaker but slightly stronger effects than lateral ventricle injections at the same doses [85]. In the anesthetized rat, intracisternal injections also produce slightly stronger effects than intrathecal injections [79, 84]. Comparison of the three levels in the awake rat also shows fourth ventricle injections of 0.3 nmol OxA to be slightly more potent than lateral ventricle injections, themselves equivalent to intrathecal injections at the upper thoracic level (Carrive, unpublished observations, Luong L PhD thesis). This indicates that orexin can act directly on lower brainstem

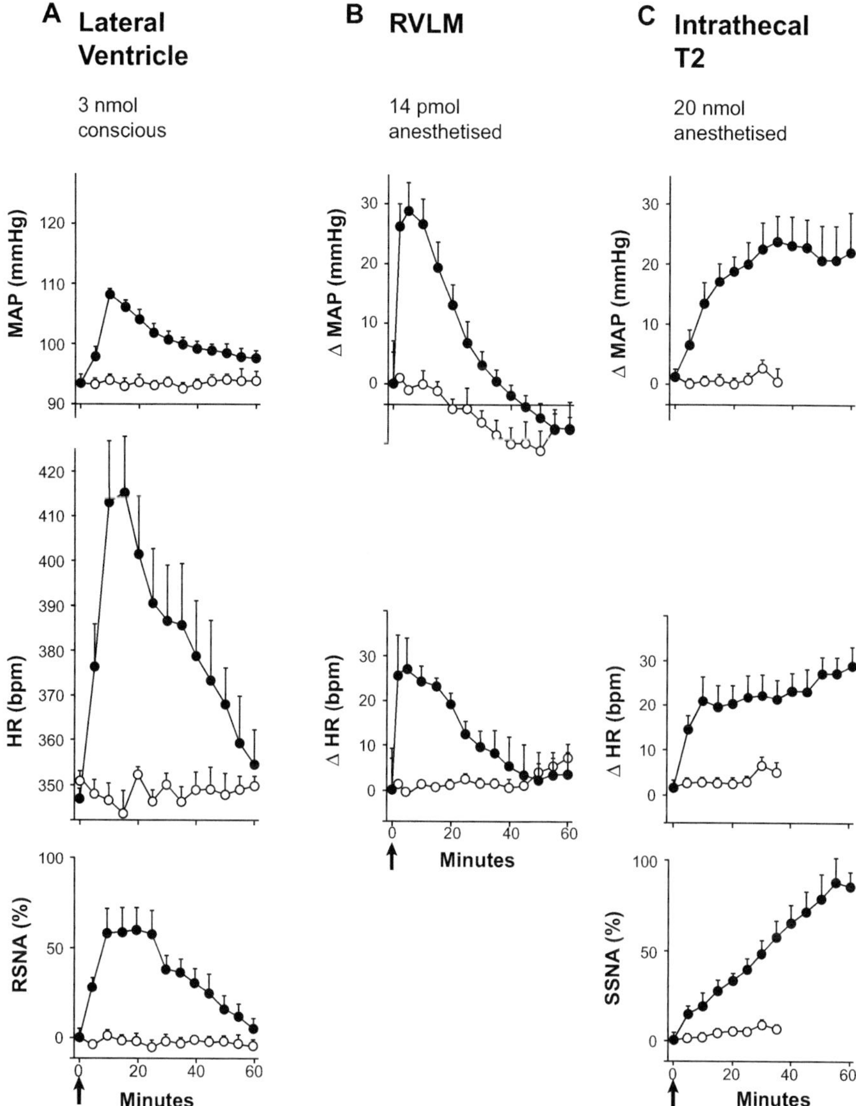

Fig. 1 Cardiovascular and sympathetic responses to central injection of orexin-A in the rat. (**A**) Lateral ventricle injection in conscious freely moving rats. (**B**) Unilateral microinjection in the rostral ventrolateral medulla (RVLM) under anesthesia. (**C**). Intrathecal injection at spinal level T2 under anesthesia. The amount of orexin-A injected varies between the three experiments (modified from Shirasaka et al. [77], Chen et al. [79], and Shahid et al. [80]with the author's permission). Mean ± SEM. *HR* heart rate, *MAP* mean arterial pressure, *RSNA* renal sympathetic nerve activity, *SSNA* splanchnic sympathetic nerve activity

centers and the spinal cord to produce potent cardiovascular effects. We can rule out an action limited to solely forebrain structures.

3.1.2 Cardiovascular Effects of Orexin Microinjected Within Regions of the Central Autonomic Network

SPN Injection of OxA and OxB on identified SPNs evokes strong depolarization due to a direct postsynaptic action [54, 84]. The response was as strong with OxA as with OxB, and the OxA effect was only reduced by 60% after pretreatment with SB334867 [54]. This led the authors to suggest that both receptor subtypes might be involved.

RVLM Chen et al. [79] were the first to report the pressor and tachycardic effects of OxA (14 pmol) in the RVLM of anesthetized rats (Fig. 1B). Machado et al. [86] showed that similar effects could be evoked in the conscious animal at very low doses (0.19 pmol). Dun et al. [87] recorded from RVLM neurons and demonstrated a direct excitatory effect of OxA and OxB. A similar study on neonate RVLM pressor neurons by Huang et al. [83] revealed equipotent depolarizing effect of OxA, OxB, and the specific OxB agonist [Ala11, D-Leu15]-OxB. These authors further showed that TCS-OX2 29 was more potent than SB334867 in reducing the effect of OxA [83]. Similar observations were reported by Shahid et al. [33] who showed that OxA (25 and 50 pmol) and [Ala11, D-Leu15]-OxB (0.75 pmol) had equipotent cardiovascular effects when injected in the RVLM and that SB334867 (1 nmol) reduced the OxA effect by half. In contrast, Xiao et al. [88] reported that RVLM OxA (100 pmol) was strongly reduced by SB408124 (100 pmol), while TCS-OX2 29 (100 pmol) had little effect in normotensive anesthetized rats [88].

Medullary raphé and RVMM When injected in the medullary raphé area and RVMM of anesthetized rats, OxA evokes clear tachycardic responses with a small nonsignificant increase in MAP at 2.5 and 12 pmol [27, 35]. However, injections in the awake animal at a higher dose (30 pmol) produce pressor and tachycardic effects [89].

Solitary Nucleus Microinjections of low doses (2.5–5 pmol) of OxA and OxB in the nucleus of the solitary tract evoke equipotent bradycardic and depressor responses [44, 90, 91]; however, at higher doses (>40 pmol) both evoke tachycardic and pressor responses with OxA more potent than OxB. Interestingly, OxA effects were abolished by SB334867 but not by an antiserum against Ox2, while OxB effects were abolished by the Ox2R antiserum but not by SB334867. This is one of the rare studies to have challenged OxA and OxB effects with both Ox1R and Ox2R blockers [91].

Nucleus Ambiguus Microinjections of OxA (2.5 pmol) evokes bradycardic effects when injected in the external part of the nucleus ambiguus, presumably by activating vagal cardiac preganglionic neurons [26]. It is not known what receptor mediates these effects although these two areas do contain Ox1R.

Forebrain OxA injections (0.05–0.5 pmol) in the subfornical organ evoke depressor and bradycardic responses mediated by a drop in sympathetic tone [92]. In contrast, injections of OxA (30 pmol) in the amygdala and periaqueductal gray evoke pressor and tachycardic responses (personal, unpublished observations).

To summarize, centrally injected orexin can evoke significant cardiovascular changes in both awake and anesthetized states. They can be evoked from multiple levels of the neuraxis but are probably strongest from lower brainstem and spinal cord. The overall effect is a pressor and tachycardic effect, although in some places the opposite can be observed (e.g., subfornical organ). The fact that the OxA effect is practically always stronger than that of OxB would suggest that both receptors are involved, although this still needs to be verified with selective antagonists of the two receptors. One problem when trying to tease out the contribution of each receptor is that in most cases, only SORA1s have been tested and in most cases SB334867, which is not the best antagonist in terms of specificity and selectivity [47, 65, 69]. In the few cases when Ox2R has been challenged following an OxA injection, or when a selective OxB agonist was injected, Ox2R-mediated effects were always stronger than what one would have expected from the effects of antagonism of Ox1R alone. This is the case, for example, in the RVLM [33, 83]. It is not clear if this is due to the lack of selectivity of the available SORAs or some form of interaction between the two receptors. In brief, these experiments suggest that both receptors are likely involved in the cardiovascular effect of orexins. Now that more and better selective SORAs are available, future experiments should always try to challenge both receptors in parallel when testing orexin effects.

3.2 *Contribution of Endogenous Orexin to Cardiovascular Responses*

Orexin microinjections have revealed the general lines of orexin cardiorespiratory effects and some of its sites of action. However, the main question is how endogenous orexin acts in the behaving animal and how it contributes to the animal's response. Coordinated release of orexin at multiple levels of the neuraxis in synchrony with activation of other non-orexinergic systems would have far more subtle effects than those of a ventricular or intracerebral injection. The approach in this case is to challenge a cardiovascular or respiratory response (centrally evoked or associated to a particular behavior or reflex) with systemically injected DORAs and SORAs. This can be done by artificially activating orexin neurons either by intracerebral manipulation or by applying external stimuli or stressors likely to activate orexin neurons.

3.2.1 Cardiovascular Responses Evoked by Central Activation of Orexin Neurons

Disinhibition of the PeF is one way of inducing an orexin-mediated response. Remarkably, although orexin neurons (1) contain other peptides such as dynorphin and use glutamate as their main neurotransmitter [12, 13] and (2) only represent a fraction of the output neurons of the perifornical area, still, 50% of the pressor response to bicuculline injection in the perifornical hypothalamus is blocked by systemic administration of the DORA almorexant (15 mg/kg, iv) [93] (Fig. 2). Interestingly, the reduction of the tachycardic response was not as marked (30%), consistent with Kayaba et al.'s [9] observations in the orexin knockout mouse after PeF injection of 0.3 mM bicuculline. It shows that the peptide plays an important role in the output of the perifornical area.

To find out which receptors are involved, Beig et al. [60] compared the effect of almorexant to that of the SORA1 ACT335827 and the SORA2 EMPA and TCS-OX2 29 (all 15 mg/kg, iv). The results showed that the SORAs produced significant reductions when given in combination, but not when given separately.

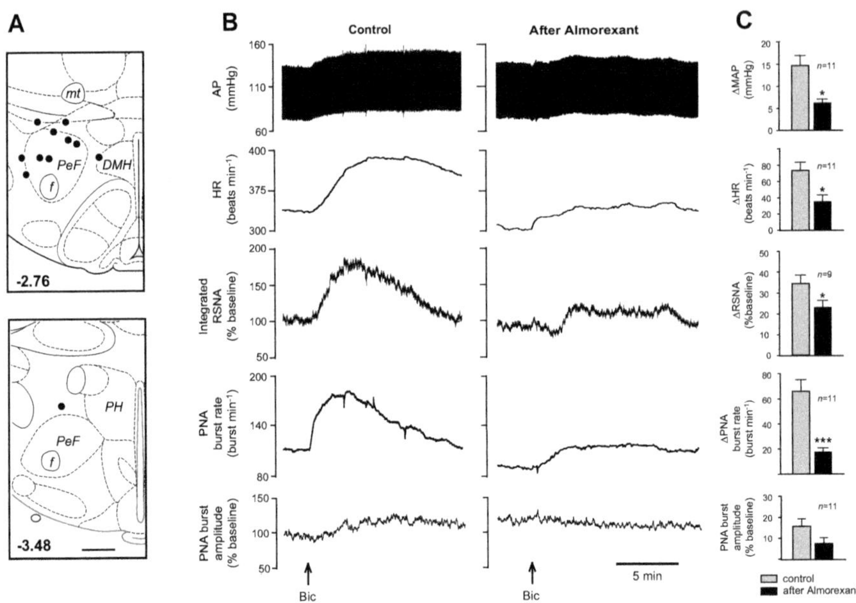

Fig. 2 Effect of systemic almorexant on the cardiorespiratory response to disinhibition of the perifornical hypothalamus in the anesthetized rat. (**A**) Location of the sites of injections at which a 10 pmol injection of the GABA-A antagonist bicuculline was made. (**B**). Cardiorespiratory effect of a 10 pmol injection of bicuculline before and after injection of almorexant (15 mg/kg iv). (**C**) Group data, mean ± SEM. *AP* arterial pressure, *DMH* dorsomedial hypothalamus, *f* fornix, *HR* heart rate, *mt* mamillothalamic track, *PeF* perifornical area, *PH* posterior hypothalamus, *PNA* phrenic nerve activity, *RSNA* renal sympathetic nerve activity. *Scale bar*: 500 μm (Modified with permission from Iigaya et al. [93])

The reductions achieved in this case were close to those of almorexant. Thus, both receptors contribute to the cardiovascular response evoked by endogenous orexin release. Most of them would presumably be located in the brainstem (rostral ventral medulla) and thoracic cord; however, part of the action may be in the perifornical area itself since almorexant microinjections in the PeF can reduce the tachycardic and pressor responses to perifornical disinhibition [94]. The presence of Ox1R and Ox2R autoreceptors on orexin neurons supports this possibility [59, 95].

Two other studies have tested centrally evoked responses with SORAs, although they were limited to SORA1s. In the rabbit, SB334867 and SB408124 (7 mg/kg, iv) markedly reduced the pressor response evoked by electrical stimulation of the dorsal hypothalamus and periaqueductal gray [96]. In conscious rats, SB334867 (10 mg/kg, iv) also reduced by about 40–50% the tachycardic, pressor, and hyperthermic response to muscimol injection in the medial preoptic area, an area that exerts a tonic inhibition on the dorsal hypothalamic area [97].

3.2.2 Cardiovascular Responses Evoked by Acute Stressors and Behavioral Challenges

Kayaba et al. [9] were the first to show an orexinergic contribution to the cardiovascular response of an integrated behavioral response, in this case social stress evoked by the resident-intruder test (Fig. 3A). In the same study, they also showed that the cardiovascular response to tail pinch was unaffected, suggesting that orexin neurons may contribute preferentially to psychological stressors than physical ones (Fig. 3B). This was tested and confirmed by Furlong et al. [98] who showed that almorexant (300 mg/kg, io) reduced the pressor and tachycardic responses of novelty and conditioned fear to context but not restraint or cold exposure (Fig. 3C–F). The pressor responses to novelty and conditioned fear, the tachycardic response to novelty, and the cardiac sympathetic response to conditioned fear were reduced by 45% or more. This led the authors to suggest that the stress responses to which orexin contributes must involve a psychological component and an interaction with the environment. Further work from the same group compared the effect of SORA1 ACT335827 and SORA2 EMPA with that of almorexant (all 100 mg/kg ip) on the cardiovascular response to novelty [95]. The results showed that the two SORAs could produce significant reductions when given alone but the effect was always stronger and closer to that of almorexant when they were given together, confirming the view of a dual contribution of orexin receptors.

A major study by Johnson et al. [99] showed that both the responses evoked by sodium lactate in panic prone rats were almost abolished (tachycardic response) or markedly reduced (pressor response) by short interfering RNA against prepro-orexin RNA. The same effect could be obtained with systemic SB334867 and to some extent SB408124 (both 30 mg/kg, ip). This Ox1R-mediated effect was recently confirmed with compound 56 (30 mg/kg, sc), a more selective SORA1 [68]; however, no SORA2 has yet been tested with this form of stress. In another study, Johnson et al. [100] demonstrated a partial but significant reduction of the

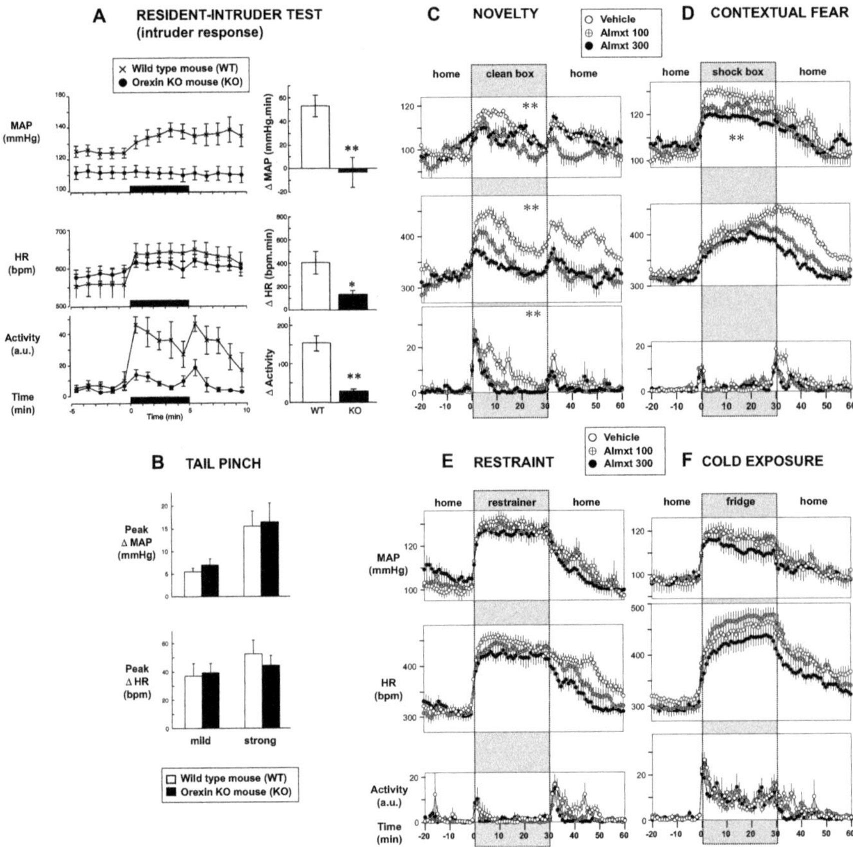

Fig. 3 Cardiovascular response to acute stressors in orexin knockout mice (**A, B**) and in rats pretreated systemically with almorexant (**C–F**). (**A**). Cardiovascular and activity responses to a 5-min forced exposure to a resident mouse (psychosocial stress) in orexin knockout mice. Histograms on the right show the same changes expressed as area under the curve over the test period. Note the marked reduction in all three parameters. *******p* < 0.01; ******p* < 0.5. (**B**). Cardiovascular response to mild or strong tail pinch in orexin knockout mice. The response is unchanged in orexin knockout mice. (**C**) Cardiovascular and activity responses to a 30-min exposure to a new environment (novelty stress) in rats pretreated 2.5 h before with either almorexant 300, 100 mg/kg, or vehicle administered orally. Note the marked reduction in all three parameters. (**D**) Cardiovascular and activity responses to a 30-min reexposure to a box in which the animal has previously received footshocks (conditioned fear to context) in rats pretreated as in (**C**). A marked reduction in MAP is observed, but not in HR. Additional experiments done in this study with atropine blockade showed that the sympathetic cardiac component was as significantly reduced as the blood pressure response. Activity remained low because the animal was in a freezing posture for most of the reexposure. (**E**). Cardiovascular and activity responses to a 30-min restraint in a Plexiglas cylinder in rats pretreated as in (**C**). There was no significant reduction in blood pressure or heart rate after almorexant with this stressor. (**F**) Cardiovascular and activity responses to a 30-min exposure in a fridge at 4°C in rats pretreated as in (**C**). No significant reduction in blood pressure or heart rate after almorexant was observed with this stressor. Mean ± SEM. *******p* < 0.01; ******p* < 0.5. *HR* heart rate, *MAP* mean arterial pressure (Modified with permission from Kayaba et al. [9] and Furlong et al. [98])

pressor and bradycardic response to hypercapnia (20% CO_2, a suffocation signal) after compound 56 (30 mg/kg, sc). Interestingly, the SORA2 JnJ10397049 and the DORA DORA-12 (both 30 mg/kg sc, see [47]) had no effect, and SB334867 only reduced the pressor response (30 mg/kg. ip). This would suggest a selective role for Ox1R in the response to hypercapnia; however, the lack of effect of DORA-12 is difficult to explain. In a third study of the same kind, Johnson et al. [101] showed that the tachycardic, but not pressor, response to the anxiogenic partial inverse benzodiazepine agonist FG7142 was nearly abolished by SB334867 (30 mg/kg ip).

Two other studies using SORA1 may be mentioned. SB334867 (10 mg/kg, iv) was shown to reduce the pressor but not tachycardic response to a moderate dose of methamphetamine [102]. In another study, ACT335827 (100–300 mg/kg, io) significantly reduced the tachycardic response to social stress, but the reduction of the pressor response was not significant [67].

To summarize, these studies show that orexin contributes to the cardiovascular response of certain forms of stress, which tend to be more emotional/psychological than physical and in which there is often an interaction with the environment. The contribution of Ox1R and Ox2R to these responses is still not clear and may depend on the type of stressor. Thus, novelty seems to engage both receptors with synergistic effects, while the response to hypercapnia may engage Ox1R only. Since both pathways converge on SPNs, the only way this could be possible is if SPNs only had Ox1R and novelty engaged Ox2R upstream to SPNs. In any case, it is possible to obtain significant reductions of cardiovascular responses with SORA1 only. From a therapeutic point of view, this indicates that SORA1 may be sufficient to reduce the cardiovascular response to some forms of stress.

3.2.3 Cardiovascular Changes Associated with Chronic Stress or Chronic Heart Failure

Orexin is involved in the response to acute stress. It may also contribute to the cardiovascular adaptations of certain forms of chronic states. Thus, Xiao et al. [88] used electric shocks over a period of 14 days to produce a stress-induced hypertensive state (+30 mmHg). Remarkably, this treatment doubled the number of orexin neurons and almost doubled the amount of Ox1R in the RVLM (Ox2R was not investigated) (Fig. 4). Consistent with this effect, unilateral microinjections of OxA in RVLM produced greater pressor and tachycardic responses in these animals. Conversely, RVLM injections of the SORA1 SB408124 (100 pmol) reduced systolic pressure and heart rate in these hypertensive animals but not in the normotensive non-stressed controls. Ox2R was involved as well since the SORA2 TCS-OX2 29 also reduced systolic pressure in these hypertensive animals. Thus, chronic stress can upregulate the orexin system, and both receptors, at least in the RVLM, contribute to the resulting hypertension. Most remarkable is the increase in orexin neurons as a result of chronic stress, which shows that orexin expression is state dependent. Similar observations have been made with rapid eye

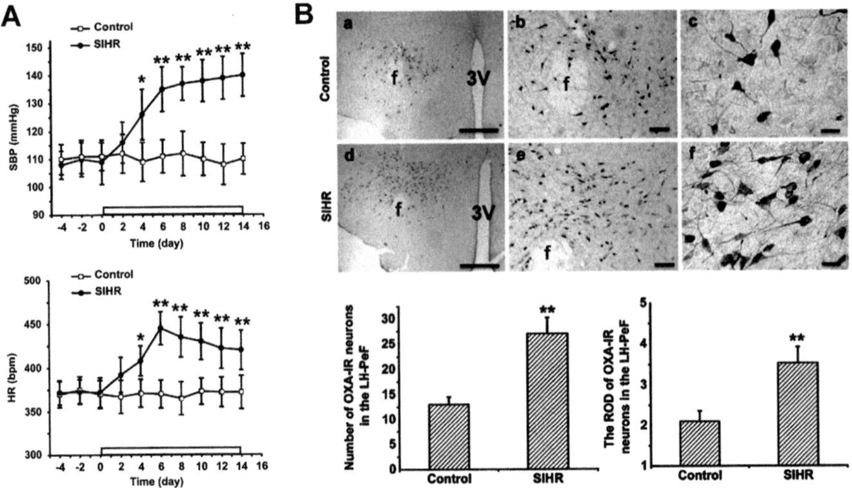

Fig. 4 Increased orexin expression in stress-induced hypertensive rats (SIHRs). (**A**) Daily footshocks for 14 days gradually increased baseline systolic blood pressure (SBP) to hypertensive levels in otherwise normotensive rats. Heart rate (HR) was also increased. Mean ± SEM. (**B**) Immunohistochemical staining of orexin in these animals reveals a doubling in the number of orexin neurons and in the relative optical density (ROD) of orexin-A, demonstrating the plasticity of the orexin system in response to stress. *3V* third ventricle, *f* fornix. ***p* < 0.01. *Scale bars* from left to right: 100, 50, and 20 μm (Reprinted with permission from Xiao et al. [88])

movement sleep deprivation (a 20% increase) [103] and with prepro-orexin mRNA 2 weeks after a single footshock (almost 100% increase, [104, 105]).

Conversely, heart failure, which is induced by coronary artery ligation and associated with an increase in sympathetic activity causes a remarkable reduction (~75%) in prepro-orexin mRNA levels 16 weeks after ligation [106]. Thus, orexin is not involved in the compensatory increase in sympathetic outflow. Its reduced expression is more likely linked to the reduced motivation, social interaction, and depressive state associated with the condition [107, 108].

3.3 Contribution of Orexin to Spontaneous Hypertension

There is now good evidence that at least some forms of spontaneous hypertension are due to an upregulation of orexin signaling. This could have important implications for our understanding and treatment of essential hypertension or primary hypertension. Although it affects 95% of hypertensive patients, the causes of essential hypertension are poorly understood [109]. There is, however, a growing recognition that it may be of neurogenic origin and that the sympathetic nervous system is critical as a trigger and driver of hypertension [110, 111]. Two animal models of essential hypertension have been studied so far, the spontaneously hypertensive rat (SHR) and the hypertensive Schlager mouse.

3.3.1 SHR Hypertension

The role of orexin in SHR hypertension was demonstrated by two different groups in two papers published the same year. One was by Lee et al. [112] and the other by Li et al. [113]. Li et al. [113] showed that almorexant (300 mg/kg, io) administration in the conscious animal normalized hypertension by approximately 50% (Fig. 5), reduced noradrenaline plasma concentration by 50% also and reduced heart rate. This was observed during wakefulness and non-REM sleep and across the circadian cycle. The same dose of almorexant had no effect on these parameters when tested in the normotensive Wistar Kyoto rat (WKY) control. The study by Lee

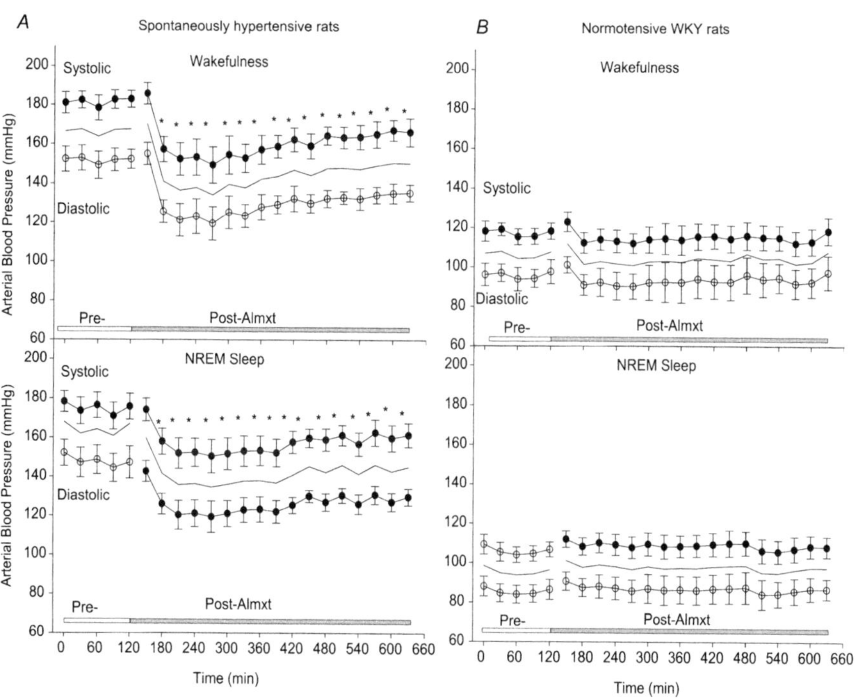

Fig. 5 Blockade of both Ox1R and Ox2R with almorexant reduces blood pressure in spontaneously hypertensive rats but not normotensive WKY rats. (A) Changes in arterial pressure after almorexant (300 mg/kg per oral) during states of wakefulness or non-rapid eye movement sleep (NREM) in the dark period. Wakefulness and NREM states were determined by EEG analysis. (A) In spontaneously hypertensive rats (SHRs, $n = 7$). (B) In normotensive WKY rats ($n = 5$). Mean ± SEM. $*p < 0.002$, compared with preinjection baseline. Note that the reduction in blood pressure was observed during both states. Similar results were obtained during the light period (Reprinted with permission from Li et al. [113])

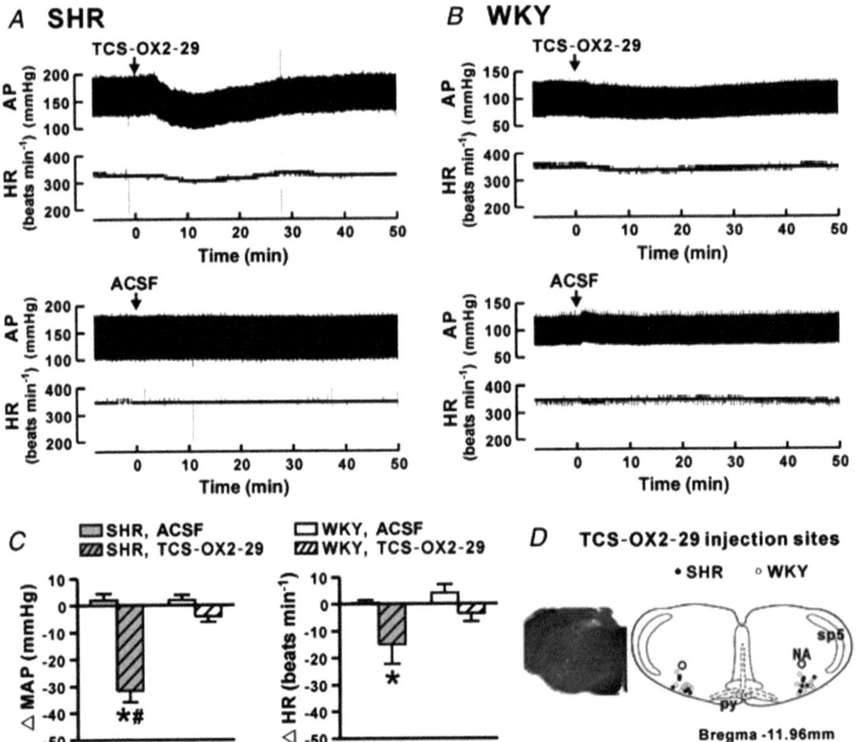

Fig. 6 Blockade of RVLM Ox2R reduces blood pressure in spontaneously hypertensive rats but not normotensive WKY rats. RVLM Ox2R were blocked with bilateral microinjections of the SORA2 TCS-OX2-29 (50 pmol) or artificial cerebrospinal fluid (ACSF). (**A**). Changes in arterial pressure (AP) and heart rate (HR) in SHRs. (**B**). Changes in WKY rats. (**C**) Group data showing the maximal change in mean arterial pressure (MAP) in each strain after TCS-OX2-29 or ACSF. *Significantly different between TCS-OX2-29 and ACSF within the same strain. #Significantly different between SHRs and WKYs for the same treatment. (**D**) Location of sites of injection. Abbreviations: *NA* nucleus ambiguus, *py* pyramidal tract, and *sp5* spinal trigeminal tract (Reprinted with permission from Lee et al. [112])

et al. [112], which was conducted in the anesthetized preparation, showed that icv injection of the SORA2 TCS-OX2 29 (10 and 30 nmol) but not the SORA1 SB334867 (100 nmol) reduced MAP and HR in the SHR. In contrast, neither drug had any effect on MAP in WKY rats. They further showed that TCS-OX2 29 injected in the RVLM (50 pmol) reduced MAP and HR in the SHR but not in the WKY (Fig. 6). These studies indicated that SHR hypertension involves overactive orexin signaling and suggested that the RVLM Ox2R plays a critical role in maintaining this state.

Lee et al. [112] also examined expression levels of Ox1R and Ox2R in the dorsal hypothalamus, paraventricular thalamus, RVLM, and nucleus of the solitary tract, areas likely to be involved in orexin actions on the cardiovascular system. They found no difference between SHR and WKY except for Ox2R in the RVLM where,

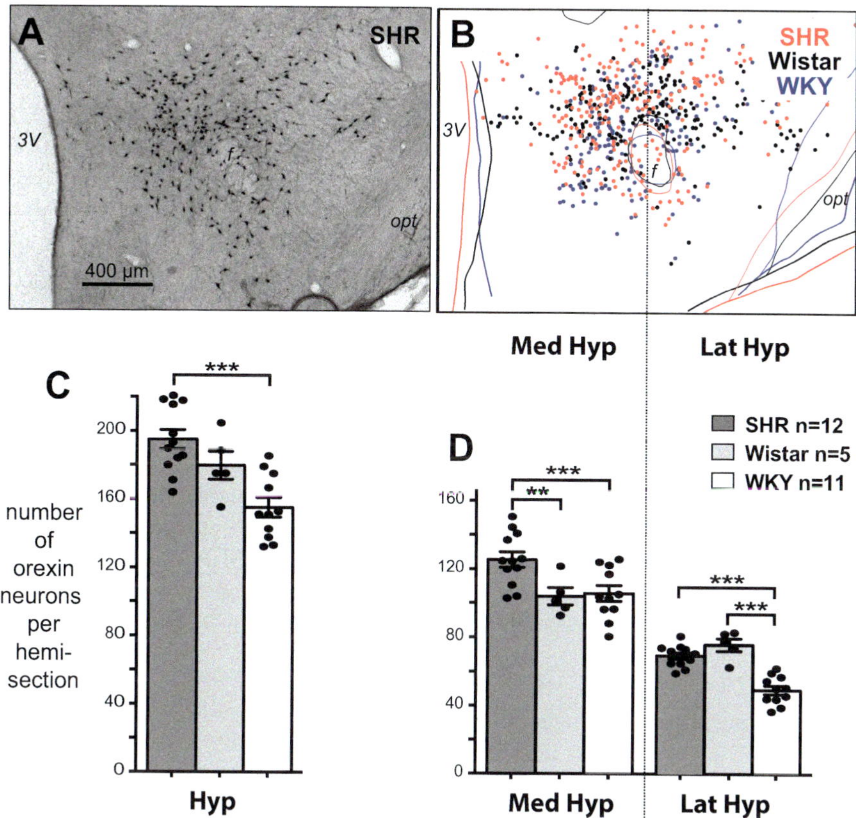

Fig. 7 Distribution of orexin neurons in the SHR compared to the Wistar and WKY rat. (**A**) Photomicrograph of a coronal section showing immunostained orexin neurons in the SHR. *3V* third ventricle, *f* fornix, *opt* optic tract. (**B**) Overlaid distribution of orexin neurons in the three strains. (**C**) Group data comparing the number of orexin neurons in the hypothalamus (Hyp) between the three strains. (**D**) Group data comparing the number of orexin neurons within the medial and lateral hypothalamus (Med Hyp and Lat Hyp, respectively) between the three strains. Note that SHR has significantly more neurons than both Wistar and WKY in the medial hypothalamus and that this difference reflects the blood pressure phenotype. Mean ± SEM. ***$p < 0.001$; **$p < 0.01$. The counts were performed on the six best hemisections per brain. (Modified with permission from Clifford et al. [114])

surprisingly, the expression in SHR was lower, not higher than in the WKY. This suggested that the upregulated signaling might be driven by an increase in orexin release at this synapse, not an increase in receptor expression. This was confirmed first by Clifford et al. [114] who showed a 20% increase in the number of orexin neurons in the medial hypothalamus when comparing SHR with Wistar and WKY rats (Fig. 7). Soon after, Lee et al. [115] confirmed this difference in the number of medial orexin neurons between SHR and WKY. Furthermore, they showed that the number of RVLM-projecting orexin neurons in the medial hypothalamus is greater in the SHR than in the WKY and that orexin concentrations in the hypothalamus and RVLM are 2–3 times higher in the SHR. These authors also showed that RVLM injections of OxA and the Ox2R agonist [Ala11, D-Leu15]-OxB had a stronger effect on blood pressure in the SHR than the WKY, indicating that the RVLM Ox2R receptor is in fact more responsive despite its level of expression being lower. Evidence is shown that the greater responsiveness is due to enhanced Ox2R-neuronal nitric oxide synthase signaling. It would appear therefore that the upregulated orexinergic signal leading to hypertension in the SHR is due in part to a greater number of orexin neurons tonically releasing more orexin on RVLM Ox2R and in part to more sensitive RVLM Ox2R.

Interestingly, SHR tend to be overactive and take longer to habituate to environmental stimuli and stressors [116–118], a behavioral phenotype that is consistent with upregulated orexin signaling. In other words, SHR is genetically programmed to be like the stress-induced hypertensive animals described by Xiao et al. [88] in which chronic stress increased the number of orexin neurons and orexin signaling in the RVLM. Further work is needed to verify a critical role of RVLM Ox2R and explore other potential targets like the SPNs. Critical questions however are what drives this upregulation of orexin signaling and does this contribute to the development of hypertension in the juvenile prehypertensive SHR? Lee et al. [115] have shown evidence that orexin expression is upregulated at 4 weeks of age, which is shortly before hypertension sets in.

3.3.2   Hypertensive Schlager Mouse

The BPH/2J Schlager mouse is a genetic model of sympathetically driven hypertension [119, 120] characterized by an abnormally high difference in blood pressure between the dark active phase and the light inactive phase. Gene microarray studies by Marquez et al. [121, 122] comparing the BPH/2J mouse with its normotensive BPN/3J counterpart revealed a more than twofold difference in the expression of the hypocretin gene (Hcrt) as well as a nearly fourfold difference between the dark active period and the light active period. When these mice were tested with intraoral or intraperitoneal injection of almorexant (300 and 100 mg/kg, respectively), a clear reduction in blood pressure similar to that of the SHR (normalization of 50%) was observed during the active period but not during the light inactive period [123] (Fig. 8). The hypertensive animals also had 29% more orexin neurons; however, this extra number was located in the lateral hypothalamus unlike in the SHR (Fig. 9). This difference is probably related to the fact that the hypertension in the Schlager mouse is mostly observed in the dark active phase, whereas in the SHR, it is the same night and day. This is an interesting difference that could reveal some important functional organization among orexin neurons.

Apart from these subtle differences, the BPH/2J Schlager mouse is, like the SHR, more responsive to environmental stimuli than the normotensive BPN/3J, which is consistent with an upregulated orexinergic system [119]. Enhanced expression of the Hcrt gene (the gene that codes for the prepro-orexin precursor) is also observed in the juvenile BPH/2J at the prehypertensive state [121, 122] as

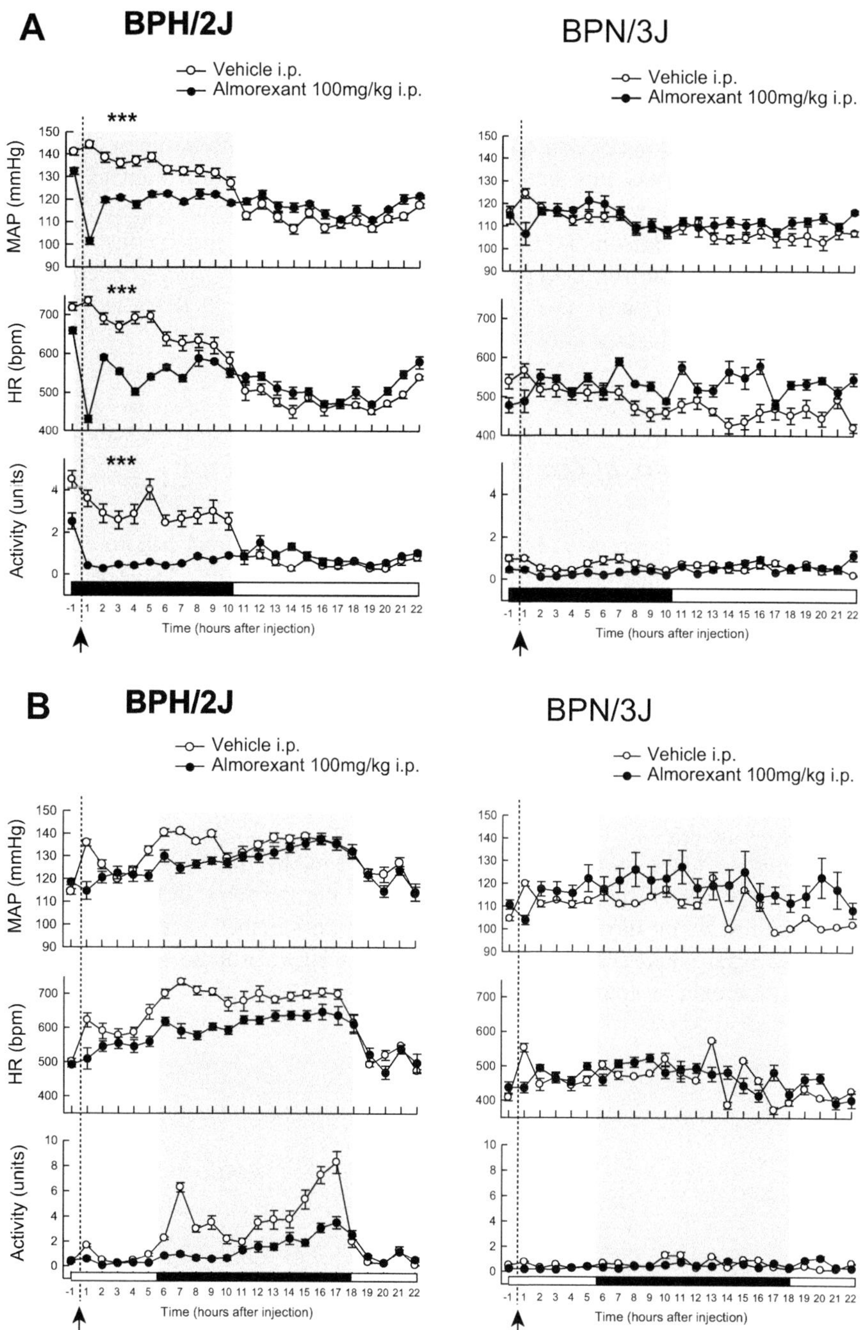

Fig. 8 Effects of blockade of both Ox1R and Ox2R with almorexant (100 mg/kg, ip) on mean arterial pressure (MAP), heart rate (HR), and activity in hypertensive BPH/2J and normotensive BPN/3J Schlager mice. (**A**) The injection was made 2 h after the beginning of the dark active period (*shaded area*). Note the marked reduction in all three parameters during the dark period after almorexant in the BPH/2J but not the BPN/3J. ***$p < 0.001$. (**B**) The injection is done in the

observed in the SHR. It will be interesting to compare the Schlager mouse with the SHR, both being neurogenic forms of hypertension with an overactive orexinergic system.

One important question that needs to be addressed now, in light of this research, is whether some forms of neurogenic hypertension are due to an upregulation of orexin signaling. If it occurs in rats and mice, it is likely to occur in humans as well, whether it is spontaneous or stress induced. It won't take long to find out now that suvorexant (alias Belsomra), the first DORA to be approved by the FDA, is available for clinical use. Suvorexant is currently prescribed for the treatment of insomnia, but it may also have antihypertensive properties. DORAs and SORAs may therefore form a new class of antihypertensive drugs.

3.4 Contribution of Orexin to Basal Blood Pressure

Blockade of orexin receptors in a healthy normotensive animal has no effect on blood pressure and heart rate if the animal is at rest [95, 98, 99, 113]. This is to be expected because there is little activity of orexin neurons at rest and therefore no activation of orexin receptors. Thus, orexin receptor antagonists such as suvorexant should not reduce blood pressure below its resting baseline, making them relatively safe from a cardiovascular point of view.

One would expect to see the same thing in orexin knockout mice, but it is not the case. In a study by Kayaba et al. [9], the blood pressure of orexin knockout mice was 15 mm Hg lower than in wild-type controls, and the difference was the same throughout the day-night cycle. In contrast, in a study by Bastianini et al. [75], the blood pressure in the orexin knockout mouse was actually 10 mm Hg higher than in the wild-type control in the light inactive phase but the same in the dark phase. Thus, in this case the day-night variations were reduced in the knockout mouse. It is not clear what could account for the differences between these two studies and those with orexin receptor blockade.

Fig. 8 (continued) last 5 h of the light inactive period. Except for the reduction of the response to handling in the hour following the injection, almorexant caused no significant reduction during the light inactive period in the BPH/2J. Mean ± SEM (Modified with permission from Jackson et al. [123])

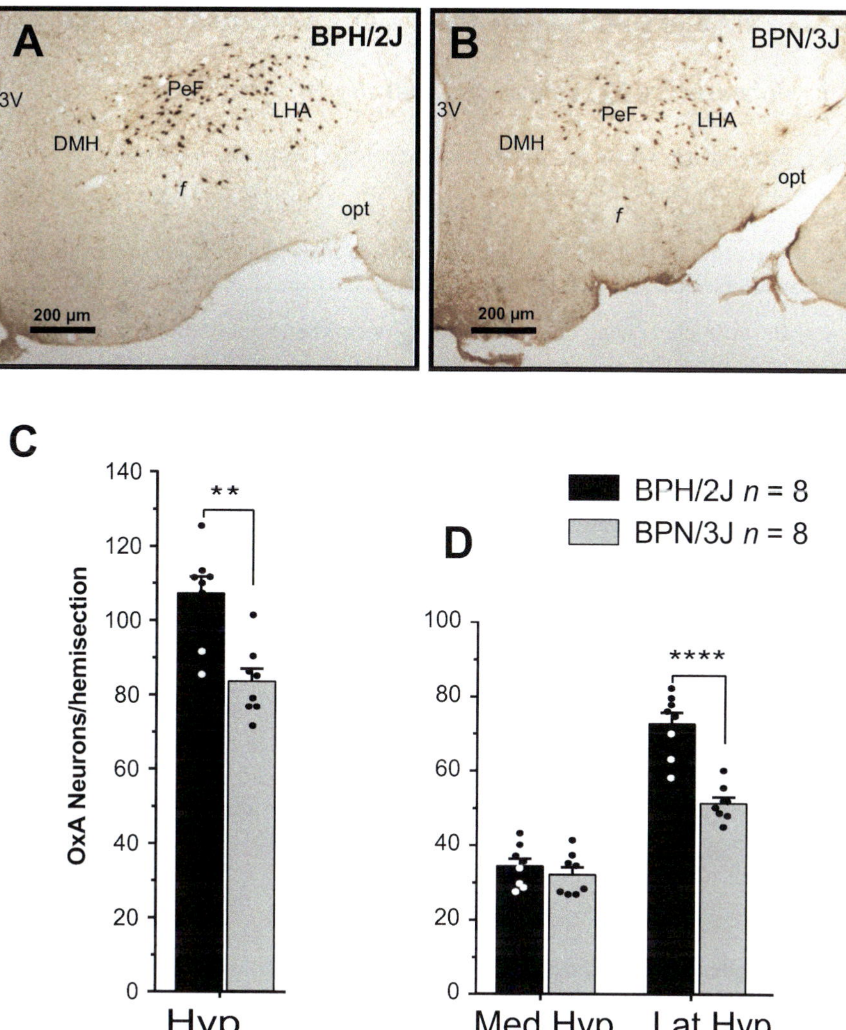

Fig. 9 Distribution of orexin neurons in the hypertensive BPH/2J compared with the normotensive BPN/3J Schlager mouse. (**A**, **B**) Photomicrographs of immunostained orexin neurons in the BPH/2J and BPN/3J, respectively. *3V* third ventricle, *DMH* dorsomedial hypothalamus, *f* fornix, *LH* lateral hypothalamus, *PeF* perifornical hypothalamus, *opt* optic tract. (**C**) Group data comparing the number of orexin neurons in the hypothalamus between the two strains. (**D**) Group data comparing the number of orexin neurons within the medial and lateral hypothalamus (Med Hyp and Lat Hyp, respectively) between the three strains. Note that BPH/2J have significantly more neurons than BPN/3J in the lateral hypothalamus, unlike SHR which have more neurons in the medial hypothalamus compared with normotensive rats. Mean ± SEM. ***$p < 0.001$; **$p < 0.01$. The counts were performed on the six best hemisections per brain (Modified with permission from Jackson et al. [123])

4 Orexin and Central Modulation of Respiratory Function

4.1 Respiratory Responses to Centrally Injected Orexin

4.1.1 Respiratory Effects of Orexin Microinjected in the Ventricles or Subarachnoid Space

The first reports on respiratory effects of orexin appeared a few years after those on cardiovascular effects [36, 85].

Lateral Ventricle Zhang et al. [85] showed in urethane-anesthetized mice that icv injection of OxA (0.03–3 nmol) induced significant increases in both tidal volume and respiratory frequency with moderate increases in blood pressure and heart rate at higher doses (0.3–3 nmol). Respiratory activation by OxA (3 nmol) and OxB (3 nmol) was later confirmed using un-anesthetized freely moving mice [124]. SB334867 (30 nmol) did not affect basal respiration indicating modulatory role of orexins in respiratory regulation [124].

Cisterna Magna Zhang et al. [85] showed that intracisternal injection of OxA (0.03–3 nmol) increased blood pressure, heart rate, and respiratory tidal volume without affecting respiratory frequency.

Intrathecal Young et al. [36] showed that intrathecal (cervical C3–C5 region where phrenic motoneurons are located) perfusion of OxA (40–200 μg/ mL × 5 μL = 0.06–0.3 nmol) in anesthetized rats resulted in a significant increase in phrenic nerve amplitude without any effect on its burst rate. Later, Shahid et al. [80] showed a prominent increase in phrenic nerve amplitude and moderate decrease in its firing rate resulting in an overall increase in minute ventilation activity by OxA (5–20 nmol). Changes in phrenic nerve activity are considered to reflect changes in central respiratory drive because they used anesthetized, vagotomized, and artificially ventilated rats to avoid possible feedback effects from changes in ventilation.

The respiratory effects of orexin at these three different levels of injection appear to be more complex than the cardiovascular effect. OxA increased tidal volume or phrenic amplitude at all three levels of injection, whereas respiratory frequency was increased by intracerebroventricular injection, unchanged by intracisternal injection, and decreased by intrathecal injection. This may reflect differences in the distribution of orexin receptors and/or difference in the availability of injected OxA among multiple respiratory centers. Respiratory rhythm (rate) is predominantly determined in the pre-Bötzinger complex, whereas respiratory amplitude is modulated by multiple sites including the trapezoid nucleus, ventral respiratory group neurons in the medulla oblongata, and phrenic motor nucleus in the spinal cord.

4.1.2 Respiratory Effects of Orexin Microinjected Within Regions of the Central Respiratory Network

Pre-Bötzinger Complex Microinjection of OxA into the pre-Bötzinger complex in anesthetized rats increased the amplitude of the diaphragmatic electromyogram (i.e., an increase in the depth of ventilation) without a change in respiratory rate [36]. However, a more recent study reported that OxB excited inspiratory and pre-inspiratory neurons located in the pre-Bötzinger region and increased respiratory rate [125].

RVLM Microinjection of OxA (12.5–100 pmol) into the RVLM induced a marked increase in phrenic amplitude together with increases in blood pressure, heart rate, and sympathetic nerve activity [33]. The effect on phrenic burst rate was minimal, which was similar to the result from intracisternal injection.

Kölliker-Fuse Nucleus Microinjection of OxB into the pontine Kölliker-Fuse nucleus in an in situ perfused brainstem preparation significantly increased the burst rate of phrenic nerve activity [126]. In addition, OxB evoked particularly prolonged pre-inspiratory discharge of the hypoglossal nerve, indicating a role in upper airway patency that is relevant to the etiology of sleep apnea.

Hypoglossal Nucleus Microinjection of OxA into the hypoglossal nucleus of decerebrated cats increased genioglossus muscle activity [127], implicating a role of orexin in regulating upper airway patency. Zhang et al. [128] confirmed a similar excitation of genioglossus muscle activity in anesthetized rats, and it was reported that this excitation was blocked by both SORA1 and SORA2.

Retrotrapezoid Nucleus In a brain slice preparation and in the presence of a cocktail of synaptic neurotransmission blockers, bath application of OxA excited pH-sensitive retrotrapezoid neurons that are implicated as a primary central CO_2 chemosensor [29].

4.2 Contribution of Endogenous Orexin to Respiratory Responses Evoked by Central Activation of Orexin Neurons

Disinhibition of the PeF in the anesthetized preparation can evoke strong respiratory effects with increases in frequency and amplitude. After systemic almorexant (15 mg/kg iv), this increase in frequency was reduced by 70% which is more than the reduction in the associated pressor and tachycardic response (50 and 30%, respectively, see above) [93] (Fig. 2). Interestingly, the increase in amplitude was not significantly reduced by almorexant. This shows that orexin makes an important contribution to the change in breathing frequency evoked from this part of the hypothalamus. A similar observation was made by Kayaba et al. [9] in the orexin knockout mouse, at least after low and moderate doses of bicuculline.

4.3 Contribution of Endogenous Orexin to Respiration and Respiratory Chemoreflex

It is well known that respiration, blood pressure, and body temperature fluctuate with a ~24-h rhythm (circadian rhythm) with nadirs occurring during nighttime in humans. In sharp contrast to humans, mice and rats sleep for a short duration (an episode of sleep lasts for 10–30 min) many times during both daytime and nighttime. These animals are called nocturnal because their total wake time is longer during the nighttime than the daytime and not because they are continuously awake during the nighttime as humans are during the daytime. In these fragmented sleepers, respiration, blood pressure, and body temperature decrease while they are sleeping regardless of the time of day. Therefore, the state of vigilance is a strong determinant of these autonomic parameters.

The reduced metabolic demand during sleep cannot explain the diminished blood pressure and minute ventilation that occurs at that time because the partial pressure of arterial CO_2 increases during sleep [129]. During non-rapid eye movement (NREM) sleep, the rhythm and amplitude of the ventilation and heartbeats are stable and regular. Sleep-related neuronal mechanisms may actively suppress ventilation because minute ventilation decreases during sleep, even in a hypercapnic environment. During NREM sleep, the tonus of the upper airway muscles decreases so markedly that airway resistance increases considerably. At the same time, decreases in the contraction of the intercostal muscles and of the diaphragm are small [129]. Therefore, sleep affects the neurons that regulate the upper airway and those controlling the thorax differently. During REM sleep, there are remarkable surges and pauses in cardiorespiratory activity, although metabolic demands seem to not be remarkably changed. The general hypersensitivity of the nervous system during REM sleep can be ruled out since hypoxic and hypercapnic ventilatory responses are smaller during REM sleep than during NREM sleep. The pulmonary stretch receptor reflex and the irritant receptor reflex are also suppressed during sleep, and, hence, cough develops only after arousal from sleep [130]. Although these phenomena are well known, their underlying mechanisms remain to be elucidated.

On the other hand, orexin neurons show the following state-dependent activity [131–133]: orexin neuron activity increases just before waking, remains high during wakefulness, and increases considerably during exercise and/or heightened alertness. Thus, orexin may be the missing link between arousal/active stress-coping behaviors and the associated bodily changes that are mainly governed by the autonomic nervous system. This hypothesis is supported by anatomical and physiological evidence [134] and the fact that stressors activate orexin-containing neurons [135–141]. To test this hypothesis, Kuwaki et al. used orexin-deficient mice and examined their basal autonomic functions and responses to stressful stimuli.

4.3.1 Contribution to Basal Respiration

In both orexin knockout mice and wild-type control mice, the respiratory frequency [quiet wake (QW) > NREM ≈ REM] and tidal volume (QW > NREM > REM) showed a clear sleep-wake dependency. Fragmentation of sleep episodes in the orexin knockout mice did not affect the relationship between the state of vigilance and minute ventilation [142]. In addition, there was no difference in the circadian pattern of ventilation (both with regard to cycle length and amplitude) [143], although in the orexin knockout mice, the total sleep time was longer, and movements were fewer during the night [143]. Therefore, orexin does not appear to contribute to basal breathing when an animal is at rest and under normal air conditions. Rather, orexin seems to be an amplifier of the reflex response magnitude to stressors (see below).

4.3.2 Contribution to the Respiratory Chemoreflex

The magnitudes of the hypercapnic (5% CO_2–21% O_2 and 10% CO_2–21% O_2) and hypoxic (15% O_2) ventilatory responses in both the orexin knockout and wild-type mice also showed a clear dependence on the state of vigilance [142]. That is, the order of the magnitudes of the changes in minute ventilation was QW > NREM > REM, irrespective of the stimulus or genotype (except for hypercapnic responses in ORX-KO mice, QW ≈ NREM). Unlike the general similarity between mutants and controls, the hypercapnic responses in the orexin knockout mice during QW were significantly attenuated to approximately half of that in wild-type mice when they were evaluated as an increase in minute ventilation (Fig. 10). When evaluated as the slope of the hypercapnic chemoreflex, the hypercapnic ventilatory responses of the orexin knockout mice did not increase with arousal from sleep, although the NREM > REM relationship was the same as that in the wild-type mice (Fig. 10). The respiratory abnormality in the orexin knockout mice was reproduced by the administration of the SORA1 SB334867 to wild-type mice without affecting the state of vigilance. In addition, supplementation with orexin-A or orexin-B partially restored the hypercapnic chemoreflex in the ORX-KO mice [124].

Based on these results, we previously proposed that the hypercapnic response during sleep periods relies on unknown mechanisms that are independent of orexin and that the response is augmented by orexin during awake periods [144]. This proposal is consistent with reports that state that CO_2 activates orexin neurons in hypothalamic slice preparations [145] and in mice in vivo [146]. Wakefulness driving to respiration has been proposed to increase hypercapnic responsiveness

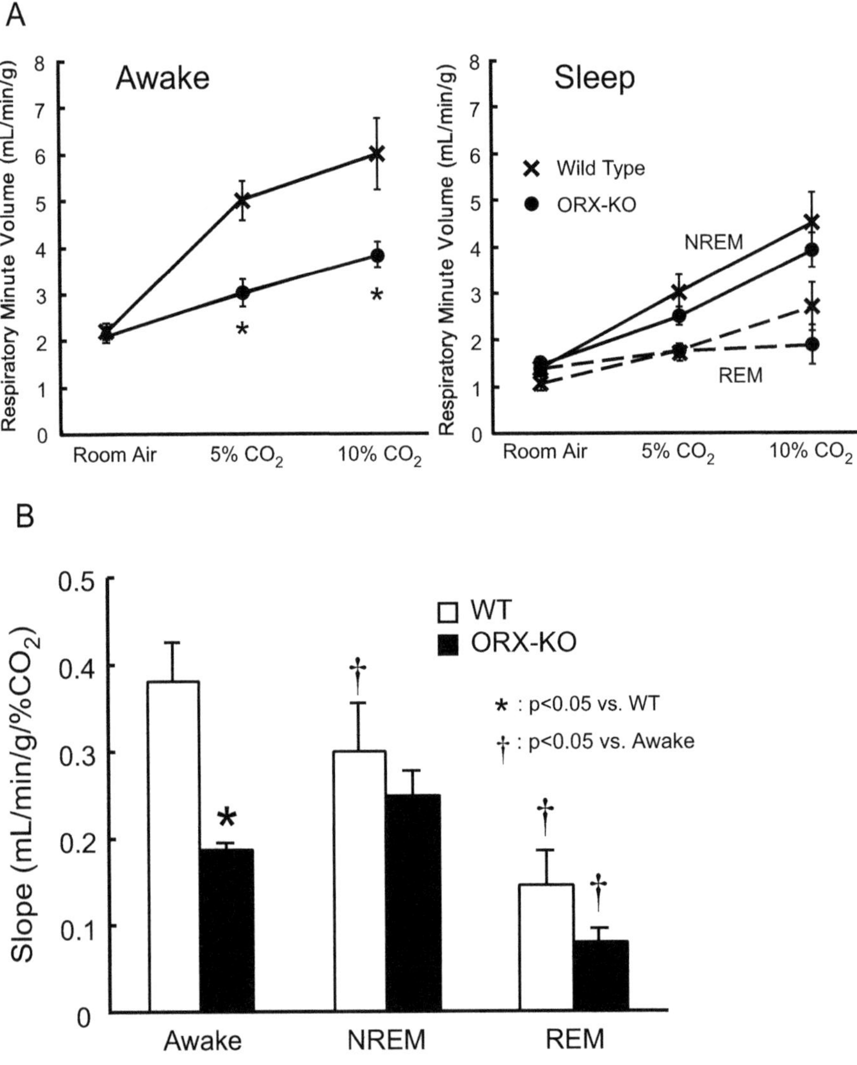

Fig. 10 Abnormality of vigilance state-dependent changes in hypercapnic ventilatory responsiveness in prepro-orexin-knockout mice. (**A**) Hypercapnic responses of respiratory minute volume in wild-type (WT) mice and prepro-orexin knockout mice (ORX-KO) during quiet awake (*left panel*) and sleep (*right panel*) periods. (**B**) Hypercapnic responses are evaluated by calculating the slope of the relationship between inspired CO_2 concentration and respiratory minute volume. Data are presented as means $\pm$ SEM of 5 WT mice and 5 ORX-KO mice. Abbreviations: *NREM* non-rapid eye movement sleep, *REM* rapid eye movement sleep (Adapted from Kuwaki and Zhang [45])

[147], and the administration of orexin has been thought to increase ventilation through its awakening effect. However, orexin knockout mice do wake without orexin, and their hypercapnic responsiveness remains unchanged from the magnitude that is seen during sleep [124], indicating that waking itself could not augment the hypercapnic responsiveness in the absence of orexin.

Nattie's group reported that the microdialysis of SB334867 into the medullary raphé nucleus of rats [148] reduced the hypercapnic ventilatory response during wakefulness in the dark (active period) but not during wakefulness in the light (inactive period). Microdialysis of SB334867 into the retrotrapezoid nucleus (RTN) of rats [149] reduced the hypercapnic ventilatory response predominantly during wakefulness in the light, which is the inactive period (it was not studied in the dark period). In fact, there are orexin-containing nerve terminals on the RTN neurons, and orexin excites them [29]. These results suggest that either the medullary raphé or RTN may be the site responsible for the orexinergic modulation of hypercapnic chemostimulation and that the modulation depends on both the state of vigilance and the circadian period [150].

A recent report showed that hypoxic, but not hypercapnic, responses were attenuated in human narcoleptic patients [151]. Unfortunately, this report did not consider the possible effects of the circadian rhythm or the dependence on the state of vigilance. Therefore, possible role of orexin in vigilance state-dependent modulation of respiratory chemoreflex in humans remains an open question.

4.4 Orexin and Sleep Apnea

In orexin knockout mice, the qualitative characteristics of sleep apneas, defined as cessation of ventilation for at least two respiratory cycles [152, 153], were similar to those observed in the wild-type mice [142]. Spontaneous and post-sigh apneas were observed during NREM sleep, and spontaneous but not post-sigh apneas were observed during REM sleep, although sighs were recorded during both NREM and REM sleep [142]. Moreover, all the apneas appeared to be of central origin because the intercostal EMG indicated that firing stopped during the apneic episodes. From a quantitative point of view, however, spontaneous apneas during both NREM and REM sleep were significantly more frequent (about 2–3 times higher) in the ORX-KO mice than in the wild-type mice, whereas the frequency of post-sigh apneas during NREM did not differ between the two types of mice (Fig. 11). Such differences were preserved when the breathing gas mixture was changed to hypoxic or hypercapnic gases [142]. We recently found more severe sleep apnea in orexin neuron-ablated mice [45]. Possible involvement of orexin in reducing sleep apnea was further confirmed by the recent observation that icv injection of Ox2R agonist, [Ala11, D-Leu15]-OxB, significantly attenuated occurrence of sleep apnea in C57BL/6 mice that usually show sleep apnea without any intervention [154]. In addition, chronic tracheal narrowing in rats increased hypothalamic orexin mRNA and protein content [155]. Administration of the DORA almorexant induced severe

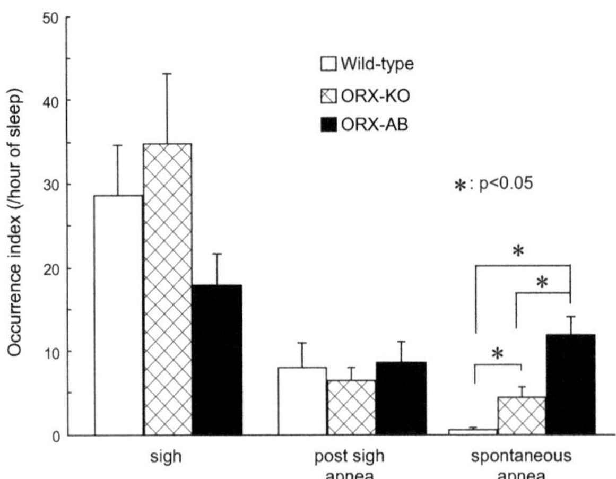

Fig. 11 Sigh and apnea occurrence indices in wild-type, prepro-orexin knockout (ORX-KO) and orexin neuron-ablated (ORX-AB) mice during NREM sleep when the animals breathed room air. Data are presented as means ± standard error of the mean (SEM) of wild-type mice (*open bar, n = 5*), ORX-KO mice (*hatched bar, n = 5*), and ORX/ATX-Tg mice (*solid bar, n = 6*) (Adapted from Kuwaki and Zhang [45])

breathing difficulties, indicating that the rise in orexin was a compensatory response.

These results indicate that orexin exerts an inhibitory effect on the genesis of spontaneous sleep apneas. Our observation is similar to a previous finding that indicates a high incidence of sleep apneas among narcolepsy patients [156, 157]. Thus, for respiratory integrity, orexin appears to be indispensable not only during wakefulness (previous section) but also during sleep periods. This proposal may appear to be in conflict with the reports stating that the spontaneous activity of the orexin neurons increases during awake periods and is decreased during sleep periods [131, 132]. However, these reports also stated that small increases in orexinergic neuronal activity were observed during the REM period and they were higher than those recorded during the NREM period. Isolated orexin neurons that are deprived of any synaptic input show spontaneous activity [23], indicating their potential firing during sleep. The magnitude of exaggeration of spontaneous sleep apnea by orexin deficiency was greater during REM sleep than during NREM [142]. Thus, the activation of orexin neurons during REM sleep may exert an inhibitory effect on the genesis of spontaneous sleep apneas.

Yamaguchi et al. [158] have recently found that intermittent (five times of 5-min duration, mimicking repetitive sleep apnea) but not sustained (25 min) hypoxia (10% O_2) efficiently activated orexin neurons (Fig. 12). In addition, intermittent hypoxia-induced long-lasting augmentation of ventilation, also called respiratory long-term facilitation, was attenuated or even disappeared in orexin-deficient mice [159, 160]. From these results, we think that activation of orexin neurons by intermittent hypoxia would inhibit further occurrence of apneas by activating the respiratory system. In fact, icv administration of orexin activated ventilation even in the anesthetized condition [36, 85]. Thus, deficiency of orexin may not be the direct cause of sleep apnea but rather a permitting factor for progressive occurrence of sleep apnea.

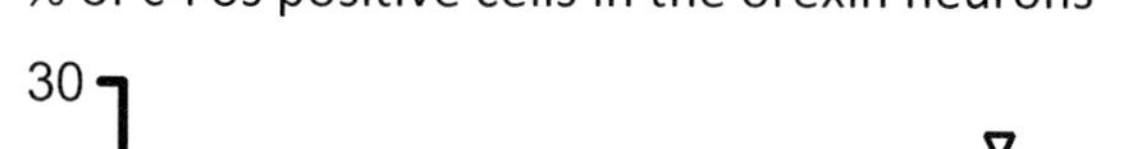

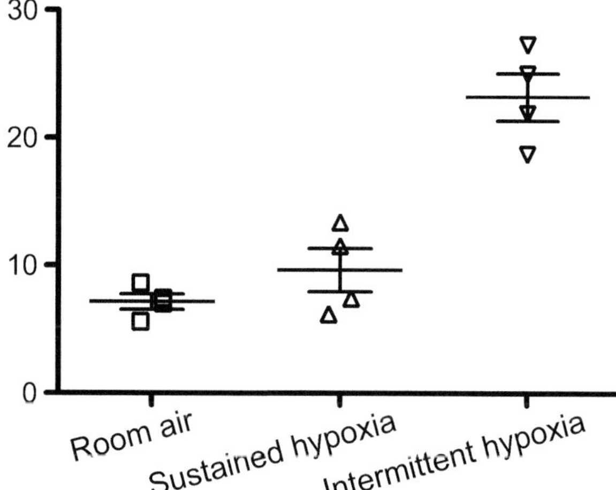

Fig. 12 Activation of orexin neurons by intermittent but not sustained hypoxia. Activation of orexin neurons was evaluated by double immunostaining for c-Fos, a marker of neuronal activation, and for orexin. (Adapted from Yamaguchi et al. [158])

We do not consider that the decreased influence of orexin can explain all human sleep apneas because most reported human apneas are of the obstructive type. Nevertheless, there are good reasons to suspect the possible involvement of orexin deficiency in some cases of human apneas, most probably in narcoleptic patients. First, orexin neurons are activated under hypoglycemia [161] and are inhibited by glucose [23, 162] through the TWIK-related acid-sensitive K^+ (TASK) subfamily of tandem-pore potassium channels [163]. Hyperglycemia is a powerful predictor of impaired breathing during sleep in both humans and animals [164, 165]. Second, orexin innervates the hypoglossal nucleus [42] that plays a major role in preventing obstructive sleep apneas [166]. Third, some obstructive sleep apnea patients showed low levels of plasma orexin [17, 167, 168], although a contradictory result [169] and no change in CSF orexin [170] have also been reported. Finally, at least during the induction phase of sleep, most apneas are thought to be of central origin in humans [129]. Narcolepsy patients have frequent sleep apneas and were thought to be of central origin [156].

4.5 *Hypnotics and Respiration*

As mentioned in the previous section, the DORA suvorexant is now available for clinical use as a hypnotic in Japan and the USA. This drug is expected to have less respiratory side effects than traditional benzodiazepine hypnotics that activate the GABAergic system in the brain. This notion may be true at least about basal respiration because orexin has a minor role in basal respiration. However, as mentioned above, orexin seems to play an important role in arousal- and stress-linked augmentation of chemoreflex and in limiting progressive occurrence of sleep apnea. Therefore, careful attention should be paid with its use in patients of lung disease even though it was safe in mild cases of chronic obstructive pulmonary disease [171].

5 Conclusion

It is now clear that the orexinergic system plays an important role in central cardiorespiratory control. It is excitatory and contributes to the cardiovascular and respiratory adjustments associated with wakefulness and motivated behavior, including some forms of stress. Orexin terminals and orexin receptors are found in all the regions known to regulate cardiovascular and respiratory function including premotor centers and even motor centers (i.e., SPN and phrenic motoneurons). Anatomical and pharmacological studies point toward a primary role for Ox1R; however, this view is biased by the fact that most studies have used Ox1R SORAs (i.e., SB334867). Ox2R is also found in most central cardiovascular centers and when challenged is often found to be as important as Ox1R. Thus, it is not clear how the two receptors contribute to orexin cardiorespiratory action, if they magnify and/ or complement each other.

Nevertheless, blockade of orexin receptors can reduce the acute cardiovascular response evoked by psychological stressors, as well as hypertension induced by chronic stress. Furthermore, blockade of orexin receptors can reduce blood pressure in two models of spontaneous hypertension, the SHR and BPH/2J Schlager mouse, raising the possibility that upregulation of orexin signaling could be a contributing factor in the development of hypertension in human. This upregulation may be due to an increase in orexin expression and a greater sensitivity of its receptors, possibly Ox2R.

It is most likely that orexin also increases ventilation during stress responses, although this has not been verified yet. Orexin does however increase the ventilatory response to hypercapnia, which is an important reflex in the regulation of breathing. Interestingly, orexin can also play a protective role in sleep apnea to maintain breathing homeostasis. In fact, it may be upregulated in this pathological condition.

The orexin system opens a new window for pharmacological intervention in the treatment of pathologies related to cardiorespiratory function. Suvorexant, the first DORA to be approved by the FDA for the treatment of insomnia, may possibly turn out to be useful for the treatment or prevention of some forms of hypertension; however, it may not be indicated in cases of obstructive sleep apnea. Exciting developments in this field can be expected over the coming years.

Acknowledgment Supported by grants from the National Health of Medical Research Council of Australia and from the Ministry of Education, Science, Culture, and Sports in Japan.

Conflict of Interest The research was conducted in the absence of any commercial or financial relationships that could be construed as a potential conflict of interest.

References

1. Carter ME, Brill J, Bonnavion P, Huguenard JR, Huerta R, de Lecea L (2012) Mechanism for hypocretin-mediated sleep-to-wake transitions. Proc Natl Acad Sci U S A 109(39):E2635–E2644. doi:10.1073/pnas.1202526109
2. Saper CB, Scammell TE, Lu J (2005) Hypothalamic regulation of sleep and circadian rhythms. Nature 437(7063):1257–1263. doi:10.1038/nature04284
3. Mahler SV, Moorman DE, Smith RJ, James MH, Aston-Jones G (2014) Motivational activation: a unifying hypothesis of orexin/hypocretin function. Nat Neurosci 17 (10):1298–1303. doi:10.1038/nn.3810
4. Sakurai T (2014) The role of orexin in motivated behaviours. Nat Rev Neurosci 15 (11):719–731. doi:10.1038/nrn3837
5. Nambu T, Sakurai T, Mizukami K, Hosoya Y, Yanagisawa M, Goto K (1999) Distribution of orexin neurons in the adult rat brain. Brain Res 827(1–2):243–260
6. Peyron C, Tighe DK, van den Pol AN, de Lecea L, Heller HC, Sutcliffe JG, Kilduff TS (1998) Neurons containing hypocretin (orexin) project to multiple neuronal systems. J Neurosci 18 (23):9996–10015
7. Abrahams VC, Hilton SM, Zbrozyna A (1960) Active muscle vasodilation produced by stimulation of the brainstem: its significance in the defence reaction. J Physiol (London) 154:491–513
8. Hilton SM (1982) The defence-arousal system and its relevance for circulatory and respiratory control. J Exp Biol 100:159–174
9. Kayaba Y, Nakamura A, Kasuya Y, Ohuchi T, Yanagisawa M, Komuro I, Fukuda Y, Kuwaki T (2003) Attenuated defense response and low basal blood pressure in orexin knockout mice. Am J Physiol 285(3):R581–R593. doi:10.1152/ajpregu.00671.2002
10. Smith OA, DeVito JL, Astley CA (1990) Neurons controlling cardiovascular responses to emotion are located in lateral hypothalamus-perifornical region. Am J Physiol 259:R943–R954
11. Carrive P (2011) Central circulatory control. Psychological stress and the defense reaction. In: Llewellyn-Smith IJ, Verberne A (eds) Central regulation of autonomic function, 2nd edn. Oxford University Press, New York, pp. 220–237
12. Rosin DL, Weston MC, Sevigny CP, Stornetta RL, Guyenet PG (2003) Hypothalamic orexin (hypocretin) neurons express vesicular glutamate transporters VGLUT1 or VGLUT2. J Comp Neurol 465(4):593–603. doi:10.1002/cne.10860
13. Chou TC, Lee CE, Lu J, Elmquist JK, Hara J, Willie JT, Beuckmann CT, Chemelli RM, Sakurai T, Yanagisawa M, Saper CB, Scammell TE (2001) Orexin (hypocretin) neurons contain dynorphin. J Neurosci 21(19):RC168
14. Muschamp JW, Hollander JA, Thompson JL, Voren G, Hassinger LC, Onvani S, Kamenecka TM, Borgland SL, Kenny PJ, Carlezon Jr WA (2014) Hypocretin (orexin) facilitates reward by attenuating the antireward effects of its cotransmitter dynorphin in ventral tegmental area. Proc Natl Acad Sci U S A 111(16):E1648–E1655
15. de Lecea L, Kilduff TS, Peyron C, Gao X, Foye PE, Danielson PE, Fukuhara C, Battenberg EL, Gautvik VT, Bartlett FS, Frankel WN, van den Pol AN, Bloom FE, Gautvik KM, Sutcliffe JG (1998) The hypocretins: hypothalamus-specific peptides with neuroexcitatory activity. Proc Natl Acad Sci U S A 95(1):322–327

16. Sawai N, Ueta Y, Nakazato M, Ozawa H (2010) Developmental and aging change of orexin-A and -B immunoreactive neurons in the male rat hypothalamus. Neurosci Lett 468(1):51–55. doi:10.1016/j.neulet.2009.10.061

17. Sakurai T, Nagata R, Yamanaka A, Kawamura H, Tsujino N, Muraki Y, Kageyama H, Kunita S, Takahashi S, Goto K, Koyama Y, Shioda S, Yanagisawa M (2005) Input of orexin/hypocretin neurons revealed by a genetically encoded tracer in mice. Neuron 46 (2):297–308. doi:10.1016/j.neuron.2005.03.010

18. Yoshida K, Mccormack S, España RA, Crocker A, Scammell TE (2006) Afferents to the orexin neurons of the rat brain. J Comp Neurol 494(5):845–861. doi:10.1002/cne.20859

19. Pitkänen A, Savander V, Ledoux JE (1997) Organization of intra-amygdaloid circuitries in the rat: an emerging framework for understanding functions of the amygdala. Trends Neurosci 20(11):517–523

20. Saper CB (2004) Central autonomic nervous system. In: Paxinos G (ed) The rat nervous system, 3rd edn. Elsevier, San Diego, pp. 761–794

21. Bochorishvili G, Nguyen T, Coates MB, Viar KE, Stornetta RL, Guyenet PG (2014) The orexinergic neurons receive synaptic input from C1 cells in rats. J Comp Neurol 522 (17):3834–3846. doi:10.1002/cne.23643

22. Karnani MM, Apergis-Schoute J, Adamantidis A, Jensen LT, de Lecea L, Fugger L, Burdakov D (2011) Activation of central orexin/hypocretin neurons by dietary amino acids. Neuron 72(4):616–629

23. Yamanaka A, Beuckmann CT, Willie JT, Hara J, Tsujino N, Mieda M, Tominaga M, Yagami K, Sugiyama F, Goto K, Yanagisawa M, Sakurai T (2003) Hypothalamic orexin neurons regulate arousal according to energy balance in mice. Neuron 38(5):701–713

24. Baldo BA, Daniel RA, Berridge CW, Kelley AE (2003) Overlapping distributions of orexin/hypocretin- and dopamine-beta-hydroxylase immunoreactive fibers in rat brain regions mediating arousal, motivation, and stress. J Comp Neurol 464(2):220–237. doi:10.1002/cne.10783

25. Berthoud H-R, Patterson LM, Sutton GM, Morrison C, Zheng H (2005) Orexin inputs to caudal raphé neurons involved in thermal, cardiovascular, and gastrointestinal regulation. Histochem Cell Biol 123(2):147–156. doi:10.1007/s00418-005-0761-x

26. Ciriello J, de Oliveira CVR (2003) Cardiac effects of hypocretin-1 in nucleus ambiguus. Am J Physiol 284(6):R1611–R1620. doi:10.1152/ajpregu.00719.2002

27. Ciriello J, Li Z, de Oliveira CVR (2003) Cardioacceleratory responses to hypocretin-1 injections into rostral ventromedial medulla. Brain Res 991(1–2):84–95

28. Geerling JC, Mettenleiter TC, Loewy AD (2003) Orexin neurons project to diverse sympathetic outflow systems. Neuroscience 122(2):541–550

29. Lazarenko RM, Stornetta RL, Bayliss DA, Guyenet PG (2011) Orexin A activates retrotrapezoid neurons in mice. Respir Physiol Neurobiol 175(2):283–287. doi:10.1016/j.resp.2010.12.003

30. Liu X, Zeng J, Zhou A, Theodorsson E, Fahrenkrug J, Reinscheid RK (2011) Molecular fingerprint of neuropeptide S-producing neurons in the mouse brain. J Comp Neurol 519 (10):1847–1866. doi:10.1002/cne.22603

31. Puskás N, Papp RS, Gallatz K, Palkovits M (2010) Interactions between orexin-immunoreactive fibers and adrenaline or noradrenaline-expressing neurons of the lower brainstem in rats and mice. Peptides 31(8):1589–1597. doi:10.1016/j.peptides.2010.04.020

32. Rosin DL, Chang DA, Guyenet PG (2006) Afferent and efferent connections of the rat retrotrapezoid nucleus. J Comp Neurol 499(1):64–89. doi:10.1002/cne.21105

33. Shahid IZ, Rahman AA, Pilowsky PM (2012) Orexin A in rat rostral ventrolateral medulla is pressor, sympatho-excitatory, increases barosensitivity and attenuates the somato-sympathetic reflex. Br J Pharmacol 165(7):2292–2303. doi:10.1111/j.1476-5381.2011.01694.x

34. Smith BN, Davis SF, van den Pol AN, Xu W (2002) Selective enhancement of excitatory synaptic activity in the rat nucleus tractus solitarius by hypocretin 2. Neuroscience 115 (3):707–714
35. Tupone D, Madden CJ, Cano G, Morrison SF (2011) An orexinergic projection from perifornical hypothalamus to raphé pallidus increases rat brown adipose tissue thermogenesis. J Neurosci 31(44):15944–15955. doi:10.1523/JNEUROSCI.3909-11.2011
36. Young JK, Wu M, Manaye KF, Kc P, Allard JS, Mack SO, Haxhiu MA (2005) Orexin stimulates breathing via medullary and spinal pathways. J Appl Physiol 98(4):1387–1395. doi:10.1152/japplphysiol.00914.2004
37. Zhang JH, Sampogna S, Morales FR, Chase MH (2004) Distribution of hypocretin (orexin) immunoreactivity in the feline pons and medulla. Brain Res 995(2):205–217
38. Zheng H, Patterson LM, Berthoud H-R (2005) Orexin-A projections to the caudal medulla and orexin-induced c-Fos expression, food intake, and autonomic function. J Comp Neurol 485(2):127–142. doi:10.1002/cne.20515
39. Date Y, Mondal MS, Matsukura S, Nakazato M (2000) Distribution of orexin-A and orexin-B (hypocretins) in the rat spinal cord. Neurosci Lett 288(2):87–90
40. Llewellyn-Smith IJ, Martin CL, Marcus JN, Yanagisawa M, Minson JB, Scammell TE (2003) Orexin-immunoreactive inputs to rat sympathetic preganglionic neurons. Neurosci Lett 351 (2):115–119
41. van den Pol AN (1999) Hypothalamic hypocretin (orexin): robust innervation of the spinal cord. J Neurosci 19(8):3171–3182
42. Fung SJ, Yamuy J, Sampogna S, Morales FR, Chase MH (2001) Hypocretin (orexin) input to trigeminal and hypoglossal motoneurons in the cat: a double-labeling immunohistochemical study. Brain Res 903(1–2):257–262
43. Volgin DV, Saghir M, Kubin L (2002) Developmental changes in the orexin 2 receptor mRNA in hypoglossal motoneurons. Neuroreport 13(4):433–436
44. de Oliveira CVR, Rosas-Arellano MP, Solano-Flores LP, Ciriello J (2003) Cardiovascular effects of hypocretin-1 in nucleus of the solitary tract. Am J Physiol 284(4):H1369–H1377. doi:10.1152/ajpheart.00877.2002
45. Kuwaki T, Zhang W (2010) Autonomic malfunctions in mice model of narcolepsy. In: Santos G, Villalba L (eds) Narcolepsy: symptoms, causes and diagnosis. Nova Science Publishers, Inc., New York, pp. 1–33
46. Kuwaki T (2011) Orexin links emotional stress to autonomic functions. Autonom Neurosci 161:20–27
47. Gotter AL, Webber AL, Coleman PJ, Renger JJ, Winrow CJ (2012) International union of basic and clinical pharmacology. LXXXVI. orexin receptor function, nomenclature and pharmacology. Pharmacol Rev 64(3):389–420. doi:10.1124/pr.111.005546
48. Leonard CS, Kukkonen JP (2014) Orexin/hypocretin receptor signalling: a functional perspective. Br J Pharmacol 171:294–313
49. Thompson MD, Xhaard H, Sakurai T, Rainero I, Kukkonen JP (2014) OX1 and OX2 orexin/hypocretin receptor pharmacogenetics. Front Neurosci 8:57. doi:10.3389/fnins.2014.00057
50. Lu XY, Bagnol D, Burke S, Akil H, Watson SJ (2000) Differential distribution and regulation of OX1 and OX2 orexin/hypocretin receptor messenger RNA in the brain upon fasting. Horm Behav 37(4):335–344. doi:10.1006/hbeh.2000.1584
51. Marcus JN, Aschkenasi CJ, Lee CE, Chemelli RM, Saper CB, Yanagisawa M, Elmquist JK (2001) Differential expression of orexin receptors 1 and 2 in the rat brain. J Comp Neurol 435 (1):6–25
52. Sunter D, Morgan I, Edwards CM, Dakin CL, Murphy KG, Gardiner J, Taheri S, Rayes E, Bloom SR (2001) Orexins: effects on behavior and localisation of orexin receptor 2 messenger ribonucleic acid in the rat brainstem. Brain Res 907(1–2):27–34
53. Trivedi P, Yu H, MacNeil DJ, Van der Ploeg LH, Guan XM (1998) Distribution of orexin receptor mRNA in the rat brain. FEBS Lett 438(1–2):71–75

54. van den Top M, Nolan MF, Lee K, Richardson PJ, Buijs RM, Davies CH, Spanswick D (2003) Orexins induce increased excitability and synchronisation of rat sympathetic preganglionic neurones. J Physiol 549(Pt 3):809–821. doi:10.1113/jphysiol.2002.033290

55. Cluderay JE, Harrison DC, Hervieu GJ (2002) Protein distribution of the orexin-2 receptor in the rat central nervous system. Regul Pept 104(1–3):131–144

56. Greco MA, Shiromani PJ (2001) Hypocretin receptor protein and mRNA expression in the dorsolateral pons of rats. Brain Res Mol Brain Res 88(1–2):176–182

57. Hervieu GJ, Cluderay JE, Harrison DC, Roberts JC, Leslie RA (2001) Gene expression and protein distribution of the orexin-1 receptor in the rat brain and spinal cord. Neuroscience 103 (3):777–797

58. Bäckberg M, Hervieu G, Wilson S, Meister B (2002) Orexin receptor-1 (OX-R1) immunoreactivity in chemically identified neurons of the hypothalamus: focus on orexin targets involved in control of food and water intake. Europ J Neurosci 15(2):315–328

59. Yamanaka A, Tabuchi S, Tsunematsu T, Fukazawa Y, Tominaga M (2010) Orexin directly excites orexin neurons through orexin 2 receptor. J Neurosci 30(38):12642–12652. doi:10. 1523/JNEUROSCI.2120-10.2010

60. Beig MI, Horiuchi J, Dampney RAL, Carrive P (2015) Both Ox1R and Ox2R orexin receptors contribute to the cardiorespiratory response evoked from the perifornical hypothalamus. Clin Exp Pharmacol Physiol 42(10):1059–1067. doi:10.1111/1440-1681.12461

61. Ibrahim BM, Abdel-Rahman AA (2015) A pivotal role for enhanced brainstem Orexin receptor 1 signaling in the central cannabinoid receptor 1-mediated pressor response in conscious rats. Brain Res 1622:51–63. doi:10.1016/j.brainres.2015.06.011

62. Kukkonen JP (2012) Recent progress in orexin/hypocretin physiology and pharmacology. Biomol Concepts 3:447–463

63. Ch'ng SS, Lawrence AJ (2015) Distribution of the orexin-1 receptor (OX1R) in the mouse forebrain and rostral brainstem: a characterisation of OX1R-eGFP mice. J Chem Neuroanat. doi:10.1016/j.jchemneu.2015.03.002

64. Darwinkel A, Stanić D, Booth LC, May CN, Lawrence AJ, Yao ST (2014) Distribution of orexin-1 receptor-green fluorescent protein- (OX1-GFP) expressing neurons in the mouse brain stem and pons: co-localization with tyrosine hydroxylase and neuronal nitric oxide synthase. Neuroscience 278:253–264. doi:10.1016/j.neuroscience.2014.08.027

65. Morairty SR, Revel FG, Malherbe P, Moreau J-L, Valladao D, Wettstein JG, Kilduff TS, Borroni E (2012) Dual hypocretin receptor antagonism is more effective for sleep promotion than antagonism of either receptor alone. PLoS One 7(7):e39131. doi:10.1371/journal.pone. 0039131

66. Scammell TE, Winrow CJ (2011) Orexin receptors: pharmacology and therapeutic opportunities. Annu Rev Pharmacol Toxicol 51:243–266. doi:10.1146/annurev-pharmtox-010510-100528

67. Steiner MA, Gatfield J, Brisbare-Roch C, Dietrich H, Treiber A, Jenck F, Boss C (2013) Discovery and characterization of ACT-335827, an orally available, brain penetrant orexin receptor type 1 selective antagonist. ChemMedChem 8(6):898–903. doi:10.1002/cmdc. 201300003

68. Bonaventure P, Yun S, Johnson PL, Shekhar A, Fitz SD, Shireman BT, Lebold TP, Nepomuceno D, Lord B, Wennerholm M, Shelton J, Carruthers N, Lovenberg T, Dugovic C (2015) A selective orexin-1 receptor antagonist attenuates stress-induced hyperarousal without hypnotic effects. J Pharmacol Exp Ther 352(3):590–601. doi:10.1124/jpet.114. 220392

69. McElhinny CJ, Lewin AH, Mascarella SW, Runyon S, Brieaddy L, Carroll FI (2012) Hydrolytic instability of the important orexin 1 receptor antagonist SB-334867: possible confounding effects on in vivo and in vitro studies. Bioorg Med Chem Lett 22 (21):6661–6664. doi:10.1016/j.bmcl.2012.08.109

70. Hirose M, Egashira S, Goto Y, Hashihayata T, Ohtake N, Iwaasa H, Hata M, Fukami T, Kanatani A, Yamada K (2003) N-acyl 6,7-dimethoxy-1,2,3,4-tetrahydroisoquinoline: the first orexin-2 receptor selective non-peptidic antagonist. Org Med Chem Lett 13(24):4497–4499
71. Malherbe P, Borroni E, Gobbi L, Knust H, Nettekoven M, Pinard E, Roche O, Rogers-Evans M, Wettstein JG, Moreau J-L (2009) Biochemical and behavioural characterization of EMPA, a novel high-affinity, selective antagonist for the OX(2) receptor. Br J Pharmacol 156 (8):1326–1341. doi:10.1111/j.1476-5381.2009.00127.x
72. Brisbare-Roch C, Dingemanse J, Koberstein R, Hoever P, Aissaoui H, Flores S, Mueller C, Nayler O, Van Gerven J, De Haas SL, Hess P, Qiu C, Buchmann S, Scherz M, Weller T, Fischli W, Clozel M, Jenck F (2007) Promotion of sleep by targeting the orexin system in rats, dogs and humans. Nat Med 13(2):150–155. doi:10.1038/nm1544
73. Malherbe P, Borroni E, Pinard E, Wettstein JG, Knoflach F (2009) Biochemical and electro-physiological characterization of almorexant, a dual orexin 1 receptor (OX1)/orexin 2 receptor (OX2) antagonist: comparison with selective OX1 and OX2 antagonists. Mol Pharmacol 76(3):618–631. doi:10.1124/mol.109.055152
74. Asahi S, Egashira S-i, Matsuda M, Iwaasa H, Kanatani A, Ohkubo M, Ihara M, Morishima H (2003) Development of an orexin 2 receptor selective agonist, [Ala(11), D-Leu(15)]orexin-B. Bioorg Med Chem Lett 13(1):111–113
75. Bastianini S, Silvani A, Berteotti C, Elghozi J-L, Franzini C, Lenzi P, Lo Martire V, Zoccoli G (2011) Sleep related changes in blood pressure in hypocretin-deficient narcoleptic mice. Sleep 34(2):213–218
76. Zhang W, Sakurai T, Fukuda Y, Kuwaki T (2006) Orexin neuron-mediated skeletal muscle vasodilation and shift of baroreflex during defense response in mice. Am J Physiol 290(6): R1654–R1663. doi:10.1152/ajpregu.00704.2005
77. Shirasaka T, Nakazato M, Matsukura S, Takasaki M, Kannan H (1999) Sympathetic and cardiovascular actions of orexins in conscious rats. Am J Physiol 277(6 Pt 2):R1780–R1785
78. Samson WK, Gosnell B, Chang JK, Resch ZT, Murphy TC (1999) Cardiovascular regulatory actions of the hypocretins in brain. Brain Res 831(1–2):248–253
79. Chen CT, Hwang LL, Chang JK, Dun NJ (2000) Pressor effects of orexins injected intracisternally and to rostral ventrolateral medulla of anesthetized rats. Am J Physiol 278 (3):R692–R697
80. Shahid IZ, Rahman AA, Pilowsky PM (2011) Intrathecal orexin A increases sympathetic outflow and respiratory drive, enhances baroreflex sensitivity and blocks the somato-sympathetic reflex. Br J Pharmacol 162(4):961–973. doi:10.1111/j.1476-5381.2010.01102.x
81. Hirota K, Kushikata T, Kudo M, Kudo T, Smart D, Matsuki A (2003) Effects of central hypocretin-1 administration on hemodynamic responses in young-adult and middle-aged rats. Brain Res 981(1–2):143–150
82. Samson WK, Bagley SL, Ferguson AV, White MM (2007) Hypocretin/orexin type 1 receptor in brain: role in cardiovascular control and the neuroendocrine response to immobilization stress. Am J Physiol 292(1):R382–R387. doi:10.1152/ajpregu.00496.2006
83. Huang S-C, Dai Y-WE, Lee Y-H, Chiou L-C, Hwang L-L (2010) Orexins depolarize rostral ventrolateral medulla neurons and increase arterial pressure and heart rate in rats mainly via orexin 2 receptors. J Pharmacol Exp Ther 334(2):522–529. doi:10.1124/jpet.110.167791
84. Antunes VR, Brailoiu GC, Kwok EH, Scruggs P, Dun NJ (2001) Orexins/hypocretins excite rat sympathetic preganglionic neurons in vivo and in vitro. Am J Physiol 281(6):R1801–R1807
85. Zhang W, Fukuda Y, Kuwaki T (2005) Respiratory and cardiovascular actions of orexin-A in mice. Neurosci Lett 385(2):131–136
86. Machado BH, Bonagamba LGH, Dun SL, Kwok EH, Dun NJ (2002) Pressor response to microinjection of orexin/hypocretin into rostral ventrolateral medulla of awake rats. Regul Pept 104:75–81
87. Dun NJ, Le Dun S, Chen CT, Hwang LL, Kwok EH, Chang JK (2000) Orexins: a role in medullary sympathetic outflow. Regul Pept. 96(1–2):65–70

88. Xiao F, Jiang M, Du D, Xia C, Wang J, Cao Y, Shen L, Zhu D (2012) Orexin A regulates cardiovascular responses in stress-induced hypertensive rats. Neuropharmacology 67C:16–24. doi:10.1016/j.neuropharm.2012.10.021

89. Luong LNL, Carrive P (2012) Orexin microinjection in the medullary raphé increases heart rate and arterial pressure but does not reduce tail skin blood flow in the awake rat. Neuroscience 202:209–217. doi:10.1016/j.neuroscience.2011.11.073

90. Ciriello J, Caverson MM, McMurray JC, Bruckschwaiger EB (2013) Co-localization of hypocretin-1 and leucine-enkephalin in hypothalamic neurons projecting to the nucleus of the solitary tract and their effect on arterial pressure. Neuroscience 250:599–613. doi:10.1016/j.neuroscience.2013.07.054

91. Shih C-D, Chuang Y-C (2007) Nitric oxide and GABA mediate bi-directional cardiovascular effects of orexin in the nucleus tractus solitarii of rats. Neuroscience 149(3):625–635. doi:10.1016/j.neuroscience.2007.07.016

92. Smith PM, Samson WK, Ferguson AV (2007) Cardiovascular actions of orexin-A in the rat subfornical organ. J Neuroendocrinol. 19(1):7–13

93. Iigaya K, Horiuchi J, McDowall LM, Lam ACB, Sediqi Y, Polson JW, Carrive P, Dampney RAL (2012) Blockade of orexin receptors with Almorexant reduces cardiorespiratory responses evoked from the hypothalamus but not baro- or chemoreceptor reflex responses. Am J Physiol 303(10):R1011–R1022. doi:10.1152/ajpregu.00263.2012

94. Stettner GM, Kubin L (2013) Antagonism of orexin receptors in the posterior hypothalamus reduces hypoglossal and cardiorespiratory excitation from the perifornical hypothalamus. J Appl Physiol 114(1):119–130. doi:10.1152/japplphysiol.00965.2012

95. Beig MI, Dampney BW, Carrive P (2014) Both Ox1r and Ox2r orexin receptors contribute to the cardiovascular and locomotor components of the novelty stress response in the rat. Neuropharmacology 89C:146–156. doi:10.1016/j.neuropharm.2014.09.012

96. Nisimaru N, Mittal C, Shirai Y, Sooksawate T, Anandaraj P, Hashikawa T, Nagao S, Arata A, Sakurai T, Yamamoto M, Ito M (2013) Orexin-neuromodulated cerebellar circuit controls redistribution of arterial blood flows for defense behavior in rabbits. Proc Natl Acad Sci U S A. doi:10.1073/pnas.1312804110

97. Rusyniak DE, Zaretsky DV, Zaretskaia MV, Dimicco JA (2011) The role of orexin-1 receptors in physiologic responses evoked by microinjection of PgE2 or muscimol into the medial preoptic area. Neurosci Lett 498(2):162–166. doi:10.1016/j.neulet.2011.05.006

98. Furlong TM, Vianna DML, Liu L, Carrive P (2009) Hypocretin/orexin contributes to the expression of some but not all forms of stress and arousal. Europ J Neurosci 30 (8):1603–1614. doi:10.1111/j.1460-9568.2009.06952.x

99. Johnson PL, Truitt W, Fitz SD, Minick PE, Dietrich A, Sanghani S, Träskman-Bendz L, Goddard AW, Brundin L, Shekhar A (2009) A key role for orexin in panic anxiety. Nat Med 16(1):111–115. doi:10.1038/nm.2075

100. Johnson PL, Federici LM, Fitz SD, Renger JJ, Shireman B, Winrow CJ, Bonaventure P, Shekhar A (2015) Orexin 1 and 2 receptor involvement in co2 -induced panic-associated behavior and autonomic responses. Depress Anxiety 32(9):671–683. doi:10.1002/da.22403

101. Johnson PL, Samuels BC, Fitz SD, Federici LM, Hammes N, Early MC, Truitt W, Lowry CA, Shekhar A (2012) Orexin 1 receptors are a novel target to modulate panic responses and the panic brain network. Physiol Behav 107(5):733–742. doi:10.1016/j.physbeh.2012.04.016

102. Rusyniak DE, Zaretsky DV, Zaretskaia MV, Durant PJ, Dimicco JA (2012) The orexin-1 receptor antagonist SB-334867 decreases sympathetic responses to a moderate dose of methamphetamine and stress. Physiol Behav 107(5):743–750. doi:10.1016/j.physbeh.2012.02.010

103. Allard JS, Tizabi Y, Shaffery JP, Manaye K (2007) Effects of rapid eye movement sleep deprivation on hypocretin neurons in the hypothalamus of a rat model of depression. Neuropeptides 41(5):329–337

104. Chen X, Li S, Kirouac GJ (2014) Blocking of corticotrophin releasing factor receptor-1 during footshock attenuates context fear but not the upregulation of prepro-orexin mRNA in rats. Pharmacol Biochem Behav 120:1–6. doi:10.1016/j.pbb.2014.01.013

105. Chen X, Wang H, Lin Z, Li S, Li Y, Bergen HT, Vrontakis ME, Kirouac GJ (2013) Orexins (hypocretins) contribute to fear and avoidance in rats exposed to a single episode of footshocks. Brain Struct Funct. doi:10.1007/s00429-013-0626-3

106. Hayward LF, Hampton EE, Ferreira LF, Christou DD, Yoo J-K, Hernandez ME, Martin EJ (2015) Chronic heart failure alters orexin and melanin concentrating hormone but not corticotrophin releasing hormone-related gene expression in the brain of male Lewis rats. Neuropeptides. doi:10.1016/j.npep.2015.06.001

107. Frey A, Popp S, Post A, Langer S, Lehmann M, Hofmann U, Sirén A-L, Hommers L, Schmitt A, Strekalova T, Ertl G, Lesch K-P, Frantz S (2014) Experimental heart failure causes depression-like behavior together with differential regulation of inflammatory and structural genes in the brain. Front Behav Neurosci 8:376. doi:10.3389/fnbeh.2014.00376

108. Schoemaker RG, Smits JF (1994) Behavioral changes following chronic myocardial infarction in rats. Physiol Behav 56(3):585–589

109. Carretero OA, Oparil S (2000) Essential hypertension. Part I: definition and etiology. Circulation 101(3):329–335

110. Fisher JP, Paton JFR (2012) The sympathetic nervous system and blood pressure in humans: implications for hypertension. J Hum Hypertens 26(8):463–475. doi:10.1038/jhh.2011.66

111. Korner P (2007) Essential hypertension and its causes: neural and non-neural mechanisms. Oxford University Press, New York

112. Lee Y-H, Dai Y-WE, Huang S-C, Li TL, Hwang L-L (2013) Blockade of central orexin 2 receptors reduces arterial pressure in spontaneously hypertensive rats. Exp Physiol. doi:10.1113/expphysiol.2013.072298

113. Li A, Hindmarch CCT, Nattie EE, Paton JFR (2013) Antagonism of orexin receptors significantly lowers blood pressure in spontaneously hypertensive rats. J Physiol. doi:10.1113/jphysiol.2013.256271

114. Clifford L, Dampney BW, Carrive P (2015) Spontaneously hypertensive rats have more orexin neurons in their medial hypothalamus than normotensive rats. Exp Physiol 100(4):388–398. doi:10.1113/expphysiol.2014.084137

115. Lee Y-H, Tsai M-C, Li TL, Dai Y-WE, Huang S-C, Hwang L-L (2015) Spontaneously hypertensive rats have more orexin neurons in the hypothalamus and enhanced orexinergic input and orexin 2 receptor-associated nitric oxide signalling in the rostral ventrolateral medulla. Exp Physiol. doi:10.1113/EP085016

116. Paré WP (1989) Stress ulcer and open-field behavior of spontaneously hypertensive, normotensive, and Wistar rats. Pavlov J Biol Sci 24(2):54–57

117. Sagvolden T (2000) Behavioral validation of the spontaneously hypertensive rat (SHR) as an animal model of attention-deficit/hyperactivity disorder (AD/HD). Neurosci Biobehav Rev 24(1):31–39

118. Sagvolden T, Pettersen MB, Larsen MC (1993) Spontaneously hypertensive rats (SHR) as a putative animal model of childhood hyperkinesis: SHR behavior compared to four other rat strains. Physiol Behav 54(6):1047–1055

119. Davern PJ, Nguyen-Huu T-P, La Greca L, Abdelkader A, Head GA (2009) Role of the sympathetic nervous system in Schlager genetically hypertensive mice. Hypertension 54(4):852–859. doi:10.1161/HYPERTENSIONAHA.109.136069

120. Schlager G (1974) Selection for blood pressure levels in mice. Genetics 76(3):537–549

121. Marques FZ, Campain AE, Davern PJ, Yang YHJ, Head GA, Morris BJ (2011) Genes influencing circadian differences in blood pressure in hypertensive mice. PLoS One 6(4):e19203. doi:10.1371/journal.pone.0019203

122. Marques FZ, Campain AE, Davern PJ, Yang YHJ, Head GA, Morris BJ (2011) Global identification of the genes and pathways differentially expressed in hypothalamus in early

and established neurogenic hypertension. Physiol Genomics 43 (12):766–771. doi:10.1152/physiolgenomics.00009.2011

123. Jackson KL, Dampney BW, Moretti J-L, Stevenson ER, Davern PJ, Carrive P, Head GA (2016) The contribution of orexin to the neurogenic hypertension in BPH/2 J mice. Hypertension 67(5):959–969

124. Deng B-S, Nakamura A, Zhang W, Yanagisawa M, Fukuda Y, Kuwaki T (2007) Contribution of orexin in hypercapnic chemoreflex: evidence from genetic and pharmacological disruption and supplementation studies in mice. J Appl Physiol 103(5):1772–1779

125. Sugita T, Sakuraba S, Kaku Y, Yoshida K, Arisakaa H, Kuwana S (2014) Orexin induces excitation of respiratory neuronal network in isolatedbrainstem spinal cord of neonatal rat. Respir Physiol Neurobiol 200:105–109

126. Dutschmann M, Kron M, Mörschel M, Gestreau C (2007) Activation of Orexin B receptors in the pontine Kölliker-Fuse nucleus modulates pre-inspiratory hypoglossal motor activity in rat. Respir Physiol Neurobiol 159(2):232–235

127. Peever JH, Lai Y-Y, Siegel JM (2003) Excitatory effects of hypocretin-1 (orexin-A) in the trigeminal motor nucleus are reversed by NMDA antagonism. J Neurophysiol 89:2591–2600

128. Zhang GH, Liu ZL, Zhang BJ, Geng WY, Song NN, Zhou W, Cao YX, Li SQ, Huang ZL, Shen LL (2014) Orexin A activates hypoglossal motoneurons and enhances genioglossus muscle activity in rats. Br J Pharmacol 171:4233–4246

129. Krieger J (2000) Respiratory physiology: breathing in normal subjects. In: Kryger MH, Roth T, Dement WC (eds) Principles and practice of sleep medicine. W.B. Saunders, Philadelphia, pp. 229–241

130. Douglas NJ (2000) Respiratory physiology: control of ventilation. In: Kryger MH, Roth T, Dement WC (eds) Principles and practice of sleep medicine. W.B. Saunders, Philadelphia, pp. 221–228

131. Lee M, Hassani O, Jones B (2005) Discharge of identified orexin/hypocretin neurons across the sleep-waking cycle. J Neurosci 25(28):6716–6720

132. Mileykovskiy B, Kiyashchenko L, Siegel J (2005) Behavioral correlates of activity in identified hypocretin/orexin neurons. Neuron 46(5):787–798

133. Takahashi K, Lin JS, Sakai K (2008) Neuronal activity of orexin and non-orexin waking-active neurons during wake–sleep states in the mouse. Neuroscience 153:860–870

134. Kuwaki T (2015) Thermoregulation under pressure: a role for orexin neurons. Temperature 2:379–391

135. Espana RA, Valentino RJ, Berridge CW (2003) Fos immunoreactivity in hypocretin-synthesizing and hypocretin-1 receptor-expressing neurons: effects of diurnal and nocturnal spontaneous waking, stress and hypocretin-1 administration. Neuroscience 121(1):201–217

136. Ida T, Nakahara K, Murakami T, Hanada R, Nakazato M, Murakami N (2000) Possible involvement of orexin in the stress reaction in rats. Biochem Biophys Res Commun 270 (1):318–323

137. Kuwaki T, Zhang W, Nakamura A (2007) State-dependent adjustment of the central autonomic regulation: role of orexin in emotional behavior and sleep/wake cycle. In: Kubo T, Kuwaki T (eds) Central mechanisms of cardiovascular regulation. Research Signport, Kerala, India, pp. 57–73

138. Watanabe S, Kuwaki T, Yanagisawa M, Fukuda Y, Shimoyama M (2005) Persistent pain and stress activate pain-inhibitory orexin pathways. Neuroreport 16(1):5–8

139. Winsky-Sommerer R, Yamanaka A, Diano S, Borok E, Roberts AJ, Sakurai T, Kilduff TS, Horvath TL, de Lecea L (2004) Interaction between the corticotropin-releasing factor system and hypocretins (orexins): a novel circuit mediating stress response. J Neurosci 24 (50):11439–11448

140. Zhang W, Sunanaga J, Takahashi Y, Mori T, Sakurai T, Kanmura Y, Kuwaki T (2010) Orexin neurons are indispensable for stress-induced thermogenesis in mice. J Physiol 588 (21):4117–4129

141. Zhu L, Onaka T, Sakurai T, Yada T (2002) Activation of orexin neurones after noxious but not conditioned fear stimuli in rats. Neuroreport 13(10):1351–1353
142. Nakamura A, Zhang W, Yanagisawa M, Fukuda Y, Kuwaki T (2007) Vigilance state-dependent attenuation of hypercapnic chemoreflex and exaggerated sleep apnea in orexin knockout mice. J Appl Physiol 102:241–248
143. Kuwaki T, Deng B-S, Nakamura A, Zhang W, Yanagisawa M, Fukuda Y (2005) Abnormal respiration in orexin knockout mice. In: Kumar A, Mallick H (eds) Proceedings of the 2nd interim congress of the world federation of sleep research and sleep medicine society. Medimond S.r.l, Bologna, Italy, pp. 69–72
144. Kuwaki T (2008) Orexinergic modulation of breathing across vigilance states. Respir Physiol Neurobiol 164:204–212
145. Williams RH, Jensen LT, Verkhratsky A, Fugger L, Burdakov D (2007) Control of hypothalamic orexin neurons by acid and CO_2. Proc Natl Acad Sci U S A 104(25):10685–10690
146. Sunanaga J, Deng B-S, Zhang W, Kanmura Y, Kuwaki T (2009) CO_2 activates orexin-containing neurons in mice. Respir Physiol Neurobiol 166(3):184–186
147. Phillipson E (1978) Control of breathing during sleep. Am Rev Respir Dis 118:909–939
148. Dias MB, Li A, Nattie EE (2010) The orexin receptor-1 (OX1R) in the rostral medullary raphé contributes to the hypercapnic chemoreflex in wakefulness, during the active period of the diurnal cycle. Respir Physiol Neurobiol 170(9):96–102
149. Dias M, Li A, Nattie E (2009) Antagonism of orexin receptor-1 in the retrotrapezoid nucleus inhibits the ventilatory response to hypercapnia predominantly in wakefulness. J Physiol 587 (9):2059–2067
150. Kuwaki T, Li A, Nattie EE (2010) State-dependent central chemoreception: a role of orexin. Respir Physiol Neurobiol 173:223–229
151. Han F, Mignot E, Wei Y, Dong S, Li J, Lin L, An P, Wang L, Wang J, He M, Gao H, Li M, Gao Z, Strohl K (2010) Ventilatory chemoresponsiveness, narcolepsy-cataplexy and human leukocyte antigen DQB1*0602 status. Eur Respir J 36(3):577–583
152. Nakamura A, Fukuda Y, Kuwaki T (2003) Sleep apnea and effect of chemostimulation on breathing instability in mice. J Appl Physiol 94(2):525–532
153. Nakamura A, Kuwaki T (2004) Sleep apnea in mice: a useful animal model for study of SIDS? Pathophysiology 10(3/4):253–257
154. Moore MW, Akladious A, Hu Y, Azzam S, Feng P, Strohl KP (2014) Effects of orexin 2 receptor activation on apnea in the C57BL/6 J mouse. Respir Physiol Neurobiol 200:118–125
155. Tarasiuk A, Levi A, Berdugo-Boura N, Yahalom A, Segev Y (2014) Role of orexin in respiratory and sleep homeostasis during upper airway obstruction in rats. Sleep 37:987–998
156. Chokroverty S (1986) Sleep apnea in narcolepsy. Sleep 9:250–253
157. Harsh J, Peszka J, Hartwig G, Mitler M (2000) Night-time sleep and daytime sleepiness in narcolepsy. J Sleep Res 9:309–316
158. Yamaguchi K, Futatsuki T, Ushikai J, Kuroki C, Minami T, Kakihana Y, Kuwaki T (2015) Intermittent but not sustained hypoxia activates orexin-containing neurons in mice. Respir Physiol Neurobiol 206:11–14
159. Terada J, Nakamura A, Zhang W, Yanagisawa M, Kuriyama T, Fukuda Y, Kuwaki T (2008) Ventilatory long-term facilitation in mice can be observed both during sleep and wake periods and depends on orexin. J Appl Physiol 104(2):499–507
160. Toyama S, Sakurai T, Tatsumi K, Kuwaki T (2009) Attenuated phrenic long-term facilitation in orexin neuron-ablated mice. Respir Physiol Neurobiol 168(3):295–302
161. Cai XJ, Evans ML, Lister CA, Leslie RA, Arch JRS, Wilson S, Williams G (2001) Hypoglycemia activates orexin neurons and selectively increases hypothalamic orexin-B levels: responses inhibited by feeding and possibly mediated by the nucleus of the solitary tract. Diabetes 50:105–112

162. Burdakov D, Gerasimenko O, Verkhratsky A (2005) Physiological changes in glucose differentially modulate the excitability of hypothalamic melanin-concentrating hormone and orexin neurons in situ. J Neurosci 25(9):2429–2433
163. Burdakov D, Jensen LT, Alexopoulos H, Williams RH, Fearon IM, O'Kelly I, Gerasimenko O, Fugger L, Verkhratsky A (2006) Tandem-pore K+ channels mediate inhibition of orexin neurons by glucose. Neuron 50(5):711–722
164. Polotsky VY, Wilson JA, Haines AS, Scharf MT, Soutiere SE, Tankersley CG, Smith PL, Schwartz AR, O'Donnell CP (2001) The impact of insulin-dependent diabetes on ventilatory control in the mouse. Am J Respir Crit Care Med 163:624–632
165. Punjabi NM, Shahar E, Rediline S, Gottlieb DJ, Givelber R, Resnick HE, Sleep Heat Health Study Investigators (2004) Sleep-disordered breathing, glucose intolerance, and insulin resistance: the Sleep Heart Health Study. Am J Epidemiol 160:521–530
166. Jordan AS, White DP (2008) Pharyngeal motor control and the pathogenesis of obstructive sleep apnea. Respir Physiol Neurobiol 160(1):1–7
167. Sakurai S, Nishijima T, Takahashi S, Yamauchi K, Arihara Z, Takahashi K (2005) Low plasma orexin-A levels were improved by continuous positive airway pressure treatment in patients with severe obstructive sleep apnea-hypopnea syndrome. Chest 127(3):731–737
168. Busquets X, Barbe F, Barcelo A, de la Pena M, Sigritz N, Mayoralas L, Ladaria A, Agusti A (2004) Decreased plasma levels of orexin-A in sleep apnea. Respiration 71(6):575–579
169. Igarashi N, Tatsumi K, Nakamura A, Sakao S, Takiguchi Y, Nishikawa T, Kuriyama T (2003) Plasma orexin-A levels in obstructive sleep apnea-hypopnea syndrome. Chest 124 (4):1381–1385
170. Kanbayashi T, Inoue Y, Kawanishi K, Takasaki H, Aizawa R, Takahashi K, Ogawa Y, Abe M, Hishikawa Y, Shimizu T (2003) CSF hypocretin measures in patients with obstructive sleep apnea. J Sleep Res 12(4):339–341
171. Sun H, Palcza J, Rosenberg R, Kryger M, Siringhaus T, Rowe J, Lines C, Wagner JA, Troyer MD (2015) Effects of suvorexant, an orexin receptor antagonist, on breathing during sleep in patients with chronic obstructive pulmonary disease. Resp Med 109:416–426

Role of the Orexin/Hypocretin System in Stress-Related Psychiatric Disorders

Morgan H. James, Erin J. Campbell, and Christopher V. Dayas

Abstract Orexins (hypocretins) are critically involved in coordinating appropriate physiological and behavioral responses to aversive and threatening stimuli. Acute stressors engage orexin neurons via direct projections from stress-sensitive brain regions. Orexin neurons, in turn, facilitate adaptive behavior via reciprocal connections as well as via direct projections to the hypophysiotropic neurons that coordinate the hypothalamic-pituitary-adrenal (HPA) axis response to stress. Consequently, hyperactivity of the orexin system is associated with increased motivated arousal and anxiety, and is emerging as a key feature of panic disorder. Accordingly, there has been significant interest in the therapeutic potential of pharmacological agents that antagonize orexin signaling at their receptors for the treatment of anxiety disorders. In contrast, disorders characterized by inappropriately low levels of motivated arousal, such as depression, generally appear to be associated with hypoactivity of the orexin system. This includes narcolepsy with cataplexy, a disorder characterized by the progressive loss of orexin neurons and increased rates of moderate/severe depression symptomology. Here, we provide a comprehensive overview of both clinical and preclinical evidence highlighting the role of orexin signaling in stress reactivity, as well as how perturbations to this system can result in dysregulated behavioral phenotypes.

M.H. James
Brain Health Institute, Rutgers University/Rutgers Biomedical and Health Sciences, Piscataway, NJ, USA
Florey Institute of Neuroscience and Mental Health, University of Melbourne, Parkville, VIC 2337, Australia

E.J. Campbell and C.V. Dayas (✉)
School of Biomedical Sciences and Pharmacy, Centre for Brain and Mental Health, University of Newcastle, Callaghan, NSW, Australia
Hunter Medical Research Institute, 1 Kookaburra Circuit, New Lambton Heights, NSW, Australia
e-mail: Christopher.Dayas@newcastle.edu.au

Curr Topics Behav Neurosci (2017) 33: 197–220
DOI 10.1007/7854_2016_56
Published Online: 12 January 2017

Keywords Anxiety • Corticosterone • Depression • Dual orexin receptor antagonists (DORAs) • HPA axis • Panic • Single orexin receptor antagonists (SORAs) • Stress

Contents

1 Introduction

Orexin-A and orexin-B ("the orexins", also known as the hypocretins [hcrt] 1 and 2, respectively) are hypothalamic neuropeptides that signal via two G-protein-coupled receptors termed orexin receptor 1 (OX_1R; hcrt receptor 1) and 2 (OX_2R; hcrt receptor 2) [1, 2]. Early studies focused on the role of the orexins in promoting food-seeking behavior, consistent with the restricted expression profile of orexin neuronal cell bodies within the lateral hypothalamic (LH) feeding area [2]. As discussed in significant detail in other chapters of this edition, the orexins have since been implicated in a range of other physiological processes, including cardiorespiratory control [3], sleep/wake cycle, and reward seeking [4, 5]. Central to each of these functions is the role of the orexins in promoting adaptive behavior in response to environmental opportunities or threats [6]. In this regard, it is not surprising that the orexin system is closely linked with stress reactivity. Indeed, significant evidence now suggests that the orexins play a key role in driving both behavioral and neuroendocrine responses to challenges that evoke hypothalamic–pituitary–adrenal (HPA) axis activation, and that dysregulation of the orexin system can lead to phenotypes associated with pathological stress reactivity.

In this chapter, we first outline findings from anatomical studies that implicate the orexins as central players in the brain's autonomic, behavioral, and endocrine responses to stress. Key to this modulatory capacity are the reciprocal connections that exist between amygdaloid, brainstem, and hypothalamic centers. Next, we describe preclinical evidence indicating that engagement of the orexin system drives behavioral and physiological responses that allow the animal to appropriately respond to a specific challenge. While under normal circumstances these responses are highly adaptive – in that they enhance the animal's likelihood of survival – chronic over- or under-activation of this system can result in maladaptive responses. These findings closely parallel opposing patterns of orexin system dysfunction observed in human anxiety versus depression. Accordingly, we discuss the latest preclinical and clinical evidence linking changes in the orexin system with disorders that are associated with elevated versus low motivated arousal. Finally, we discuss the therapeutic potential and possible caveats of targeting the orexin system for the treatment of such disorders.

2　Orexin System and Stress Circuitry

Orexin neurons are capable of integrating a variety of central and peripheral stress-related inputs, and are critical actuators in the neural circuitry of stress [7]. Since the distribution of orexin receptors and orexin afferent/efferent projections has been described in great detail elsewhere [8, 9], here, we aim to highlight the connectivity of orexin neurons with regions of the brain that control stress reactivity in an attempt to emphasize how the orexin neurons are strategically positioned to play a key role in the coordination of the central stress response. We pay particular attention to the actions of orexins at the central and peripheral branches of the HPA axis.

2.1　Orexin Receptor Distribution and Efferent Targets of Orexin Neurons

Orexin neurons project widely throughout the brain, consistent with their diverse physiological actions. Terminal orexin fibers are observed in key autonomic and respiratory sites, as well systems involved with emotion, and attention [10].

2.1.1　Arousal and Autonomic Structures

Several arousal-related regions in the brainstem receive orexinergic projections. For example, orexin-positive terminals are highly abundant in the locus coeruleus (LC) [9, 11] and dorsal raphé (DR) [9]. Orexin A and B excite noradrenergic LC neurons [11, 12] and orexin A increases norepinephrine release at LC efferent sites that are associated with attention [13]. Similarly, orexin peptides excite serotonergic neurons in the DR [14]. The effects of orexin in both LC and DR appear to be mediated predominantly by OX_1R, as OX_2R expression is very low or undetectable in these regions [8, 14, 15]. Orexin neurons also strongly innervate various cholinergic brainstem structures, including laterodorsal tegmental nucleus (LDT) and the pedunculopontine tegmental nucleus (PPT), which play a critical role in the regulation of REM sleep and wakefulness [16]. Finally, neurons in the GABAergic nucleus incertus, a region involved in arousal and stress responsivity, are innervated by orexin-positive fibers and excited by orexin-A, an effect that can be blocked by the OX_2R antagonist TCS-OX2-29 but not the OX_1R antagonist SB-334867 [17, 18].

Orexin neurons also provide input to a range of non-brainstem arousal structures. For example, the histaminergic tuberomammillary nucleus (TMN) of the hypothalamus is richly innervated by orexin neurons [9] and expresses only the OX_2R [8]. TMN neurons are excited and depolarized by both orexin-A and B [19], and infusion of orexin-A into TMN rapidly promotes wakefulness [20, 21]. The basal forebrain (BF), important for modulating arousal via widespread cholinergic, GABAergic, and glutamatergic projections to the cortex [22, 23], is also modulated by orexin neurons. Orexin-A excites BF cholinergic neurons, and causes release of acetylcholine in the cortex, which in turn promotes wakefulness [24]. Orexin-A also excites most GABAergic and glutamatergic neurons in the BF, which are also involved in promoting behavioral arousal [24, 25]. For a full review on how orexin neurons interact with arousal systems, the reader is directed to several excellent recent reviews [16, 17, 26, 27].

Regions involved in coordinating the autonomic response to stressors also receive input from orexin neurons. In particular, the rostral medullary raphé region is innervated by orexin-fibers, and local infusions of orexin-A in this region evoke significant and long lasting increases in heart rate and mean arterial pressure [28]. Orexin neurons also innervate and excite neurons in the retrotrapezoid nucleus on the ventral surface of the medulla, a key relay for hypothalamic pathways that regulate breathing [29]. The raphé pallidus expresses both OX_1R and OX_2R, and activation of this region with infusions of orexin-A into the fourth ventricle is associated with increased body temperature, heart rate, and locomotor activity [30]. Further, orexin-A has excitatory effects on neurons in the nucleus of the solitary tract [31], and infusion of orexin-A into the caudal, lateral, and medial subnuclei of the NTS elicits both depressor and bradycardia responses [32–34] and potentiates the heart rate component of the baroreflex [32, 33]. Orexin fibers also innervate the periaqueductal gray (PAG) [9], and orexin-A depresses GABAergic evoked IPSCs in this region [35]. Infusions of orexin-A into the PAG decrease formalin-induced nociceptive behaviors [36]. The extent of orexin innervation in these arousal and autonomic regions highlights the potential of orexin in modulating responses to environmental stressors through connections with descending autonomic and arousal systems.

2.1.2 Stress and Emotion Systems

Psychological and physiological stressors evoke distinct patterns of behavior that are often reflected as emotional states [37]. Orexin-containing nerve terminals and receptors are found in well-characterized stress/emotion-sensitive centers including the central nucleus of the amygdala (CeA) and the bed nucleus of the stria terminalis (BNST). The CeA and extended amygdala have well-established roles in anxiety, particularly in response to stimuli that evoke fear or threat. Interestingly, CeA neurons express both OX_1R and OX_2R [8, 38], and both orexin A and B depolarize neurons in the central medial nucleus of the CeA via OX_2R [39]. In contrast, BNST neurons almost exclusively express OX_1R [15] and orexin-A excites only a subset of BNST neurons in a slice preparation [40]. Orexin neurons also innervate other regions of the amygdala, including the basolateral amygdala, which expresses both OX_1R and OX_2R [41] and the medial amygdala which predominantly expresses OX_1R [8, 15]. Significant research has also been devoted to the role of orexinergic input to the paraventricular thalamus (PVT) [42–45], as this region displays extremely dense innervation by orexin-positive terminals [9], expresses both OX_1R and OX_2R [8], and is highly sensitive to a range of stressors [46–50]. Application of either orexin A or B excites the majority of PVT neurons in a dose-dependent manner, and the increase in firing rate in these cells is approximately three times larger for orexin B compared to orexin A [51]. Moreover, orexin B excites cortically projecting glutamatergic PVT neurons, resulting in substantial excitation in multiple layers of the medial prefrontal cortex [52]. In addition to these areas, orexin neurons innervate other stress-relevant regions, including lateral septum, prefrontal cortex, various hypothalamic regions, and the nucleus accumbens [8] (see Fig. 1 for an overview). Further work is required to characterize the expression profile of OX_1R versus OX_2R on different cell types in each of these regions.

2.2 Stress-Sensitive Afferent Inputs to Orexin Neurons

The afferent regulation of LH orexin neurons is likely to be critical to the recruitment and modulation of this system, especially given the small number of orexin-containing neurons. Sakurai et al. [54] used a molecular genetic approach to create a transgenic mouse line that expressed a fusion protein between the nontoxic C-terminal fragment of tetanus toxin (TTC, which is transported in a retrograde fashion) and GFP (TTC::GFP) exclusively in orexin neurons [54]. This TTC::GFP fusion protein was retrogradely transported to the cell bodies of interconnected neurons. Using this approach, the authors described a pattern of retrograde labeling in a range of stress-relevant regions, including BNST, medial amygdala, and C1 adrenergic neurons in the rostroventrolateral medulla (RVLM). This study also identified the existence of orexin–orexin neuron connections. Given the small number of orexin neurons and their extensive projections throughout the brain, this LH-based excitatory orexin circuit may provide a way of synchronizing and amplifying orexin neuron output. Additionally, because orexin neurons also contain glutamate [56–58] and dynorphin [59], it is possible that these circuits provide a mechanism through which other neurotransmitters and neuropeptides can regulate orexin neurons. This intra-LH action may enhance the salience of external stimuli, including stress-predictive cues. Subsequent studies using traditional anterograde

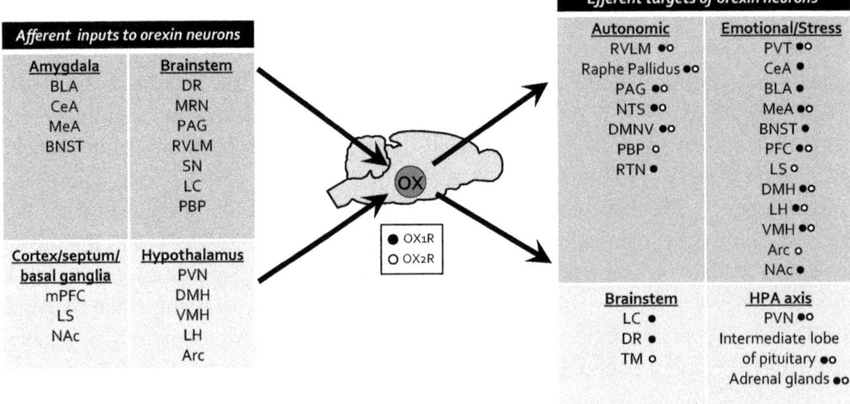

Fig. 1 Afferent inputs and efferent targets of the orexin system. A schematic illustration representing the stress-related afferent inputs to orexin neurons and efferent outputs of orexin neurons, as well as the pattern of orexin receptor distribution in each of the efferent regions. Orexin (OX), orexin receptors 1 and 2 (OX₁R, OX₂R), central amygdala (CeA), medial amygdala (MeA), bed nucleus of the stria terminalis (BNST), medial prefrontal cortex (mPFC), lateral septum (LS), dorsal raphé nucleus (DR), midbrain median raphé nucleus (MRN), periaqueductal gray (PAG), rostroventrolateral medulla (RVLM), paraventricular nucleus of the hypothalamus (PVN), lateral hypothalamus (LH), nucleus of the solitary tract (NTS), dorsal motor nucleus of the vagus (DMNV), retrotrapezoid nucleus (RTN), locus coeruleus (LC), histaminergic tuberomammillary nucleus (TM), paraventricular thalamus (PVT), basolateral amygdala (BLA), dorsomedial hypothalamus (DMH), ventromedial hypothalamus (VMH), arcuate nucleus (Arc), nucleus accumbens shell (NAc), substantia nigra (SN), parabrachial nucleus of the ventral tegmental area (PBP). *References*: Date et al. [53], Marcus et al. [8], Peyron et al. [9], Sakurai et al. [54], Trivedi et al. [15], and Yoshida et al. [55]

and retrograde neural tract-tracing techniques have confirmed many of the afferent systems identified by Sakurai and colleagues, and have also reported projections from stress-relevant regions including medial prefrontal cortex, lateral septum, central amygdala, DR, and PAG [55, 60].

An overview of all stress-related regions shown to project to LH orexin neurons is shown in Fig. 1. Perhaps the most striking collective observation from these tracing studies is the fact that many of the regions that innervate the orexin field also receive reciprocal input from orexin neurons (see above). This bidirectional relationship with stress-related regions strongly suggests that the orexin neurons are central to the integration of stress signals and initiation of adaptive behavioral and physiological responses [6, 61].

2.3 Orexin Regulation of the HPA Axis

Corticotrophin releasing factor (CRF) neurons in the medial parvocellular division of the paraventricular nucleus of the hypothalamus (mpPVN) constitute the apex of the HPA axis. Both physiological and psychological stressors recruit CRF neurons, resulting in CRF being released into the hypophyseal portal system. CRF binds to corticotroph cells in the anterior pituitary gland causing them to synthesize and release adrenocorticotropic hormone (ACTH) into the systemic blood circulation. ACTH subsequently promotes the production and release of glucocorticoids from the adrenal cortex, the most important being cortisol in humans or corticosterone in rats. Cortisol/corticosterone act to break down energy stores to release glucose, suppress inflammation and trigger behavioral arousal (for review, see [62]).

2.3.1 Expression of Orexins and Orexin Receptors in the Central Branch of the HPA Axis

Anatomical evidence suggests that the orexin system can influence mpPVN CRF neurons. For example, both OX_1R and OX_2R are expressed in the PVN of rat and mouse, although there is some inconsistency regarding the relative expression of each receptor between studies [15, 38, 63]. Both OX_1R and OX_2R are strongly expressed in the intermediate lobe of the pituitary, whereas the expression of OX_1R is greater than OX_2R in the anterior lobe [53]. Orexin-positive fibers are also observed in the PVN and median eminence, where CRF is released into the hypophyseal portal system [53]. Additionally, intracerebroventricular (ICV) infusions of the orexins activate CRF cells and increase CRF mRNA levels in the mpPVN. This orexin-induced stimulation of CRF results in HPA-axis activation, as reflected by increased plasma ACTH and corticosterone levels – an effect that can be blocked by CRF receptor antagonists [64, 65]. However, it is unclear if these actions are direct or through other stress-sensitive sites that initiate HPA axis activity. Further, orexin-A stimulates the release of CRF when applied to hypothalamic explants [64] and causes depolarization and an increase in spike frequency in PVN cells from mpPVN [65].

2.3.2 Effect of Orexins on the Peripheral Branch of the HPA Axis

Orexin peptides may act directly on the adrenal glands to modulate glucocorticoid release. Both OX_1R and OX_2R have been detected in rat adrenal homogenates [66–68] as well as in freshly dispersed and cultured human adrenocortical cells [69]. Studies that have attempted to characterize the relative expression of each receptor in adrenal subregions in rats have shown that OX_2R are expressed in abundance in the zona glomerulosa (ZG) and reticularis (ZR) [70, 71], whereas OX_1R levels in all regions are either undetectable or very low relative to OX_2R [72, 73]. In contrast, experiments using human tissue have shown that OX_1R and OX_2R are both expressed equally in adrenocortical cells [69, 74], zona fasiculata (ZF), ZR and adrenal medulla [75] whereas only OX_1R is expressed in ZG [76]. Regardless, it appears that orexin is able to act directly on the adrenals to stimulate glucocorticoid release, as application of orexin-A (but not orexin-B) increases basal corticosterone/cortisol secretion from dispersed or cultured rat and human ZF/ZR cells [74, 76], and this effect is blocked by OX_1R-specific antibodies [69].

2.3.3 Orexin Antagonists and HPA-Axis Control

ICV application of orexin peptides increases plasma ACTH levels, actions that are blocked by pretreatment with orexin receptor antagonists [77]. In contrast, orexin receptor antagonists have no effect on basal HPA tone. For example, ICV infusions of the orexin-1 receptor antagonist SB-408124, or oral treatment with the dual orexin receptor antagonist (DORA) almorexant, have no effect on basal plasma corticosterone levels [78, 79]. We interpret these findings as evidence that orexin neurons are not engaged under basal conditions. In contrast, in response to environmental threats, or experimental approaches that promote orexin system engagement, orexins act to promote HPA axis function and facilitate adaptive responding. These findings are likely to be significant when considering clinical approaches that target the orexin system (discussed below), given the importance of basal HPA axis tone for waking and normal metabolic functions.

2.3.4 Optogenetic Stimulation of Orexin Neurons Activates Stress Regions and HPA Axis

A recent study by Bonnavion et al. [7] provided direct evidence of orexin-induced control of HPA axis function. Using a lentiviral approach, these authors expressed channel rhodopsin 2 (ChR2) in hypothalamic orexin neurons. Photostimulation of orexin::ChR2 neurons increased Fos expression in the PVN, as well as other stress-related regions innervated by orexin cells, including PVT, DR, and LC. Activation of orexin neurons also increased plasma corticosterone levels that peaked at 20 min post-photostimulation, and this was blocked by systemic administration of the OX_1R antagonist SB-334867. Increases in corticosterone levels were contingent on the number and frequency of light pulses applied; at least 18,000 light pulses or 30 min stimulation (20 Hz, 10 s pulse with 10 s inter-pulse intervals) were necessary to increase corticosterone levels, suggesting that prolonged intense stimulation of orexin neurons is necessary to produce HPA axis activation.

3 Orexin and Stress Reactivity and Behavior

3.1 Effects of Stress Exposure on Orexin Cell Activity

Consistent with the role for orexins in motivated behavior, exposure to threatening stimuli potently activates the orexin system. Psychological stressors, including exposure to fear-associated contexts [80, 81], novel environments [81], acute foot shock [82–84], and the forced swim test [85], increase levels of prepro-orexin mRNA or Fos-protein, a marker of activity [86], in orexin neurons. Physiological stressors, including intraperitoneal (i.p.) injections [87], morphine withdrawal [88], CO_2 stress [89], acute sleep deprivation [90], and food restriction [91], also increase Fos in orexin neurons. Pharmacological agents known to engender stress behavior, such as the anxiogenic drug FG-7142 [61], yohimbine [18], caffeine [61], and sodium lactate [92] also potently activate orexin neurons. Finally, increasing evidence suggests that painful stimuli, including formalin injections, recruit orexin function [93]. There is some evidence that stressors selectively recruit medial populations of orexin neurons, whereas rewarding stimuli selectively recruit lateral orexin neurons, however findings from studies investigating this potential dichotomy have been mixed [6, 18, 26, 61, 84, 94, 95].

3.2 Orexin Modulation of Stress Behavior

Activation of orexin neurons by threatening or aversive stimuli facilitates the engagement of adaptive behaviors. Consistent with orexin neuron recruitment under these conditions, exogenous administration of orexin peptides into the lateral ventricles promotes the expression of innate stress/emotional behaviors, including grooming, face washing, burrowing, searching and increased general locomotor behavior [96, 97]. Central administration of orexin-A, and in some cases orexin-B, also evokes anxiety-like behavior on classical behavioral assays such as the elevated plus maze and the light-dark test in rats [98] and mice [99, 100].

These effects can be recapitulated by infusions of orexins directly into stress-sensitive brain regions such as the PVT and BNST (see Sect. 2.1), which increases stress-related behavior in the open field and social interaction tests [40, 101]. Recent studies have also used optogenetics to assess orexin signaling in anxiety-like behavior. For example, Bhatnager and colleagues showed that optogenetic stimulation of orexin neurons reduced time spent in the interaction zone in a social interaction test [102], whereas Bonnavion et al. [7] reported increased freezing and reduced exploration in an open field test following extended photostimulation of orexin neurons. Interestingly, in both studies, no changes in behavior were observed when orexin neurons were photostimulated while the animal was in their home cage, consistent with the interpretation that orexin neurons are primarily involved in modulating state dependent behaviors [6].

4 Orexins and Stress-Related Psychiatric Disease

4.1 Evidence of Orexin Dysregulation in Human Psychiatric Disease

Dysregulated orexin signaling has been observed in various neuropsychiatric disease states, including anxiety and depression. For example, Johnson and colleagues [92] observed that orexin levels in CSF were elevated in patients with panic anxiety compared to control subjects. Interestingly, this study also showed that when panic anxiety was comorbid with major depressive disorder (MDD), orexin CSF levels tended to be *lower* than control patients. Several other studies have reported evidence of reduced orexin functioning in depressive populations. Thus, Brundin and colleagues reported that patients with MDD have reduced levels of orexin-A relative to healthy controls, and that orexin levels were predictive (negatively correlated) with depressive symptomology [103–105]. These authors also reported that in patients who had attempted suicide, orexin CSF levels were significantly higher at 6- and 12-month evaluations, as compared to the time of the attempt [103]. Further, Salomon and colleagues [106] reported reduced diurnal variation in orexin CSF levels in MDD patients, relative to healthy controls, consistent with sleep–wake and arousal disruption in depression. Similarly, Rotter et al. [107] observed that orexin mRNA levels in whole blood were negatively correlated with Hamilton Depression rating scores. Finally, [108] reported that orexin-A peptide levels in the amygdala were reduced in subjects with flat or negative affect. For a summary of these clinical findings, refer to Fig. 2.

4.2 Evidence of Orexin Dysregulation in Animal Models of Psychiatric Disease

4.2.1 Panic and Anxiety-Like Behavior

Animal models often involve the use of psychological, physiological, or pharmacological stressors to evoke a state of anxiety-like behavior. Using CO_2 gas challenge or administration of the anxiety provoking agent sodium lactate, Johnson, Shekhar, and colleagues [61, 92] have shown that orexin neurons in the dorsomedial/perifornical regions are critical to eliciting coordinated behavioral and physiological panic-like responses. Further, as outlined above, the expression of anxiety-like behavior is generally associated with an overall increase in orexin system activity (see Fig. 2). In support, the spontaneously hypertensive rat, which is often used as a model for anxiety and attention deficit/hyperactivity disorder [138, 139], exhibits increased numbers of orexin neurons in the hypothalamus compared to normotensive controls [140, 141].

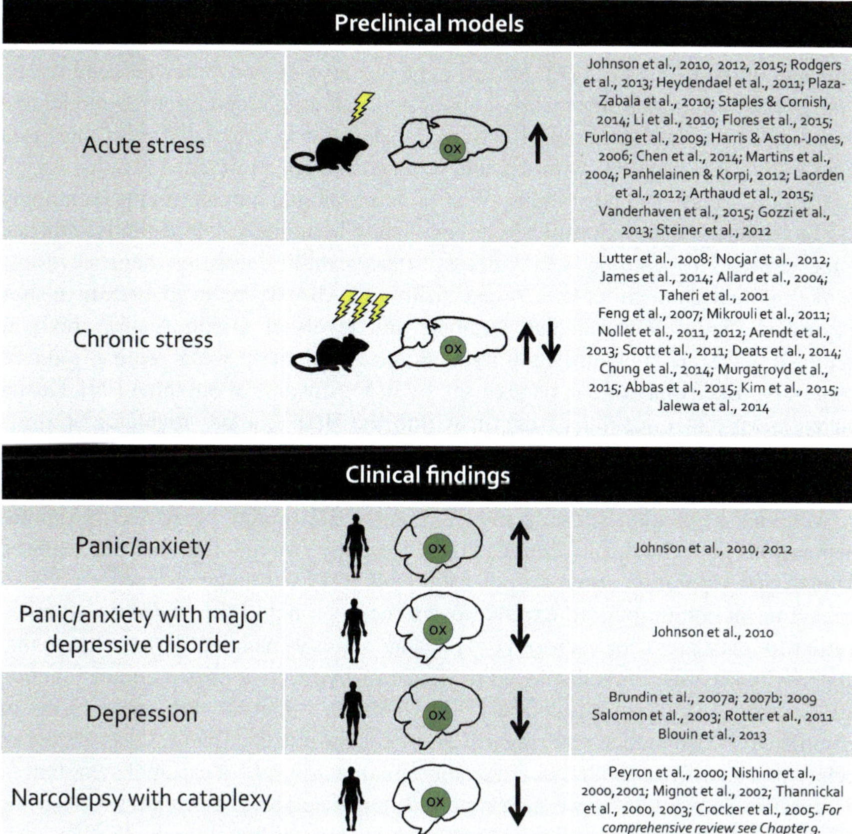

Fig. 2 Orexin system function in preclinical models of stress and clinical neuropsychiatric disease states. Preclinical studies indicate that exposure to acute stressors (physiological and psychological) activates orexin neurons and that this is associated with the expression of panic, anxiety, and stress-related behaviors, as well as elevated HPA axis activation [10, 44, 61, 80–82, 84, 85, 87–90, 92, 101, 109–115]. Repeated or chronic stress leads to either an increase *or* decrease in orexin system function, depending on the type and intensity of the stressor, and is generally associated with a behavioral phenotype of low motivated arousal, or depressive-like behavior [48, 99, 116–130]. These findings are generally reflected in the clinical literature, where patients suffering panic disorder exhibit increased CSF OXA levels, whereas patients with depression, or disorders associated with depression such as narcolepsy/cataplexy, have low CSF levels of OXA, relative to healthy controls ([10, 61, 92, 103–108, 131–137])

4.2.2 Depression-Like Behavior

The Wistar Kyoto rat is a selectively bred strain that expresses hormonal and behavioral profiles that resemble those observed in depressive patients such as increased behavioral despair and altered corticosterone responses to psychological stressors [142]. These rats exhibit a wide range of orexin-related deficiencies relative to the Wistar strain, including lower numbers of hypothalamic orexin neurons, reduced orexin soma size, reduced levels of prepro-orexin mRNA levels, and lower orexin-A immunoreactivity in various regions including the hypothalamus and amygdala [117, 124]. Recent evidence also demonstrates reduced orexin system function in rats chronically housed under dim light conditions, a model that is thought to mimic seasonal affective disorder and is associated with increased immobility in the forced swim test and reduced sucrose preference [129].

Chronic stress is a known antecedent of depression in humans and is commonly used to induce a depression-like behavioral state in animals. Interestingly, rats and mice exposed to chronic social defeat stress exhibit increased depressive-like behavior in tests such as the forced swim test, display reduced orexin mRNA expression, reduced orexin cell numbers and levels of orexin A and orexin B peptides [120, 122]. Consistent with an orexin deficit, social defeat induced depressive-like behavior can be reversed by ICV infusions of orexin-A [99]. Others have shown that infusions of orexin-A into the BNST, a key regulator of stress behavior, can recover depression-like behavior in rats exposed to chronic restraint stress [99].

Maternal separation stress, a model of early life trauma or neglect, promotes depression-like behavior in animals [143]. Recently, we reported that rats exposed to maternal separation stress on postnatal days 2–14 exhibited reduced motivated behavior, including reduced exploration in the open field [48], and reduced motivation to lever press for sucrose [144]. Using activity mapping, it was shown that maternally separated rats displayed evidence of hypoactive orexin circuit function after restraint stress in adulthood [48]. These data suggest that some forms of chronic stress can induce rewiring of LH orexin circuits that manifests as abnormal behavioral stress reactivity and reduced motivated arousal – a cardinal symptom of depression. Interestingly, we also reported that a period of exercise reversed the hypoactivity observed in orexin circuits in male (but not female) rats. Further work is required to determine the specific effects of exercise on LH circuits (orexin neurons, inhibitory and excitatory interneurons), and to identify neural targets for augmenting orexin neuron activity.

It is important to acknowledge that not all preclinical studies examining depressive-like behaviors report a downregulation of orexin system function. For example, the Flinders Sensitive Line of rat, which exhibits a depressive-like phenotype, displays increased numbers of orexin neurons compared to controls [121]. Similarly, studies using genetically modified mice lacking OX_1R show reduced depressive-like behavior in the forced swim test [116, 123]. Non-genetic models of depression can also increase orexin system function. For example, animal models of chronic restraint stress-induced depression result in increased orexin mRNA levels in the basolateral amygdala (BLA) and orexin siRNA injections into the BLA, resulting in a local knockdown of orexin mRNA, produce antidepressant-like behavioral effects [119]. Similarly, the chronic unpredictable mild stress model of depression in mice resulted in increased orexin neuron activity and this was reversed with 6 weeks of fluoxetine treatment, a commonly prescribed medication for depression in humans [126]. Finally, repeated subcutaneous corticosterone injections increased the number of orexin neurons in the hypothalamus which was associated with the expression of depressive-like behavior [118].

The reasons for these disparate findings are currently unclear. One possibility is that differences in the category of stressor, the intensity or duration of stress exposure dictates the effects of repeated stress on orexin function. Regardless, collectively these studies point to dysregulation of orexin system function associated with the expression of depression-like behavior.

5 Orexin Antagonists as Modulators of Stress-Induced Behaviors

The DORA almorexant (Belsomra) was recently approved by the FDA and Pharmaceuticals and Medical Devices Agency for use in humans for the treatment of insomnia, and other orexin antagonists are in various stages of preclinical and clinical trials [145–149]. Accordingly, there is significant interest in the use of these compounds in the treatment of other disorders including stress-related psychiatric diseases. In support of this idea, a significant body of preclinical evidence indicates that acute orexin receptor antagonism is highly effective at ameliorating the behavioral responses to anxiety-provoking stimuli. The selective OX_1R antagonist, SB-334867 is effective at reducing anxiety-like behavior following different stressors, including cat odor exposure [111, 112, 114], acute nicotine treatment [111], sodium lactate injections [92], and fear-associated contexts [82]. These behavioral effects of SB-334867 are associated with concomitant reductions in the activation of PVN neurons [110, 114] and reduced ACTH release [110]. Only a small number of studies have investigated the effects of alternative OX_1R antagonists, including one study that reported, using fMRI in rats, that treatment with GSK1059865 prevented increased activity in fronto-hippocampal and amygdala regions in response to the pharmacological stressor yohimbine [109]. Another study demonstrated that systemic administration of the OX_1R antagonist ACT335827 reduces the pressor and tachycardic responses to novelty stress [150]. Some studies have investigated the anxiolytic properties of DORAs, including almorexant, which is effective at attenuating fear potentiated startle responses [113] and generalized avoidance behavior [151], as well as autonomic responses to novelty stress [150]. More research is required to ascertain whether single- or dual-orexin receptor antagonists (SORAs, DORAs, respectively) provide optimal anxiolytic effects.

Together, these preclinical findings are generally supportive of the potential for orexin receptor antagonists for the treatment of anxiety-related disorders. It is important to note, however, that the potential therapeutic benefits of these medications for stress disorders be weighed against the potential that these drugs may interfere with other physiological systems, including sleep/wake cycle and reward function [152]. This is particularly relevant given findings of suppressed orexin system function in human patients and animals displaying low motivated arousal. Furthermore, there is a distinct lack of studies assessing the effects of long term administration of orexin receptor antagonists on depressive and reward-based functions. Given the data highlighted, it is critical that this knowledge gap is addressed before this treatment approach is considered.

6 Conclusions

Orexin neurons are critical for promoting adaptive behavior in response to stressful and anxiety-provoking stimuli. Orexin neurons are recruited by both psychological and physiological stressors and this activation is tightly linked with the expression of coping behaviors. This is achieved via strong reciprocal interactions between the orexin neurons and key stress-sensitive regions of the brain, as well as arousal and autonomic centers. Orexin neurons also signal directly to both the central (PVN, pituitary) and peripheral (adrenal) branches of the HPA axis, and thereby also play a key role in coordinating the neuroendocrine response to stress.

Under acute stress, orexin neurons act to promote adaptive behavior – such as freezing in a threatening environment or increasing the time spent in safe and enclosed zones in an unfamiliar territory. In the case of chronic or repeated stress, however, orexin neuron function can become dysregulated, manifesting as maladaptive behaviors. While further research is required to understand the precise relationship between different patterns of orexin function and particular behavioral phenotypes, the majority of evidence to date points to a relationship between hyperactive orexin neurons and anxiety/panic phenotypes, whereas hypoactivity of orexin neurons is associated with low motivated arousal and depression-like behavior under some circumstances. This is supported by the clinical observations that CSF orexin-A levels are increased in patients with panic disorder, but decreased in patients with MDD and disorders associated with depressive symptomology, such as narcolepsy/cataplexy. It is important to acknowledge, however, that the spectrum of symptoms experienced by patients with stress-related diseases such as anxiety and depression, as well as the high comorbidity of these disorders, means that this dichotomy is likely a vast oversimplification. Individual vulnerability to stress is also likely to be a key factor determining the specific relationship between orexin cell activity and the expression of stress-related behaviors.

Orexin receptor antagonists may represent a potential therapeutic option for the treatment of anxiety-related disorders. Indeed, significant preclinical evidence now suggests that orexin antagonists are effective at reducing both behavioral and neuroendocrine responsivity to acute stressors and anxiogenic stimuli. However, a number of important questions remain to be answered. Critically, significant research is needed to fully understand how chronic stress alters the orexin system. Currently, it is unclear whether the category of stressor, or the intensity or duration of stress determines whether the orexin system is upregulated or suppressed. A greater understanding is also required regarding the contribution of OX_1R versus OX_2R to stress behavior, with respect to the use of SORAs versus DORAs as potential therapeutics. Finally, with the recent approval of DORAs for the clinical treatment of insomnia, it will be important that the consequences of long-term orexin receptor antagonist administration are closely monitored. Assuming these key issues can be adequately addressed, the orexin system offers a strong potential target for the treatment of stress-related illness.

Acknowledgments This chapter was supported by project grant 510778 from the National Health and Medical Research Council (NHMRC) of Australia and NHMRC CJ Martin Fellowship (Australia) 1072706 to M.H.J.

References

1. de Lecea L, Kilduff TS, Peyron C, Gao X, Foye PE, Danielson PE, Fukuhara C, Battenberg EL, Gautvik VT, Bartlett FS 2nd, Frankel WN, van den Pol AN, Bloom FE, Gautvik KM, Sutcliffe JG (1998) The hypocretins: hypothalamus-specific peptides with neuroexcitatory activity. Proc Natl Acad Sci U S A 95:322–327
2. Sakurai T et al (1998) Orexins and orexin receptors: a family of hypothalamic neuropeptides and G protein-coupled receptors that regulate feeding behavior. Cell 92:573–585
3. Carrive P, Kuwaki P (2016) Orexin and central modulation of cardiovascular and respiratory function. Curr Top Behav Neurosci. doi:10.1007/7854_2016_46
4. Walker LC, Lawrence AJ (2016) The role of orexins/hypocretins in alcohol use and abuse. Curr Top Behav Neurosci. doi:10.1007/7854_2016_55
5. James MH et al (2016) Role of the orexin/hypocretin system in stress-related psychiatric disorders. Curr Top Behav Neurosci. 10.1007/7854_2016_56
6. Mahler SV, Moorman DE, Smith RJ, James MH, Aston-Jones G (2014) Motivational activation: a unifying hypothesis of orexin/hypocretin function. Nat Neurosci 17:1298–1303
7. Bonnavion P, Jackson AC, Carter ME, de Lecea L (2015) Antagonistic interplay between hypocretin and leptin in the lateral hypothalamus regulates stress responses. Nat Commun 6:6266
8. Marcus JN, Aschkenasi CJ, Lee CE, Chemelli RM, Saper CB, Yanagisawa M, Elmquist JK (2001) Differential expression of orexin receptors 1 and 2 in the rat brain. J Comp Neurol 435:6–25
9. Peyron C, Tighe DK, van den Pol AN, de Lecea L, Heller HC, Sutcliffe JG, Kilduff TS (1998) Neurons containing hypocretin (orexin) project to multiple neuronal systems. J Neurosci 18:9996–10015
10. Johnson PL, Molosh A, Fitz SD, Truitt WA, Shekhar A (2012) Orexin, stress, and anxiety/ panic states. Prog Brain Res 198:133–161
11. Horvath TL, Peyron C, Diano S, Ivanov A, Aston-Jones G, Kilduff TS, van Den Pol AN (1999) Hypocretin (orexin) activation and synaptic innervation of the locus coeruleus nor-adrenergic system. J Comp Neurol 415:145–159
12. Hagan JJ et al (1999) Orexin A activates locus coeruleus cell firing and increases arousal in the rat. Proc Natl Acad Sci U S A 96:10911–10916
13. Walling SG, Nutt DJ, Lalies MD, Harley CW (2004) Orexin-A infusion in the locus ceruleus triggers norepinephrine (NE) release and NE-induced long-term potentiation in the dentate gyrus. J Neurosci 24:7421–7426
14. Brown RE, Sergeeva OA, Eriksson KS, Haas HL (2002) Convergent excitation of dorsal raphe serotonin neurons by multiple arousal systems (orexin/hypocretin, histamine and noradrenaline). J Neurosci 22:8850–8859

15. Trivedi P, Yu H, MacNeil DJ, Van der Ploeg LH, Guan XM (1998) Distribution of orexin receptor mRNA in the rat brain. FEBS Lett 438:71–75
16. Tsujino N, Sakurai T (2013) Role of orexin in modulating arousal, feeding, and motivation. Front Behav Neurosci 7:28
17. Blasiak A, Siwiec M, Grabowiecka A, Blasiak T, Czerw A, Blasiak E, Kania A, Rajfur Z, Lewandowski MH, Gundlach AL (2015) Excitatory orexinergic innervation of rat nucleus incertus – implications for ascending arousal, motivation and feeding control. Neuropharmacology 99:432–447
18. Kastman HE, Blasiak A, Walker L, Siwiec M, Krstew EV, Gundlach AL, Lawrence AJ (2016) Nucleus incertus Orexin2 receptors mediate alcohol seeking in rats. Neuropharmacology 110(Part A):82–91
19. Bayer L, Eggermann E, Serafin M, Saint-Mleux B, Machard D, Jones B, Muhlethaler M (2001) Orexins (hypocretins) directly excite tuberomammillary neurons. Eur J Neurosci 14:1571–1575
20. Eriksson KS, Sergeeva O, Brown RE, Haas HL (2001) Orexin/hypocretin excites the histaminergic neurons of the tuberomammillary nucleus. J Neurosci 21:9273–9279
21. Huang ZL, Qu WM, Li WD, Mochizuki T, Eguchi N, Watanabe T, Urade Y, Hayaishi O (2001) Arousal effect of orexin A depends on activation of the histaminergic system. Proc Natl Acad Sci U S A 98:9965–9970
22. Henny P, Jones BE (2008) Projections from basal forebrain to prefrontal cortex comprise cholinergic, GABAergic and glutamatergic inputs to pyramidal cells or interneurons. Eur J Neurosci 27:654–670
23. Rye DB, Wainer BH, Mesulam MM, Mufson EJ, Saper CB (1984) Cortical projections arising from the basal forebrain: a study of cholinergic and noncholinergic components employing combined retrograde tracing and immunohistochemical localization of choline acetyltransferase. Neuroscience 13:627–643
24. Arrigoni E, Mochizuki T, Scammell TE (2010) Activation of the basal forebrain by the orexin/hypocretin neurones. Acta Physiol 198:223–235
25. Fadel J, Frederick-Duus D (2008) Orexin/hypocretin modulation of the basal forebrain cholinergic system: insights from in vivo microdialysis studies. Pharmacol Biochem Behav 90:156–162
26. Alexandre C, Andermann ML, Scammell TE (2013) Control of arousal by the orexin neurons. Curr Opin Neurobiol 23:752–759
27. de Lecea L (2015) Optogenetic control of hypocretin (orexin) neurons and arousal circuits. Curr Top Behav Neurosci 25:367–378
28. Luong LN, Carrive P (2012) Orexin microinjection in the medullary raphe increases heart rate and arterial pressure but does not reduce tail skin blood flow in the awake rat. Neuroscience 202:209–217
29. Lazarenko RM, Stornetta RL, Bayliss DA, Guyenet PG (2011) Orexin A activates retrotrapezoid neurons in mice. Respir Physiol Neurobiol 175:283–287
30. Zheng H, Patterson LM, Berthoud HR (2005) Orexin-A projections to the caudal medulla and orexin-induced c-Fos expression, food intake, and autonomic function. J Comp Neurol 485:127–142
31. Yang B, Samson WK, Ferguson AV (2003) Excitatory effects of orexin-A on nucleus tractus solitarius neurons are mediated by phospholipase C and protein kinase C. J Neurosci 23:6215–6222
32. de Oliveira CV, Rosas-Arellano MP, Solano-Flores LP, Ciriello J (2003) Cardiovascular effects of hypocretin-1 in nucleus of the solitary tract. Am J Physiol Heart Circ Physiol 284:H1369–H1377
33. de Oliveira CV, Rosas-Arellano MP, Solano-Flores LP, Babic T, Li Z, Ciriello J (2003) Estrogen alters the bradycardia response to hypocretin-1 in the nucleus tractus solitarius of the ovariectomized female. Brain Res 978:14–23

34. Shih CD, Chuang YC (2007) Nitric oxide and GABA mediate bi-directional cardiovascular effects of orexin in the nucleus tractus solitarii of rats. Neuroscience 149:625–635
35. Ho YC, Lee HJ, Tung LW, Liao YY, Fu SY, Teng SF, Liao HT, Mackie K, Chiou LC (2011) Activation of orexin 1 receptors in the periaqueductal gray of male rats leads to antinociception via retrograde endocannabinoid (2-arachidonoylglycerol)-induced disinhibition. J Neurosci 31:14600–14610
36. Azhdari Zarmehri H, Semnanian S, Fathollahi Y, Erami E, Khakpay R, Azizi H, Rohampour K (2011) Intra-periaqueductal gray matter microinjection of orexin-A decreases formalin-induced nociceptive behaviors in adult male rats. J Pain 12:280–287
37. Dayas CV, Buller KM, Crane JW, Xu Y, Day TA (2001) Stressor categorization: acute physical and psychological stressors elicit distinctive recruitment patterns in the amygdala and in medullary noradrenergic cell groups. Eur J Neurosci 14:1143–1152
38. Cluderay JE, Harrison DC, Hervieu GJ (2002) Protein distribution of the orexin-2 receptor in the rat central nervous system. Regul Pept 104:131–144
39. Bisetti A, Cvetkovic V, Serafin M, Bayer L, Machard D, Jones BE, Muhlethaler M (2006) Excitatory action of hypocretin/orexin on neurons of the central medial amygdala. Neuroscience 142:999–1004
40. Lungwitz EA, Molosh A, Johnson PL, Harvey BP, Dirks RC, Dietrich A, Minick P, Shekhar A, Truitt WA (2012) Orexin-A induces anxiety-like behavior through interactions with glutamatergic receptors in the bed nucleus of the stria terminalis of rats. Physiol Behav 107:726–732
41. Arendt DH, Hassell J, Li H, Achua JK, Guarnieri DJ, Dileone RJ, Ronan PJ, Summers CH (2014) Anxiolytic function of the orexin 2/hypocretin A receptor in the basolateral amygdala. Psychoneuroendocrinology 40:17–26
42. James MH, Yeoh JW, Graham BA, Dayas CV (2012) Insights for developing pharmacological treatments for psychostimulant relapse targeting hypothalamic peptide systems. J Addict Res Ther S4:008
43. James MH, Charnley JL, Levi EM, Jones E, Yeoh JW, Smith DW, Dayas CV (2011) Orexin-1 receptor signalling within the ventral tegmental area, but not the paraventricular thalamus, is critical to regulating cue-induced reinstatement of cocaine-seeking. Int J Neuropsychopharmacol 14:684–690
44. Li Y, Li S, Wei C, Wang H, Sui N, Kirouac GJ (2010) Orexins in the paraventricular nucleus of the thalamus mediate anxiety-like responses in rats. Psychopharmacology (Berl) 212:251–265
45. Li Y, Wang H, Qi K, Chen X, Li S, Sui N, Kirouac GJ (2011) Orexins in the midline thalamus are involved in the expression of conditioned place aversion to morphine withdrawal. Physiol Behav 102:42–50
46. Baffi JS, Palkovits M (2000) Fine topography of brain areas activated by cold stress. A fos immunohistochemical study in rats. Neuroendocrinology 72:102–113
47. James MH, Dayas CV (2013) What about me…? The PVT: a role for the paraventricular thalamus (PVT) in drug-seeking behavior. Front Behav Neurosci 7:18
48. James MH, Campbell EJ, Walker FR, Smith DW, Richardson HN, Hodgson DM, Dayas CV (2014) Exercise reverses the effects of early life stress on orexin cell reactivity in male but not female rats. Front Behav Neurosci 8:244
49. Senba E, Matsunaga K, Tohyama M, Noguchi K (1993) Stress-induced c-fos expression in the rat brain: activation mechanism of sympathetic pathway. Brain Res Bull 31:329–344
50. Yeoh JW, James MH, Graham BA, Dayas CV (2014) Electrophysiological characteristics of paraventricular thalamic (PVT) neurons in response to cocaine and cocaine- and amphetamine-regulated transcript (CART). Front Behav Neurosci 8:280
51. Ishibashi M, Takano S, Yanagida H, Takatsuna M, Nakajima K, Oomura Y, Wayner MJ, Sasaki K (2005) Effects of orexins/hypocretins on neuronal activity in the paraventricular nucleus of the thalamus in rats in vitro. Peptides 26:471–481

52. Huang H, Ghosh P, van den Pol AN (2006) Prefrontal cortex-projecting glutamatergic thalamic paraventricular nucleus-excited by hypocretin: a feedforward circuit that may enhance cognitive arousal. J Neurophysiol 95:1656–1668

53. Date Y, Mondal MS, Matsukura S, Ueta Y, Yamashita H, Kaiya H, Kangawa K, Nakazato M (2000) Distribution of orexin/hypocretin in the rat median eminence and pituitary. Brain Res Mol Brain Res 76:1–6

54. Sakurai T, Nagata R, Yamanaka A, Kawamura H, Tsujino N, Muraki Y, Kageyama H, Kunita S, Takahashi S, Goto K, Koyama Y, Shioda S, Yanagisawa M (2005) Input of orexin/hypocretin neurons revealed by a genetically encoded tracer in mice. Neuron 46:297–308

55. Yoshida K, McCormack S, Espana RA, Crocker A, Scammell TE (2006) Afferents to the orexin neurons of the rat brain. J Comp Neurol 494:845–861

56. Schone C, Burdakov D (2012) Glutamate and GABA as rapid effectors of hypothalamic "peptidergic" neurons. Front Behav Neurosci 6:81

57. Schone C, Apergis-Schoute J, Sakurai T, Adamantidis A, Burdakov D (2014) Coreleased orexin and glutamate evoke nonredundant spike outputs and computations in histamine neurons. Cell Rep 7:697–704

58. Schone C, Cao ZF, Apergis-Schoute J, Adamantidis A, Sakurai T, Burdakov D (2012) Optogenetic probing of fast glutamatergic transmission from hypocretin/orexin to histamine neurons in situ. J Neurosci 32:12437–12443

59. Chou TC, Lee CE, Lu J, Elmquist JK, Hara J, Willie JT, Beuckmann CT, Chemelli RM, Sakurai T, Yanagisawa M, Saper CB, Scammell TE (2001) Orexin (hypocretin) neurons contain dynorphin. J Neurosci 21:RC168

60. Sartor GC, Aston-Jones GS (2012) A septal-hypothalamic pathway drives orexin neurons, which is necessary for conditioned cocaine preference. J Neurosci 32:4623–4631

61. Johnson PL, Samuels BC, Fitz SD, Federici LM, Hammes N, Early MC, Truitt W, Lowry CA, Shekhar A (2012) Orexin 1 receptors are a novel target to modulate panic responses and the panic brain network. Physiol Behav 107:733–742

62. Herman JP, Figueiredo H, Mueller NK, Ulrich-Lai Y, Ostrander MM, Choi DC, Cullinan WE (2003) Central mechanisms of stress integration: hierarchical circuitry controlling hypothalamo-pituitary-adrenocortical responsiveness. Front Neuroendocr 24:151–180

63. Hervieu GJ, Cluderay JE, Harrison DC, Roberts JC, Leslie RA (2001) Gene expression and protein distribution of the orexin-1 receptor in the rat brain and spinal cord. Neuroscience 103:777–797

64. Russell SH, Small CJ, Dakin CL, Abbott CR, Morgan DG, Ghatei MA, Bloom SR (2001) The central effects of orexin-A in the hypothalamic-pituitary-adrenal axis in vivo and in vitro in male rats. J Neuroendocrinol 13:561–566

65. Samson WK, Taylor MM, Follwell M, Ferguson AV (2002) Orexin actions in hypothalamic paraventricular nucleus: physiological consequences and cellular correlates. Regul Pept 104:97–103

66. Lopez M, Seoane L, Senaris RM, Dieguez C (2001) Prepro-orexin mRNA levels in the rat hypothalamus, and orexin receptors mRNA levels in the rat hypothalamus and adrenal gland are not influenced by the thyroid status. Neurosci Lett 300:171–175

67. Lopez M, Senaris R, Gallego R, Garcia-Caballero T, Lago F, Seoane L, Casanueva F, Dieguez C (1999) Orexin receptors are expressed in the adrenal medulla of the rat. Endocrinology 140:5991–5994

68. Malendowicz LK, Jedrzejczak N, Belloni AS, Trejter M, Hochol A, Nussdorfer GG (2001) Effects of orexins A and B on the secretory and proliferative activity of immature and regenerating rat adrenal glands. Histol Histopathol 16:713–717

69. Ziolkowska A, Spinazzi R, Albertin G, Nowak M, Malendowicz LK, Tortorella C, Nussdorfer GG (2005) Orexins stimulate glucocorticoid secretion from cultured rat and human adrenocortical cells, exclusively acting via the OX1 receptor. J Steroid Biochem Mol Biol 96:423–429

70. Johren O, Bruggemann N, Dendorfer A, Dominiak P (2003) Gonadal steroids differentially regulate the messenger ribonucleic acid expression of pituitary orexin type 1 receptors and adrenal orexin type 2 receptors. Endocrinology 144:1219–1225
71. Johren O, Neidert SJ, Kummer M, Dendorfer A, Dominiak P (2001) Prepro-orexin and orexin receptor mRNAs are differentially expressed in peripheral tissues of male and female rats. Endocrinology 142:3324–3331
72. Kummer M, Neidert SJ, Johren O, Dominiak P (2001) Orexin (hypocretin) gene expression in rat ependymal cells. Neuroreport 12:2117–2120
73. Nanmoku T, Isobe K, Sakurai T, Yamanaka A, Takekoshi K, Kawakami Y, Ishii K, Goto K, Nakai T (2000) Orexins suppress catecholamine synthesis and secretion in cultured PC12 cells. Biochem Biophys Res Commun 274:310–315
74. Spinazzi R, Ziolkowska A, Neri G, Nowak M, Rebuffat P, Nussdorfer GG, Andreis PG, Malendowicz LK (2005) Orexins modulate the growth of cultured rat adrenocortical cells, acting through type 1 and type 2 receptors coupled to the MAPK p42/p44- and p38-dependent cascades. Int J Mol Med 15:847–852
75. Mazzocchi G, Malendowicz LK, Gottardo L, Aragona F, Nussdorfer GG (2001) Orexin A stimulates cortisol secretion from human adrenocortical cells through activation of the adenylate cyclase-dependent signaling cascade. J Clin Endocrinol Metab 86:778–782
76. Mazzocchi G, Malendowicz LK, Aragona F, Rebuffat P, Gottardo L, Nussdorfer GG (2001) Human pheochromocytomas express orexin receptor type 2 gene and display an in vitro secretory response to orexins A and B. J Clin Endocrinol Metab 86:4818–4821
77. Chang H, Saito T, Ohiwa N, Tateoka M, Deocaris CC, Fujikawa T, Soya H (2007) Inhibitory effects of an orexin-2 receptor antagonist on orexin A- and stress-induced ACTH responses in conscious rats. Neurosci Res 57:462–466
78. Girault EM, Foppen E, Ackermans MT, Fliers E, Kalsbeek A (2013) Central administration of an orexin receptor 1 antagonist prevents the stimulatory effect of Olanzapine on endogenous glucose production. Brain Res 1527:238–245
79. Steiner MA, Sciarretta C, Brisbare-Roch C, Strasser DS, Studer R, Jenck F (2013) Examining the role of endogenous orexins in hypothalamus-pituitary-adrenal axis endocrine function using transient dual orexin receptor antagonism in the rat. Psychoneuroendocrinology 38:560–571
80. Flores A, Saravia R, Maldonado R, Berrendero F (2015) Orexins and fear: implications for the treatment of anxiety disorders. Trends Neurosci 38:550–559
81. Furlong TM, Vianna DM, Liu L, Carrive P (2009) Hypocretin/orexin contributes to the expression of some but not all forms of stress and arousal. Eur J Neurosci 30:1603–1614
82. Chen X, Li S, Kirouac GJ (2014) Blocking of corticotrophin releasing factor receptor-1 during footshock attenuates context fear but not the upregulation of prepro-orexin mRNA in rats. Pharmacol Biochem Behav 120:1–6
83. Chen X, Wang H, Lin Z, Li S, Li Y, Bergen HT, Vrontakis ME, Kirouac GJ (2014) Orexins (hypocretins) contribute to fear and avoidance in rats exposed to a single episode of footshocks. Brain Struct Funct 219:2103–2118
84. Harris GC, Aston-Jones G (2006) Arousal and reward: a dichotomy in orexin function. Trends Neurosci 29:571–577
85. Martins PJ, D'Almeida V, Pedrazzoli M, Lin L, Mignot E, Tufik S (2004) Increased hypocretin-1 (orexin-a) levels in cerebrospinal fluid of rats after short-term forced activity. Regul Pept 117:155–158
86. Morgan JI, Curran T (1991) Stimulus-transcription coupling in the nervous system: involvement of the inducible proto-oncogenes fos and jun. Annu Rev Neurosci 14:421–451
87. Panhelainen AE, Korpi ER (2012) Evidence for a role of inhibition of orexinergic neurons in the anxiolytic and sedative effects of diazepam: a c-Fos study. Pharmacol Biochem Behav 101:115–124

88. Laorden ML, Ferenczi S, Pinter-Kubler B, Gonzalez-Martin LL, Lasheras MC, Kovacs KJ, Milanes MV, Nunez C (2012) Hypothalamic orexin-a neurons are involved in the response of the brain stress system to morphine withdrawal. PLoS One 7:e36871

89. Johnson PL, Federici LM, Fitz SD, Renger JJ, Shireman B, Winrow CJ, Bonaventure P, Shekhar A (2015) Orexin 1 and 2 receptor involvement in Co2-induced panic-associated behavior and autonomic responses. Depress Anxiety 32:671–683

90. Arthaud S, Varin C, Gay N, Libourel PA, Chauveau F, Fort P, Luppi PH, Peyron C (2015) Paradoxical (REM) sleep deprivation in mice using the small-platforms-over-water method: polysomnographic analyses and melanin-concentrating hormone and hypocretin/orexin neuronal activation before, during and after deprivation. J Sleep Res 24:309–319

91. Kurose T, Ueta Y, Yamamoto Y, Serino R, Ozaki Y, Saito J, Nagata S, Yamashita H (2002) Effects of restricted feeding on the activity of hypothalamic Orexin (OX)-A containing neurons and OX2 receptor mRNA level in the paraventricular nucleus of rats. Regul Pept 104:145–151

92. Johnson PL, Truitt W, Fitz SD, Minick PE, Dietrich A, Sanghani S, Traskman-Bendz L, Goddard AW, Brundin L, Shekhar A (2010) A key role for orexin in panic anxiety. Nat Med 16:111–115

93. Campbell EJ, Watters SM, Zouikr I, Hodgson DM, Dayas CV (2015) Recruitment of hypothalamic orexin neurons after formalin injections in adult male rats exposed to a neonatal immune challenge. Front Neurosci 9:65

94. Berridge CW, Espana RA, Vittoz NM (2010) Hypocretin/orexin in arousal and stress. Brain Res 1314:91–102

95. Moorman DE, James MH, Kilroy EA, Aston-Jones G (2016) Orexin/hypocretin neuron activation is correlated with alcohol seeking and preference in a topographically specific manner. Eur J Neurosci 43:710–720

96. Ida T, Nakahara K, Katayama T, Murakami N, Nakazato M (1999) Effect of lateral cerebroventricular injection of the appetite-stimulating neuropeptide, orexin and neuropeptide Y, on the various behavioral activities of rats. Brain Res 821:526–529

97. Ida T, Nakahara K, Murakami T, Hanada R, Nakazato M, Murakami N (2000) Possible involvement of orexin in the stress reaction in rats. Biochem Biophys Res Commun 270:318–323

98. Suzuki M, Beuckmann CT, Shikata K, Ogura H, Sawai T (2005) Orexin-A (hypocretin-1) is possibly involved in generation of anxiety-like behavior. Brain Res 1044:116–121

99. Chung HS, Kim JG, Kim JW, Kim HW, Yoon BJ (2014) Orexin administration to mice that underwent chronic stress produces bimodal effects on emotion-related behaviors. Regul Pept 194–195:16–22

100. Palotai M, Telegdy G, Jászberényi M (2014) Orexin A-induced anxiety-like behavior is mediated through GABA-ergic, α- and β-adrenergic neurotransmissions in mice. Peptides 57:129–134

101. Li Y, Li S, Wei C, Wang H, Sui N, Kirouac GJ (2010) Changes in emotional behavior produced by orexin microinjections in the paraventricular nucleus of the thalamus. Pharmacol Biochem Behav 95:121–128

102. Heydendael W, Sengupta A, Beck S, Bhatnagar S (2014) Optogenetic examination identifies a context-specific role for orexins/hypocretins in anxiety-related behavior. Physiol Behav 130:182–190

103. Brundin L, Bjorkqvist M, Traskman-Bendz L, Petersen A (2009) Increased orexin levels in the cerebrospinal fluid the first year after a suicide attempt. J Affect Disord 113:179–182

104. Brundin L, Bjorkqvist M, Petersen A, Traskman-Bendz L (2007) Reduced orexin levels in the cerebrospinal fluid of suicidal patients with major depressive disorder. Eur Neuropsychopharmacol 17(9):573–579

105. Brundin L, Petersen A, Bjorkqvist M, Traskman-Bendz L (2007) Orexin and psychiatric symptoms in suicide attempters. J Affect Disord 100(1–3):259–263

106. Salomon RM, Ripley B, Kennedy JS, Johnson B, Schmidt D, Zeitzer JM, Nishino S, Mignot E (2003) Diurnal variation of cerebrospinal fluid hypocretin-1 (Orexin-A) levels in control and depressed subjects. Biol Psychiatry 54:96–104

107. Rotter A, Asemann R, Decker A, Kornhuber J, Biermann T (2011) Orexin expression and promoter-methylation in peripheral blood of patients suffering from major depressive disorder. J Affect Disord 131:186–192

108. Blouin AM, Fried I, Wilson CL, Staba RJ, Behnke EJ, Lam HA, Maidment NT, Karlsson KAE, Lapierre JL, Siegel JM (2013) Human hypocretin and melanin-concentrating hormone levels are linked to emotion and social interaction. Nat Commun 4:1547

109. Gozzi A, Lepore S, Vicentini E, Merlo-Pich E, Bifone A (2013) Differential effect of orexin-1 and CRF-1 antagonism on stress circuits: a fMRI study in the rat with the pharmacological stressor Yohimbine. Neuropsychopharmacology 38:2120–2130

110. Heydendael W, Sharma K, Iyer V, Luz S, Piel D, Beck S, Bhatnagar S (2011) Orexins/hypocretins act in the posterior paraventricular thalamic nucleus during repeated stress to regulate facilitation to novel stress. Endocrinology 152:4738–4752

111. Plaza-Zabala A, Martin-Garcia E, de Lecea L, Maldonado R, Berrendero F (2010) Hypocretins regulate the anxiogenic-like effects of nicotine and induce reinstatement of nicotine-seeking behavior. J Neurosci 30:2300–2310

112. Staples LG, Cornish JL (2014) The orexin-1 receptor antagonist SB-334867 attenuates anxiety in rats exposed to cat odor but not the elevated plus maze: an investigation of Trial 1 and Trial 2 effects. Horm Behav 65:294–300

113. Steiner MA, Lecourt H, Jenck F (2012) The brain orexin system and almorexant in fear-conditioned startle reactions in the rat. Psychopharmacology (Berl) 223:465–475

114. Vanderhaven MW, Cornish JL, Staples LG (2015) The orexin-1 receptor antagonist SB-334867 decreases anxiety-like behavior and c-Fos expression in the hypothalamus of rats exposed to cat odor. Behav Brain Res 278:563–568

115. Rodgers RJ, Wright FL, Snow NF, Taylor LJ (2013) Orexin-1 receptor antagonism fails to reduce anxiety-like behaviour in either plus-maze-naive or plus-maze-experienced mice. Behav Brain Res 243:213–219

116. Abbas MG, Shoji H, Soya S, Hondo M, Miyakawa T, Sakurai T (2015) Comprehensive behavioral analysis of male Ox1r (-/-) mice showed implication of orexin receptor-1 in mood, anxiety, and social behavior. Front Behav Neurosci 9:324

117. Allard JS, Tizabi Y, Shaffery JP, Trouth CO, Manaye K (2004) Stereological analysis of the hypothalamic hypocretin/orexin neurons in an animal model of depression. Neuropeptides 38:311–315

118. Jalewa J, Wong-Lin K, McGinnity TM, Prasad G, Holscher C (2014) Increased number of orexin/hypocretin neurons with high and prolonged external stress-induced depression. Behav Brain Res 272:196–204

119. Kim TK, Kim JE, Park JY, Lee JE, Choi J, Kim H, Lee EH, Kim SW, Lee JK, Kang HS, Han PL (2015) Antidepressant effects of exercise are produced via suppression of hypocretin/orexin and melanin-concentrating hormone in the basolateral amygdala. Neurobiol Dis 79:59–69

120. Lutter M, Krishnan V, Russo SJ, Jung S, McClung CA, Nestler EJ (2008) Orexin signaling mediates the antidepressant-like effect of calorie restriction. J Neurosci 28:3071–3075

121. Mikrouli E, Wortwein G, Soylu R, Mathe AA, Petersen A (2011) Increased numbers of orexin/hypocretin neurons in a genetic rat depression model. Neuropeptides 45:401–406

122. Nocjar C, Zhang J, Feng P, Panksepp J (2012) The social defeat animal model of depression shows diminished levels of orexin in mesocortical regions of the dopamine system, and of dynorphin and orexin in the hypothalamus. Neuroscience 218:138–153

123. Scott MM, Marcus JN, Pettersen A, Birnbaum SG, Mochizuki T, Scammell TE, Nestler EJ, Elmquist JK, Lutter M (2011) Hcrtr1 and 2 signaling differentially regulates depression-like behaviors. Behav Brain Res 222:289–294

124. Taheri S, Gardiner J, Hafizi S, Murphy K, Dakin C, Seal L, Small C, Ghatei M, Bloom S (2001) Orexin A immunoreactivity and preproorexin mRNA in the brain of Zucker and WKY rats. Neuroreport 12:459–464
125. Feng P, Vurbic D, Wu Z, Strohl KP (2007) Brain orexins and wake regulation in rats exposed to maternal deprivation. Brain Res 1154:163–172
126. Nollet M, Gaillard P, Minier F, Tanti A, Belzung C, Leman S (2011) Activation of orexin neurons in dorsomedial/perifornical hypothalamus and antidepressant reversal in a rodent model of depression. Neuropharmacology 61(1–2):336–346
127. Nollet M, Gaillard P, Tanti A, Girault V, Belzung C, Leman S (2012) Neurogenesis-independent antidepressant-like effects on behavior and stress axis response of a dual orexin receptor antagonist in a rodent model of depression. Neuropsychopharmacology 37 (10):2210–2221
128. Arendt DH, Ronan PJ, Oliver KD, Callahan LB, Summers TR, Summers CH (2013) Depressive behavior and activation of the orexin/hypocretin system. Behav Neurosci 127(1):86–94
129. Deats SP, Adidharma W, Lonstein JS, Yan L (2014) Attenuated orexinergic signaling underlies depression-like responses induced by daytime light deficiency. Neuroscience 272:252–260
130. Murgatroyd CA, Pena CJ, Podda G, Nestler EJ, Nephew BC (2015) Early life social stress induced changes in depression and anxiety associated neural pathways which are correlated with impaired maternal care. Neuropeptides 52:103–111
131. Peyron C et al (2000) A mutation in a case of early onset narcolepsy and a generalized absence of hypocretin peptides in human narcoleptic brains. Nat Med 6(9):991–997
132. Nishino S, Ripley B, Overeem S, Lammers GJ, Mignot E (2000) Hypocretin (orexin) deficiency in human narcolepsy. Lancet 355(9197):39–40
133. Nishino S et al (2001) Low cerebrospinal fluid hypocretin (Orexin) and altered energy homeostasis in human narcolepsy. Ann Neurol 50(3):381–388
134. Mignot E et al (2002) The role of cerebrospinal fluid hypocretin measurement in the diagnosis of narcolepsy and other hypersomnias. Arch Neurol 59(10):1553–1562
135. Thannickal TC et al (2000) Reduced number of hypocretin neurons in human narcolepsy. Neuron 27(3):469–474
136. Thannickal TC, Siegel JM, Nienhuis R, Moore RY (2003) Pattern of hypocretin (orexin) soma and axon loss, and gliosis, in human narcolepsy. Brain Pathol 13(3):340–351
137. Crocker A et al (2005) Concomitant loss of dynorphin, NARP, and orexin in narcolepsy. Neurology 65(8):1184–1188
138. Ramos A, Kangerski AL, Basso PF, Da Silva Santos JE, Assreuy J, Vendruscolo LF, Takahashi RN (2002) Evaluation of Lewis and SHR rat strains as a genetic model for the study of anxiety and pain. Behav Brain Res 129:113–123
139. Sagvolden T, Russell VA, Aase H, Johansen EB, Farshbaf M (2005) Rodent models of attention-deficit/hyperactivity disorder. Biol Psychiatry 57:1239–1247
140. Clifford L, Dampney BW, Carrive P (2015) Spontaneously hypertensive rats have more orexin neurons in their medial hypothalamus than normotensive rats. Exp Physiol 100:388–398
141. Lee Y-H, Tsai M-C, Li T-L, Dai Y-WE, Huang S-C, Hwang L-L (2015) Spontaneously hypertensive rats have more orexin neurons in the hypothalamus and enhanced orexinergic input and orexin 2 receptor-associated nitric oxide signalling in the rostral ventrolateral medulla. Exp Physiol 100:993–1007
142. Will CC, Aird F, Redei EE (2003) Selectively bred Wistar-Kyoto rats: an animal model of depression and hyper-responsiveness to antidepressants. Mol Psychiatry 8:925–932
143. Winslow JT, Insel TR (1991) The infant rat separation paradigm: a novel test for novel anxiolytics. Trends Pharmacol Sci 12(11):402–404
144. Campbell EJ, Mitchell CS, Adams CA, Yeoh JW, Hodgson DM, Graham BA, Dayas CV (2016) Chemogenetic activation of the lateral hypothalamus reverses early life stress-induced deficits in motivational drive (in preparation)

145. Boss C, Roch C (2015) Recent trends in orexin research – 2010 to 2015. Bioorg Med Chem Lett 25:2875–2887
146. Howland RH (2014) Suvorexant: a novel therapy for the treatment of insomnia. J Psychosoc Nurs Ment Health Serv 52:23–26
147. Jacobson LH, Callander GE, Hoyer D (2014) Suvorexant for the treatment of insomnia. Expert Rev Clin Pharmacol 7:711–730
148. Khoo SY, Brown RM (2014) Orexin/hypocretin based pharmacotherapies for the treatment of addiction: DORA or SORA? CNS Drugs 28:713–730
149. Michelson D, Snyder E, Paradis E, Chengan-Liu M, Snavely DB, Hutzelmann J, Walsh JK, Krystal AD, Benca RM, Cohn M, Lines C, Roth T, Herring WJ (2014) Safety and efficacy of suvorexant during 1-year treatment of insomnia with subsequent abrupt treatment discontinuation: a phase 3 randomised, double-blind, placebo-controlled trial. Lancet Neurol 13:461–471
150. Beig MI, Dampney BW, Carrive P (2015) Both Ox1r and Ox2r orexin receptors contribute to the cardiovascular and locomotor components of the novelty stress response in the rat. Neuropharmacology 89:146–156
151. Viviani D, Haegler P, Jenck F, Steiner MA (2015) Orexin neuropeptides contribute to the development and persistence of generalized avoidance behavior in the rat. Psychopharmacology (Berl) 232:1383–1393
152. Yeoh JW, Campbell EJ, James MH, Graham BA, Dayas CV (2014) Orexin antagonists for neuropsychiatric disease: progress and potential pitfalls. Front Neurosci 8:36

The Role of Orexins/Hypocretins in Alcohol Use and Abuse

Leigh C. Walker and Andrew J. Lawrence

Abstract Addiction is a chronic relapsing disorder characterized by compulsive drug seeking and drug taking despite negative consequences. Alcohol abuse and addiction have major social and economic consequences and cause significant morbidity and mortality worldwide. Currently available therapeutics are inadequate, outlining the need for alternative treatments. Detailed knowledge of the neurocircuitry and brain chemistry responsible for aberrant behavior patterns should enable the development of novel pharmacotherapies to treat addiction. Therefore it is important to expand our knowledge and understanding of the neural pathways and mechanisms involved in alcohol seeking and abuse. The orexin (hypocretin) neuropeptide system is an attractive target, given the recent FDA and PMDA approval of suvorexant for the treatment of insomnia. Orexin is synthesized exclusively in neurons located in the lateral (LH), perifornical (PEF), and dorsal medial (DMH) hypothalamus. These neurons project widely throughout the neuraxis with regulatory roles in a wide range of behavioral and physiological responses, including sleep–wake cycle neuroendocrine regulation, anxiety, feeding behavior, and reward seeking. Here we summarize the literature to date, which have evaluated the interplay between alcohol and the orexin system.

Keywords Addiction • Alcohol • Hypocretin • Orexin

Contents

L.C. Walker and A.J. Lawrence (✉)
The Florey Institute of Neuroscience and Mental Health, Parkville, VIC 3052, Australia

Florey Department of Neuroscience and Mental Health, The University of Melbourne, Parkville, VIC 3010, Australia
e-mail: Andrew.Lawrence@florey.edu.au

© Springer International Publishing AG 2016
Curr Topics Behav Neurosci (2017) 33: 221–246
DOI 10.1007/7854_2016_55
Published Online: 30 November 2016

1   Introduction

Excessive alcohol intake, dependence, and abuse are serious health conditions affecting the physiological and emotional health of millions of individuals. Alcohol use disorders (AUDs) have an annual prevalence of ~10% and account for approximately 3.8% of deaths and 4.6% of disease and injury burden in developed countries [1]. It is estimated that ~15% of individuals who try alcoholic beverages become alcohol-dependent [2]. This represents a significant number of people, given the large number who use alcohol as a legal drug. Despite the need for effective therapeutics for the treatment of this disorder, relatively few medications exist. Three medications have been approved by the US Food and Drug Administration (FDA) for the treatment of alcoholism: disulfiram (Antabuse™), naltrexone (ReVia™), and acamprosate (Campral™). However, all three suffer from limited efficacy, poor patient compliance, and high relapse rates [3–5]. Recent research has focused on the repositioning of other FDA approved drugs in the treatment of AUDs as this represents a fast and economically feasible approach for drug development. For example, the γ-aminobutyric acid (GABA)$_B$ receptor agonist, baclofen (Lioresal™), approved in the treatment of spasticity, and gabapentin enacarbil (HORIZANT™), currently prescribed in the treatment of pain and epilepsy, are both instances where drugs are being repurposed for the treatment of AUDs. In addition, new strategies are also in development and/or clinical trials; for example, CRF$_1$ receptor antagonists (see https://clinicaltrials.gov).

The orexin/hypocretin (OX) system presents a novel potential therapeutic target for preventing relapse to AUDs. The recent FDA and Pharmaceuticals and Medical Devices Agency (PMDA) approval of the dual OX receptor antagonist (DORA) suvorexant (Belsomra™), for the treatment of insomnia, and other potential DORAs and single OX receptor antagonists (SORAs) in preclinical and clinical development provides promising prospects for the treatment of AUDs [6]. Early animal studies of intracranial self-stimulation (ICSS) revealed that rodents would robustly self-administer electrical current when the electrodes were placed within the lateral hypothalamus (LH) [7, 8]. Furthermore, electrical stimulation of the LH led rats to drink alcohol at intoxicating levels [7]. On the other hand, ibotenic acid lesioning of the LH diminished electrical self-stimulation [8], suggesting a role for the LH in alcohol reinforcing behavior. In the late 1990s, the OX peptide was isolated by two independent research groups [9, 10]. Shortly after its discovery there was a surge of focus on this system, with reports that dysfunction of the OX system was strongly associated with narcoleptic symptoms in animals [11, 12]. Subsequent work in humans verified that narcoleptic patients have little OX in their cerebrospinal fluid (CSF) and lack most or all OX neurons [13, 14]. Furthermore, human narcoleptics, who have few or little OX neurons, rarely exhibit stimulant abuse and seeking despite chronic treatment with stimulants [15].

The notion that OX-containing neurons of the hypothalamus were involved in drug seeking gained momentum with observations that these neurons innervate structures implicated in responses to drug abuse and arousal [16–18]. The Aston-Jones laboratory provided evidence of OX involvement in drug seeking. Chemical stimulation of the LH OX neurons reinstated a Pavlovian place preference in mice, which was attenuated by pretreatment with a selective OX_1 receptor antagonist [19]. Using the self-administration paradigms, Boutrel and colleagues extended these findings to show central infusions of OX precipitated reinstatement of cocaine seeking in rats [20]. In the alcohol field, Lawrence and colleagues provided the first evidence linking the OX system to ethanol self-administration and cue-induced reinstatement of alcohol seeking [21]. In the past decade of research, many insights have been generated in determining the role and mechanisms by which alcohol interacts with the OX system.

2 Orexin System Modulation by Ethanol Consumption

Studies into the effects of long-term chronic ethanol exposure on the modulation of the OX system have yielded somewhat contradictory results. On the one hand chronic ethanol consumption (70 days, consuming 5 g/kg/day, two-bottle choice paradigm) induced a threefold increase in the area of expression of prepro-orexin mRNA in the LH of inbred alcohol preferring (iP) rats, compared with ethanol-naïve iP rats. This effect was specific to the LH, with no differences observed in the dorsal medial hypothalamus (DMH) [21]. In agreement, voluntary ethanol intake in the intermittent access two-bottle choice paradigm led to a significant increase in OX mRNA of the perifornical area of the hypothalamus (PEF) and LH in Long–Evans rats as measured by quantitative real-time polymerase chain reaction (qRT-PCR) [22]. Alcohol consumption also increased OX mRNA expression in the hypothalamus using a zebra fish model [23]. However, a study by Morganstern and colleagues found conflicting results with Sprague-Dawley (SD) rats. Animals chronically consuming 0.75 g/kg/day of 2% ethanol, or 1.0 and 2.5 g/kg/day of 9% ethanol showed a decrease in levels of OX mRNA in the PEF measured by qRT-PCR [24]. Moreover, binge-like ethanol or sucrose consumption decreased OX-like-immunoreactivity (IR) in the mouse LH. In the PEF, OX-like-IR was also reduced after three sessions of ethanol bingeing [25]. Several methodological differences between these studies may account for the contradictory findings. These include differences in total ethanol intake, different duration of ethanol intake, rodent strain, behavioral paradigm, molecular techniques, and the end point measured. The study by Morganstern et al. [24] also examined the effects of acute ethanol consumption on OX mRNA and peptide levels in the PEF and LH. A significant increase of OX was noted by qPCR and in situ hybridization (ISH) after oral gavage of 0.75 g/kg, but not 2.5 g/kg ethanol. However, immunohistochemical (IHC) analysis of the LH showed no corresponding changes in peptide levels following 0.75 g/kg ethanol administration, but an increase in peptide following the 2.5 g/kg dose. No changes were seen in the PEF [24].

Ethanol-naïve iP and NP rats showed no difference in basal expression of preproorexin mRNA [21], suggesting that this does not markedly contribute to the alcohol preferring nature of these animals. However, High Saccharin intake (HiS) rats have greater basal expression of OX in the LH and PEF than Low Saccharin intake (LoS) rats [26]. HiS rats (compared with LoS) also exhibit increased ethanol intake [27], as well as intake of other drugs [28]. However, this difference may be more closely linked to their propensity for foods rich in fat and sugars [29]. Rats with high novelty-induced activity show both high ethanol consumption and elevated OX mRNA expression in the PEF/LH, while expression of OX was reduced in rats with high triglycerides despite similar ethanol intake [30]. Four repetitive, 2-h daily episodes of saccharin, but not ethanol, binge-like drinking blunted OX_1 receptor mRNA expression in the LH of C57BL/6J mice [31].

Recently a study by Moorman and colleagues examined OX neuronal activation following the intermittent access two-bottle choice paradigm. Ethanol preference was correlated with OX activated neurons of the LH and PEF, but not DMH in SD rats after 1 week of ethanol exposure. Furthermore, a positive correlation between the number of licks made on the ethanol bottle and percentage activation of LH and PEF OX neurons was also seen [32]. Neuronal activation has also been examined in some extrahypothalamic regions following ethanol exposure. In male Long–Evans rats, ethanol consumption leads to an increase in Fos-IR in anterior paraventricular nucleus of the thalamus (aPVT), but not posterior PVT (pPVT). This increase occurred in rats that consumed ethanol voluntarily in the intermittent access two-bottle choice paradigm, and rats administered ethanol via an oral gavage, indicating that this site specific activation was due to ethanol exposure. Furthermore, increased OX_2 receptor but not OX_1 receptor mRNA was observed in the aPVT, but not pPVT, and acute ethanol exposure increased Fos/OX_2 double labeling, but not Fos/OX_1 double labeling, specifically in the aPVT [24].

Overall, acute ethanol exposure via oral gavage seems to increase OX expression in the hypothalamus of rats. However, the timing of this increase appears to be dependent on dose [24]. In contrast, an acute binge-like drinking episode decreases OX-A in the LH of C57BL/6J mice [25]. Chronic ethanol exposure increases OX expression within the hypothalamus and activates OX neurons in the LH and PEF, which in turn may act as positive feedback to drive ethanol intake. Ethanol consumption may also alter OX receptor expression in hypothalamic (and target) regions but further elucidation is required.

3 Orexin and Ethanol Self-Administration

A study by Lawrence et al. [21] was the first to show the involvement of OX_1 receptor in ethanol self-administration. Using an OX_1-selective antagonist 1-(2-methylbenzoxazol-6-yl)-3-[1,5]naphthyridin-4-yl urea (SB-334867, 20 mg/kg; intraperitoneal, i.p.), this study showed an attenuation in operant self-administration of ethanol in male iP rats [21]. These data were reproduced in male Long–Evans rats, with the same antagonist shown to attenuate ethanol self-administration [33]. A different OX_1 receptor antagonist, 1-(6,8-difluoro-2-methylquinolin-4-yl)-3-[4-(dimethylamino)phenyl]urea (SB-408124), had no impact on ethanol self-administration in male Wistar rats at 1, 10, or 30 mg/kg (subcutaneous, s.c.) [34]. However, the authors noted the low levels of ethanol intake during self-administration and raised the possibility that activity at OX_1 receptor may only be recruited during high levels of ethanol intake, or only involved during high motivation to consume ethanol. This is in line with a study by Moorman and Aston-Jones in which 30 mg/kg (i.p.) SB-334867 reduced both ethanol consumption and preference in SD rats using the two-bottle choice paradigm. This effect was only observed in rats that expressed high preference for ethanol, while no effect was seen in those rats with a low baseline preference [35]. Further studies in male iP rats showed that low doses of SB-334867 could attenuate self-administration of ethanol [36]. In female iP rats, 10 and 30 mg/kg (i.p.) of SB-334867 reduced ethanol intake in the two-bottle choice paradigm [37].

Contrasting results have been found with the effect of OX_1 receptor antagonists on sucrose administration. In one study, SB-334867 10, 15, or 20 mg/kg did not attenuate sucrose self-administration [33]. Meanwhile, another study found that 30 mg/kg, but not 10 or 20 mg/kg, of SB-334867 decreased responding for both sucrose pellets in rats fed ad libitum and food restricted rats [38]. Furthermore in male iP rats while 5 mg/kg of SB-334867 attenuated self-administration of sucrose, a significantly greater effect on ethanol responding compared to sucrose was observed at this same dose [39]. This study showed a greater role for OX_1 receptors in the motivational properties of ethanol, as the same low dose of SB-334867 attenuated responding on a progressive ratio schedule for ethanol but not sucrose [39]. Contrastingly, 10 or 30 mg/kg SB-334867 did not significantly reduce breakpoint or ethanol consumption in a progressive ratio paradigm in female iP rats [37]. Therefore, it seems that the involvement of OX_1 signaling in reward consumption may be more tuned into ethanol compared to natural rewards.

The "drinking in the dark" (DID) paradigm can induce binge-like ethanol-drinking model that promotes high levels of drinking in mice and produces pharmacologically relevant blood ethanol concentrations [40]. Utilizing this paradigm, binge consumption of ethanol is reduced in C57BL/6J mice after pretreatment with 30 mg/kg (i.p.) SB-334867. However, at this dose a reduction in sucrose intake was also observed [37]. Another study using lower doses found administration of 5 or 10 mg/kg (i.p.) SB-334867 attenuated binge drinking in C57BL/6J mice; however, at the 10 mg/kg dose an attenuation in saccharin seeking was also noted, although locomotor impairments were ruled out as a potential confound [25]. Intracerebroventricular (i.c.v.)

administration of 3 µg/µl SB-334867 blunted ethanol, but not saccharin binge-like drinking in the DID paradigm [31]. Overall OX_1 receptor antagonism seems to reduce ethanol self-administration and voluntary intake of ethanol in rodent models. These effects may be due, in part, to the ability of these antagonists to reduce the reinforcing value of alcohol, as shown by evidence that they both decrease the breakpoint on a progressive ratio schedule of responding for alcohol [36, 39]. Note also the relatively wide dose ranges employed, between different studies. This possibly relates to different formulations of SB-334867, some of which suffer from hydrolytic instability [41], a factor that may account for some of the negative findings.

Compared to OX_1, less research has examined the role of OX_2 receptors in ethanol self-administration. An OX_2 receptor antagonist, *N*-(2,4-dibromophenyl)-*N'*-[(4S,5S)-2,2-dimethyl-4-phenyl-1,3-dioxan-5-yl]-urea (JNJ 10397049; s.c.), dose dependently reduced ethanol, but not saccharin, self-administration [34]. This antagonist did not induce motor impairments, rescue ethanol-induced motor impairments, nor impact withdrawal symptoms and had no effect on basal or ethanol-induced dopamine (DA) release in the nucleus accumbens (NAc). However, this same study found no effect with an OX_1 receptor antagonist SB-408124, using the same paradigms [34], which contradicts general consensus. Central (i.c.v.) administration of the selective OX_2 receptor antagonist (2*S*)-1-(3,4-dihydro-6,7-dimethoxy-2(1*H*)-isoquinolinyl)-3,3-dimethyl-2-[(4-pyridinylmethyl)amino]-1-butanone hydrochloride (TCS-OX2-29; 100 and 300 µg) reduced self-administration of ethanol with no impact on sucrose self-administration at the doses tested in male iP rats [42]. Importantly TCS-OX2-29 did not cause any sedative effects [42]. Another study using the OX_2 receptor antagonist *N*-biphenyl-2-yl-4-fluoro-*N*-(1H-imidazol-2-ylmethyl) benzene-sulfonamide HCl (LSN2424100) showed 60 mg/kg reduced binge-like ethanol and sucrose intake using the DID paradigm in C57BL/6J mice [37]. In female iP rats, 30 mg/kg LSN2424100 was also able to reduce the breakpoint and ethanol consumption in the progressive ratio paradigm in female iP rats. However, this same dose did not attenuate voluntary ethanol consumption in female iP rats during the two-bottle choice paradigm [37]. This suggests there may be some gender differences and highlights the importance of studies using both sexes.

In a zebra fish model of voluntary ethanol-gelatin consumption, OX-A administration (i.c.v.) increased both ethanol and food intake [43]. The DORA almorexant can decrease operant self-administration of both 20% ethanol and 5% sucrose in male Long–Evans rats at 15 mg/kg (i.p.) with no significant effect on locomotor behavior [36]. In female iP rats, higher doses of 60 and 100 mg/kg almorexant reduced both ethanol and water consumption in the two-bottle choice paradigm, while doses of 10, 30, and 60 mg/kg reduced breakpoint and ethanol consumption in a progressive ratio test. Previous studies using doses of almorexant up to 300 mg/kg (administered orally) did not observe reduced motor performances in rats [44]. However, in sleep studies 30 mg/kg almorexant can decrease alertness, while 10 mg/kg does not [45]. In C57BL/6J mice, 100 mg/kg almorexant reduced ethanol intake, without reducing sucrose intake using the DID paradigm [37] (Table 1).

Table 1 Role of orexin receptors in alcohol self-administration

Compound	Dose	Paradigm	Species	Findings	
OX₁ antagonist					
SB-334867	20 mg/kg (i.p.)	Operant self-administration	Male iP rat	Reduced ethanol responding	[21]
	10, 15, and 20 mg/kg (i.p.)	Operant self-administration	Male Long–Evans rat	Reduced ethanol responding; no effect on sucrose responding	[33]
	5 or 10 mg/kg (i.p.)	Operant self-administration	Male iP rat	Reduced ethanol and sucrose responding	[39]
	30 mg/kg (i.p.)	Two-bottle choice	Male SD rat	Reduced consumption and preference (only in high preferring)	[35]
	30 mg/kg (i.p.)	Two-bottle choice	Female iP rat	Reduced ethanol consumption	[37]
	3 or 6 ng (intra-NAc core)	Two-bottle choice	Male Wistar rat	Reduced ethanol and food consumption at 1 h, but not 2–24 h	[50]
	10 nmol (intra-PVT)	Intermittent access two-bottle choice	Male Long–Evans rat	No effect on ethanol consumption	[22]
	5 mg/kg (i.p.)	Progressive ratio	Male iP rat	Reduced breakpoint of ethanol responding, no effect on sucrose	[39]
	10 and 30 mg/kg (i.p.)	Progressive ratio	Female iP rat	No effect on breakpoint for ethanol responding	[37]
	30 mg/kg (i.p.)	Drinking in the dark	Male C57BL/6J mice	Reduced consumption of ethanol and sucrose	[37]
	5 and 10 mg/kg (i.p.)	Drinking in the dark	Male C57BL/6J mice	Reduced consumption of ethanol. 10 mg/kg also reduced saccharin consumption	[25]
	1 or 3 µg/µl (i.c.v.)	Drinking in the dark	Male C57BL/6J mice	Reduced consumption of ethanol, no effect on saccharin or food consumption	[31]
SB-408124	1, 10, and 30 mg/kg (s.c.)	Operant self-administration	Male Wistar rat	No effect on ethanol responding	[34]
OX₂ antagonist					
JNJ-10397049	3 and 10 mg/kg (s.c.)	Operant self-administration	Male Wistar rat	Reduction of ethanol responding	[34]
TCS-OX2-29	100 and 300 µg (i.c.v.)	Operant self-administration	Male iP rat	Reduced ethanol responding, no effect on sucrose	[42]
	100 µg (intra-NAc core)	Operant self-administration	Male iP rat	Reduced ethanol responding	[42]
	100 µg (intra-NAcSh)	Operant self-administration	Male iP rat	No effect on ethanol responding	[42]
	10 nmol (intra-PVT)	Intermittent access two-bottle choice	Male Long–Evans rat	aPVT, not pPVT reduced ethanol consumption	[22]

(continued)

Table 1 (continued)

Compound	Dose	Paradigm	Species	Findings	
LSN-2424100	30 mg/kg (i.p.)	Two-bottle choice	Female iP rat	No change in ethanol consumption	[37]
	30 mg/kg (i.p.)	Progressive ratio	Female iP rat	Reduced breakpoint and ethanol consumption	[37]
	60 mg/kg (i.p.)	Drinking in the dark	C57BL/6J mice	Reduced ethanol and sucrose consumption	[37]
DORA					
Almorexant	Intra-VTA	Operant self-administration	Male Long–Evans rat	Reduced ethanol responding, no effect on sucrose	[36]
	15 mg/kg (i.p.)	Operant self-administration	Male Long–Evans rat	Reduced ethanol and sucrose responding	[36]
	60 and 100 mg/kg (i.p.)	Two-bottle choice	Female iP rat	Reduced ethanol and water consumption	[37]
	10, 30, and 60 mg/kg (i.p.)	Progressive ratio	Female iP rat	Reduced breakpoint and consumption	[37]
	100 mg/kg (i.p.)	Drinking in the dark	C57BL/6J mice	Reduced ethanol consumption, no effect on sucrose	[37]
OX agonist					
OX-A	i.c.v.	Voluntary intake	Zebra fish	Increased ethanol and food consumption	[43]
	0.9 nmol Intra-PVN	Two-bottle choice	Male SD rat	Increased ethanol consumption, no effect on water or food	[46]
	0.9 nmol Intra-LH	Two-bottle choice	Male SD rat	Increased ethanol consumption, no effect on food or water	[46]
	0.9 nmol Intra-NAc	Two-bottle choice	Male SD rat	No effect on ethanol consumption	[46]
	0.9 nmol Intra-PVN	Two-bottle choice	Male SD rat	Increased ethanol consumption, no effect on water or food	[52]
OX-A + OX-B	1 nmol Intra-aPVT	Intermittent access two-bottle choice	Male Long–Evans rat	Increased ethanol consumption, no effect on sucrose consumption	[22]
	1 nmol Intra-pPVT	Intermittent access two-bottle choice	Male Long–Evans rat	No effect on ethanol consumption, increased sucrose consumption	[22]

The abovementioned data relate to studies using systemic or i.c.v. administration of drugs to manipulate the OX system. While of value, they do not reveal specific anatomic loci of action where OX may be acting in the brain to regulate self-administration of alcohol. Microinjections of OX-A (0.9 nmol) into the paraventricular nucleus of the hypothalamus (PVN) and LH increase ethanol self-administration, but not food or water consumption in SD rats [47, 52]. OX-A or OX-B microinjection into the NAc core (NAcc) increased ethanol and food consumption in male Wistar rats [48, 50]. In contrast, nonspecific intra-NAc microinjections of OX-A had no effect on ethanol intake [46]. The OX-A-induced increase in ethanol consumption is thought to be due to an increase in the number of drinking bouts, not the amount of ethanol consumed per drink [52]. Administration of OX-A or OX-B into the aPVT, but not pPVT, increased ethanol consumption in male Long–Evans rats. In contrast, OX-A administration into the pPVT, but not the aPVT, increased sucrose intake. Furthermore, administration of an OX_2 receptor antagonist, TCS-OX2-29 (10 nmol), but not the OX_1 receptor antagonist SB-334867 (10 nmol), into the aPVT reduced ethanol intake; the same antagonists microinjected into the pPVT had no effect [22].

Intra-ventral tegmental area (VTA) infusions, but not intra-substantia nigra (SN) infusions, of the DORA almorexant reduced ethanol self-administration without affecting sucrose self-administration in male Long–Evans rats. Using in vitro electrophysiological techniques, this study showed OX-A to increase firing in VTA neurons, and this enhancement was prevented by almorexant [36]. The effect of TCS-OX2-29, an OX_2 receptor antagonist, was region specific within the NAc. 100 µg TCS-OX2-29 administered intra-NAcc but not intra-NAc shell (NAcSh) reduced ethanol self-administration [42]. Furthermore, 3 or 6 ng OX_1 receptor antagonist SB-334867 microinjected into the NAcc also reduced ethanol intake in a two-bottle choice paradigm [50]. While the hypothalamic sites where OX regulates ethanol intake have been identified, there is still a need to fully evaluate potential extrahypothalamic sites of action and their interactions.

4 Orexin and Cue-Induced Relapse

The first evidence that OX plays a role in alcohol seeking was established with the demonstration that 20 mg/kg of the OX_1 receptor antagonist SB-334867 could prevent a cue-induced reinstatement of ethanol seeking in male iP rats [21]. An additional study by Jupp et al. [39] also found that SB-334867 prevented cue-induced reinstatement of alcohol seeking in male iP rats. This latter study evaluated the effect of SB-334867 both immediately after extinction and also following extinction plus 5 months protracted abstinence and found a similar attenuation of cue-induced reinstatement [39]. Despite long-term abstinence, it seems that the OX system is still involved in the integration of the salience of cues that previously signaled the availability of ethanol. SB-334867 is more effective at reducing cue-induced reinstatement of alcohol seeking compared to that for a natural reward. Thus, SB-334867 (1–10 mg/kg i.p.) attenuated cue-induced reinstatement of ethanol seeking in a dose-dependent manner. This effect was specific to ethanol, with no changes in cue-induced responding to a natural reward (SuperSac – 3% glucose and 0.125% saccharin) [48].

However, in a study by Dhaher et al. [51] the effect of 10 and 20 mg/kg (i.p.) SB-334867 on spontaneous recovery of ethanol seeking was examined in female iP rats. Spontaneous recovery is the recovery of responding in the absence of the previously trained reward, following a period of rest after extinction. This paradigm tested four consecutive days of ethanol seeking following 2 weeks abstinence since extinction training. No attenuation of renewal was seen when the OX_1 receptor antagonist SB-334867 was administered prior to the first session at either dose. However, when this test was repeated in the presence of an ethanol reward upon responding, 10 and 20 mg/kg (i.p.) SB-334867 attenuated relapse [51]. In contrast to OX_1 receptors, OX_2 receptors apparently have no role in cue-induced reinstatement of alcohol seeking, with i.c.v. administration (100 or 300 µg) of the selective OX_2 receptor antagonist TCS-OX2-29 unable to prevent cue-induced reinstatement of ethanol seeking in iP rats [42].

A number of studies have investigated the neural substrates and mechanisms underlying cue-induced relapse to alcohol seeking. There are a number of potential nuclei where OX receptor activation/antagonism may regulate ethanol seeking. Within the hypothalamus, cue-induced reinstatement of alcohol seeking is associated with the activation of OX-containing neurons. Significant increases in Fos expression in OX cells within the DMH and PEF/LH following cue-induced reinstatement of alcohol seeking were found in Wistar rats [52]. However, a more recent study found no correlation between active lever responses and percentage OX neuronal activation in any subregion of the hypothalamus in male SD rats following cue-induced reinstatement of alcohol seeking [32]. This finding is interesting, since OX antagonists can reduce cue-induced reinstatement of alcohol seeking [21, 39, 53, 54]. Although few studies have evaluated the OX activation following drug seeking, these data were in line with another study by Mahler and Aston-Jones [49], who showed no correlation between cue-induced reinstatement of cocaine seeking and OX /Fos-IR [49]. Notably, a study examining cue-induced relapse to nicotine seeking in mice found similar activation pattern to that found by Dayas et al. [52], with activation within the LH and PEF (although this was not directly correlated to degree of reinstatement). The authors argue that their cue-induced reinstatement data are in line with the previous studies examining OX neuronal activation following context-induced reinstatement, and that the persistent contextual cues are sufficient to drive activation, while brief discrete cues may not be sufficient [32].

Cue-induced reinstatement of alcohol seeking immediately after extinction increased Fos expression in the infralimbic (IL), prelimbic (PrL), orbitofrontal (OFC), and piriform cortices, NAc, bed nucleus of the stria terminalis (BNST), central nucleus of the amygdala (CeA), basolateral amygdala (BLA), LH, and DMH. Delayed reinstatement of alcohol seeking (5 months protracted abstinence after extinction) further elevated Fos induction in these regions. SB-334867 attenuated immediate reinstatement induced Fos in the NAc, PrL, and OFC, while reducing delayed reinstatement induced Fos in the piriform cortex, OFC, and PrL [39]. This shows that reinstatement following extinction and protracted abstinence is characterized by enhanced prefrontal Fos expression, analogous to increased cue reactivity in human imaging studies of AUD patients [55]. Furthermore, the OX system is seemingly involved in both of these behaviors, with SB-334867 attenuating Fos induction following cue-induced reinstatement. At both time points, the PrL was implicated as a possible target of OX innervation where cortical OX_1 receptors may modulate cue driven alcohol seeking. SB-334867 microinjected into the PrL (3 µg/ side) attenuated cue-induced reinstatement of alcohol seeking in male iP rats. This dose of SB-334867 microinjected into the PrL did not have any effect on sucrose seeking [53]. A similar effect was also noted with SB-334867 microinjected into the VTA, suggesting multiple axes where OX_1 receptors may exert control over cue driven alcohol seeking.

5 Orexin and Extinction/Context-Induced Relapse

The renewal model can be described as the recovery of an extinguished behavior that is dependent on context. In this model, animals are trained to self-administer a drug in one environment (A), the behaviors are extinguished in another environment (B), and responding is returned when the animal is returned to the original context (A). Using this model, Hamlin and colleagues showed ABA renewal of alcoholic beer seeking to induce Fos expression in the BLA, ventral NAcSh, and LH [56]. However, another in vivo study using an OX antisense morpholino to knockdown OX expression within the LH had no effect on reacquisition or ABA renewal of extinguished alcoholic beer seeking. Although the morpholino sufficiently decreased OX protein expression, alcohol reacquisition and renewal were only diminished when both OX and another hypothalamic peptide melanin concentrating hormone (MCH) pre-protein levels were reduced [57]. Although this shows a role for the LH in reacquisition and ABA renewal of alcoholic beer seeking, these data suggest that OX on its own may not be sufficient in the execution of this behavior [57].

Bilateral reversible inactivation of LH prevented context-induced reinstatement of beer and sucrose seeking in rats [58]. To elucidate the pathway mediating this effect, neuronal retrograde tracing was employed which demonstrated that NAcSh-LH projections were activated during context-induced reinstatement of beer seeking [58]. Millan et al. [54] further examined the role of the NAcSh projections to the LH. Reversible inhibition of the ventral, but not dorsal, NAcSh facilitated extinction of beer seeking (i.e., prevented reinstatement) in male Long–Evans rats. This same study showed that this region was not involved in the acquisition of extinction, but in the maintenance of this behavior. Moreover, reversible inhibition of the ventral NAcSh (and presumably disinhibition of OX neurons) increased Fos/OX-like-IR in the PEF and LH, but not DMH [54]. Furthermore, in the LH there was a significant positive correlation between beer seeking and renewal-associated Fos expression in OX positive cells [56] and a positive correlation of lever presses and OX neuron activation was observed in the LH and DMH, but not PEF following ABA renewal of alcohol seeking [32].

Although these studies implicate a role for OX and an NAcSh projection to the LH, there is some disagreement between the specific loci of these two studies. Marchant et al. [58] suggest that this is the role of the dorsal NAcSh, contrastingly Millan et al. [54] suggest that this is integral to the ventral NAcSh. It is important to note however that methodological differences especially those of a contextual nature could contribute to these apparent discrepancies.

6 Orexin and Stress-Induced Relapse

A feature of human addiction is the propensity to relapse. Stress has long been associated with an increased risk of drug relapse with stress during abstinence a predictor of outcomes for alcoholics, among other drugs of abuse [59]. OX neurons are highly responsive to stress stimuli and activate the HPA axis. Central administration of both OX-A or -B leads to increased plasma concentrations of ACTH and corticosterone in vivo via a CRF receptor dependent mechanism [60, 61]. CRF has mutual interactions with OX neurons, with anatomical evidence revealing reciprocal projections from CRF neurons to orexinergic neurons and various stressors to activate OX transcription via a CRF-dependent manner [62].

Although OX is implicated in stress-related behaviors, relatively little research has examined the role of OX in stress-induced relapse of alcohol seeking. Using other drugs of abuse, systemic administration of SB-334867 has been shown to attenuate footshock-induced reinstatement of cocaine seeking [20]. Furthermore, the CRF1/2 receptor antagonist D-Phe-CRF12-41 attenuated OX-induced reinstatement of cocaine seeking [20]. These results suggest that OX and CRF systems interact to regulate cocaine seeking. However, another study by Wang et al. [63] found that while footshock-induced reinstatement of cocaine seeking involved the CRF system, intra-VTA microinjection of OX to induce reinstatement of cocaine seeking did not [63]. Furthermore, the OX_1 antagonist SB-334867 did not modify the CRF-dependent stress-induced reinstatement of nicotine seeking [64]. These results suggest that the mechanism by which OX and CRF interact to induce reinstatement of drug seeking may be dependent on anatomic loci and/or drug of abuse.

In the alcohol field, Richards et al. [33] showed that the OX_1 receptor antagonist SB-334867 attenuated yohimbine-induced reinstatement of alcohol and sucrose seeking [33]. Furthermore, OX-A can excite neurons of the nucleus incertus (NI), predominantly via an OX_2 receptor mediated pathway [65]. Indeed, microinjections of TCS-OX2-29 (100 µg/side) but not SB-334867 (3 µg/side) directly into the NI attenuated yohimbine-induced reinstatement of alcohol seeking in iP rats. These effects were specific to the NI, with administration of TCS-OX2-29 or SB-334867 immediately adjacent to the NI, or within the fourth ventricle having no influence on this behavior. In line with these findings, whole cell patch clamp studies found that OX-A (600 nM) depolarized the majority of NI neurons. Bath application of TCS-OX2-29 (10 µM) but not SB-334867 (10 µM) prevented this depolarization [66]. These data suggest that stress activates an excitatory OX input to NI that contributes to the reinstatement of alcohol seeking, predominantly via OX_2 receptor signaling. The relaxin-3 containing neurons of the NI have been implicated in ethanol self-administration and alcohol seeking in rats and mice [67, 68] and provide a novel potential locus where motivation and reward circuits appear to interface.

It is important to note that the exact mechanisms involved in yohimbine-induced reinstatement of operant responding are not fully understood. A recent study by

Chen et al. [69] has shown that yohimbine reinstates lever pressing both in animals that were previously trained for lever pressing resulting in cue presentation regardless of reward delivery, and therefore may be involved in cue reactivity, not necessarily acting as a stressor model [69]. It is however noteworthy that yohimbine-induced reinstatement of ethanol seeking can be prevented by treatments that do not impact upon yohimbine-induced reinstatement of sucrose seeking, and reinstatement is specific to the alcohol-paired lever [67]. This suggests that cue reactivity is not the only factor behind a yohimbine-induced reinstatement of reward seeking. Moreover, clinical evidence suggests that yohimbine induces anxiety and the subjective measures of alcohol intoxication in healthy humans [70], as well as craving in human alcoholics [71]. Therefore, based on this property yohimbine has been suggested to represent an ideal cross-species probe for translational research in addiction [72]. Yohimbine is widely used in animal studies [73] and produces a robust and stable reinstatement of alcohol seeking in rats [74] and other species [72]. Both yohimbine- and footshock-induced reinstatement of alcohol seeking induce Fos and CRF expression in a similar distribution [75] and are dependent upon extrahypothalamic CRF_1 receptor signaling [74, 76], suggesting yohimbine precipitates reinstatement in a stress-like manner. Furthermore, the OX_1 receptor antagonist SB-334867 is able to attenuate both footshock- and yohimbine-induced reinstatement of drug seeking [20, 33, 77].

Interestingly, a recent clinical study evaluated the usefulness of OX blood concentration levels as a marker for relapse susceptibility. This study found that at the beginning of the study, more severely alcohol-dependent patients had significantly greater OX blood concentration than individuals with moderate addiction severity. However, after 4 weeks of abstinence, the peptide blood concentration was similar in both groups of alcoholic patients, and at a comparable level to the control group [78].

7 Orexin and Conditioned Place Preference

The first published evidence that implicated OX neurons in reward utilized the conditioned place preference (CPP) paradigm. In this study, conditioned animals that showed a preference for a reward (morphine, cocaine, food) paired chamber also had elevated Fos expression in OX neurons in the LH [19]. Extensive research has been employed using this model with other drugs of abuse; however, alcohol CPP studies remain underexplored and somewhat ambiguous. SB-334867 (30 mg/kg i.p.) attenuated a weak CPP to ethanol in DBA/2 mice; however, using a different protocol, resulting in a strong CPP, the same dose of SB-334867 had no effect. Importantly, no differences in locomotor activity were noted in DBA/2 mice treated with up to 30 mg/kg SB-334867 [79]. Another study by Shoblock et al. [34] showed that another OX_1 receptor antagonist SB-408124 had no effect on acquisition of CPP or ethanol-induced CPP in DBA/2 mice. The authors also tested the OX_2 receptor antagonist JNJ-10397049 and found 10 mg/kg (s.c.) before ethanol conditioning resulted in a failure to acquire a significant CPP. Furthermore, animals pretreated with JNJ-10397049 failed to reinstate following a 1 g/kg priming injection of ethanol [34].

8 Orexin and Withdrawal

Excessive alcohol consumption over a prolonged period of time disrupts the neurochemical balance of the brain. When alcohol consumption is reduced drastically or stopped completely, this can lead to a dysfunctional neurophysiological state and constellation of clinical signs and symptoms (alcohol withdrawal syndrome). Symptoms can range from mild (tremors, anxiety, restlessness, and nausea) to more serious and complicated (seizures and delirium tremens (DTs)). These signs and symptoms are thought to be compensatory responses representing the brain's attempt to reestablish homeostasis following continuous alcohol exposure [80, 81]. The OX system has been implicated in withdrawal from several drugs of abuse. Prepro-orexin knockout mice and WT mice pretreated with SB-334867 had attenuated physical symptoms of morphine withdrawal [82, 83]. Pretreatment with SB-334867, but not with OX2 antagonist TCS-OX2-29, attenuated the physical signs of nicotine withdrawal in mice [64].

In alcohol withdrawal studies, pretreatment with the OX_1 receptor antagonist SB-408124 (3, 10, or 30 mg/kg s.c.) and OX_2 receptor antagonist JNJ-10397049 (1, 3, or 10 mg/kg s.c.) did not impact acute withdrawal symptoms of C57BL/6J mice. This study used a visual rating scale to examine tremors, piloerection, muscle rigidity, and vocalization. Mice displayed signs of physical withdrawal when alcohol was withheld, and diazepam but not OX receptor antagonists attenuated these signs of withdrawal [34]. However, data from alcohol-dependent subjects implicates OX in affective dysregulation that accompanies alcohol withdrawal [84]. Indeed, significant associations between OX-A levels and severity of alcohol withdrawal scores were observed in alcohol-dependent subjects. Furthermore, OX mRNA levels are higher at the beginning of withdrawal and seem to decrease during the course of withdrawal [85].

9 Orexins and Locomotor Activity/Behavioral Sensitization

OX-A and OX-B administration increase locomotor activity [86]. Pretreatment with the DORA almorexant (100 or 200 mg/kg; p.o.) prevented the stimulatory effects of central OX-A in C57BL/6J mice. However, these doses and lower (50 mg/kg; p.o.) were sufficient in reducing baseline locomotor activity [87]. OX_1 receptor and OX_2 receptor deficient mice show no differences in baseline locomotor activity to wild-type mice [87]; however, this may in part be due to compensatory mechanisms during development. In C57BL/6J mice, 3 μg/μl (i.c.v.) or 10 mg/kg (i.p.) SB-334867 did not impact locomotor behavior [25], and 30 mg/kg SB-334867 in DBA/2 mice had no effect on locomotion [79]. OX-A (3 μg i.c.v.) increased locomotion in wild-type and OX_1 receptor deficient mice, but not mice lacking the OX_2 receptor or both OX receptor subtypes [87], suggesting that OX-A induced locomotion in mice may be mediated via the OX_2. The OX_1 antagonist SB-334867 (15 or 20 mg/kg i.p.) did not have any effect on distance traveled or stereotypic counts in male Long–Evans rats [33]. Furthermore, no effects on locomotion, feeding, or grooming were seen at doses lower than 20 mg/kg SB-334867; however, 30 mg/kg SB-334867 can result in nonspecific behavioral effects, including reduction in feeding, grooming, and locomotion, and an increase in resting [88]. The OX_2 receptor antagonist TCS-OX2-29 had no effect on locomotion when administered centrally (100 or 300 μg) or microinjected into the NAc (100 μg) of male iP rats [42]. Similarly, the OX_2 receptor antagonist JNJ-10397049 (1, 3, or 10 mg/kg s.c.) did not induce motor impairments in male Wistar rats [34].

The OX system does not seem to be involved in the acute motor effects induced by ethanol consumption. In male Wistar rats, ethanol (0.32, 1, and 1.5 g/kg, i.p.) dose dependently decreased rotarod performance and grip strength, while almorexant (30, 100, and 300 mg/kg, per os; p.o.) did not. In combination with ethanol, almorexant did not interfere with grip strength or rotarod performance, showing no action on motor coordination or on ethanol-induced sedation [44]. Indeed, in humans, no potentiation of alcohol impairment was seen by almorexant [89]. The OX_2 receptor antagonist JNJ-10397049 (1, 3, or 10 mg/kg s.c.) did not rescue ethanol-induced motor impairments in male Wistar rats [34]. However, central administration of 3 μg/μl of SB-334867 increased the latency to recover the righting reflex after a sedative dose of ethanol [31].

Behavioral sensitization, defined as a progressive increase in the psychomotor stimulant effects elicited by repeated exposure to drugs of abuse, has been proposed to occur as a result of drug-induced neural plasticity that regulates the attribution of salience to stimuli [90]. Pretreatment with the OX_1 receptor antagonist SB-334867 decreased ethanol-induced locomotion in mice [79] and rats [33], but not spontaneous activity in mice [79]. The OX_2 receptor antagonist JNJ-10397049 (1–10 mg/kg s.c.) attenuated ethanol-induced hyperactivity in C57BL/6J mice. However another OX_1 receptor antagonist SB-408124 failed to attenuate alcohol-induced hyperactivity [34]. Furthermore, in male Swiss albino mice, 7 days administration of 2.2 g/kg ethanol (i.p.) caused behavioral sensitization along with a trend towards increased OX/Fos-like-IR. Therefore, OX neurons of the LH seem to be recruited in the development of behavioral sensitization to ethanol, although other systems are likely involved. Administration of 20 mg/kg, but not 10 mg/kg, SB-334867 was sufficient to block the sensitized response to ethanol challenge in chronically treated mice [91]. This suggests that OX_1 signaling is required for the expression of ethanol sensitization and suggests that OX may participate in the neuroadaptations induced by chronic ethanol consumption/treatment. Studies investigating the involvement of OX in amphetamine sensitization suggest a role for both OX receptors [92]. SB-334867 prevents both the development and expression of amphetamine-induced behavioral sensitization, specifically with involvement of the VTA [93–95]. However, further elucidation in the mechanistic differences between drugs of abuse and the loci involved is required.

SB-334867 treatment (20 mg/kg i.p.) following early postnatal alcohol exposure decreased levels of activity compared with those treated with vehicle. Reductions in both alcohol-induced motor activity and stereotypy were observed, with no effect on motor activity in control rats [96]. This implicates OX in the hyperactivity observed following prenatal ethanol exposure. Interestingly, prenatal ethanol exposure stimulates neurogenesis of OX neurons in the PEF/LH of SD rats prenatally exposed to either 1 or 3 g/kg ethanol [97]. In line with rodent studies, a recent study in Zebra fish showed embryonic ethanol exposure to increase ethanol consumption, neurogenesis of hypothalamic neurons and OX expression, and locomotor and exploratory behaviors [43]. With the notion that increased OX expression contributes to an increase in ethanol consumption, this may contribute to the increased consumption of, and preference for, alcohol in the offspring produced by prenatal ethanol exposure.

10 Potential Mechanisms

OX neurons within the LH are susceptible to modulation by other peptide systems and may interact with them in combination to affect ethanol consumption/seeking. For example, Prasad and McNally [57] have shown that OX on its own may not be sufficient for renewal of alcohol seeking. They propose that it may work in concert with other neuropeptides, such as MCH, to induce this behavior. OX is implicated in the modulation of the mesolimbic dopamine system (VTA and NAc) [17, 92, 98–100]. Recently dopamine receptor subtypes within the LH have been implicated in altering ethanol intake. Microinjection of a dopamine D1 receptor agonist 6-chloro-1-phenyl-2,3,4,5-tetrahydro-1H-3-benzazepine-7,8-diol (SKF-81297; 10.8 nmol/side) or D2 receptor antagonist N-[(1-ethylpyrrolidin-2-yl)methyl]-2-methoxy-5-sulfamoylbenzamide (sulpiride; 23.4 nmol/side) into the PEF/LH increased ethanol intake. Conversely microinjection of a D1 receptor antagonist 7-chloro-3-methyl-1-phenyl-1,2,4,5-tetrahydro-3-benzazepin-8-ol (SCH-23390; 15.4 nmol/side) or D2 receptor agonist (5aR,9aR)-6-propyl-5a,7,8,9,9a,10-hexahydro-5H-pyrido[2,3-g] quinazolin-2-amine (quinelorane; 6.2 nmol/side) reduced ethanol intake in male SD rats. Furthermore, this study showed the receptor antagonists to alter OX mRNA expression, with the D1 receptor agonist increasing OX mRNA expression and the D2 receptor agonist decreasing the expression [101]. Furthermore, ABA renewal of alcoholic beer seeking increases Fos expression in the LH, via a D1 receptor dependent mechanism [56]. However, modulation of the LH dopamine receptors had no impact upon MCH expression [104].

When injected into the LH, two compounds which mimic the effects of glutamate N-methyl-D-aspartate (NMDA) and α-amino-3-hydroxy-5-methyl-4-isoxazolepropionic acid (AMPA) both significantly increased ethanol intake while having no effect on chow or water intake. Glutamate receptor antagonists had the opposite effect, significantly reducing ethanol consumption. This was specific to the LH. Intra-LH injection of NMDA significantly stimulated expression of OX in both the LH and PEF while reducing MCH in the zona incerta (ZI), whereas AMPA increased OX only in the LH and had no effect on MCH [101]. Intra-LH injection of neuropeptide S (NPS) induces a cue-induced reinstatement of alcohol seeking, which is prevented by the OX_1 receptor antagonist SB-334867, suggesting an orexinergic contribution [102]. Furthermore, this was replicated by Ubaldi et al. [103], who showed that NPS facilitated cue-induced reinstatement of ethanol seeking in male Wistar rats. This study blocked NPS facilitated cue-induced reinstatement with SB-334867 microinjections into the BNST and PVN, but not locus coeruleus (LC) or VTA. Furthermore, intra-BNST or PVN OX-A micro-injection also enhanced ethanol seeking, but to a lesser extent than intrahypothalamic NPS administration [103].

Substance P (SP) in the PVT mediates the stimulatory effect of ethanol consumption by OX [104]. These effects were ethanol specific, with SP administration into the aPVT reducing and pPVT increasing sucrose intake. This study found that SP (5 nmol) administration into the aPVT, but not the pPVT or dorsal third ventricle (d3v), increased ethanol consumption in an intermittent access two-bottle choice paradigm, which could be attenuated by pretreatment with two different SP receptor (neurokinin 1 receptor/tachykinin receptor 1; NK1R) antagonists, (2S,3S)-3-[[3,5-bis(trifluoromethyl)phenyl]-methoxy]-2-phenyl-piper-idine hydrochloride (L-733060; 5 nmol) and (3aR,7aR)-octahydro-2-[1-imino-2-(2-methoxyphenyl)ethyl]-7,7-diphenyl-4H-isoindol (RP-67580; 5 nmol). Further-more, OX-A (1 nmol) and OX-B (1 nmol) administration stimulated SP mRNA and peptide expression, specifically in the aPVT, and local aPVT administration of L733060 (5 nmol) attenuated OX-B induced ethanol intake [104]. Overall this shows a variety of peptide systems that act upon or alongside the orexinergic system facilitate alcohol seeking in rodent models.

11 Discussion

OX_1 receptors are implicated in operant self-administration [21, 35, 36], home-cage alcohol consumption [22, 31, 35, 37], and cue- and stress-induced reinstatement of ethanol seeking [21, 33]. All these studies used the same OX_1 antagonist SB-334867, but results span different rodent species, different strains, and different genders. However, see Shoblock et al. [34] for an example of an OX_1 receptor antagonist (SB-408124) failing to attenuate self-administration, and Dhaher et al. [51] for an example of SB-334867 failing to attenuate Pavlovian spontaneous recovery. In contrast, although the OX_2 receptor antagonists JNJ-10397049 and TCS-OX2-29 reduce operant self-administration [34, 42], results using the OX_2 antagonist LSN-242100 in home-cage ethanol consumption paradigms vary depending on the rodent species [37]. The OX_2 receptor does not appear to be involved in cue-induced reinstatement in male iP rats [42], but may have some loci specific involvement in stress-induced reinstatement [66]. The DORA almorexant reduced ethanol consumption in operant responding paradigm [36], home-cage consumption, and binge-like drinking of ethanol [37]. Overall it appears that OX_1 and OX_2 receptors are involved in the reinforcing effects of alcohol. However, the effects of OX on ethanol intake depend on both the receptor and the brain region with which it interacts. The dichotomy in orexinergic involvement may be related to differences in signaling between OX_1 and OX_2 receptors. The two receptors have different distribution throughout the mesocorticolimbic circuitry [105], different G-coupled protein signaling [106], and transmitter selectivity [10].

Some evidence suggests that there may be a functional dichotomy between LH and PEF/DMH OX neurons in regard to arousal and reward seeking [107]. While the LH OX neurons have a proposed role in reward seeking, the PEF and DMH are implicated in arousal- and stress-related behaviors [107]. This putative dichotomy has been suggested for multiple drugs of abuse and natural rewards [19, 108, 109]. In alcohol studies, this dichotomy was reported for renewal of alcoholic beer seeking [56]; however, no region-specific activation was noted in cue-induced ethanol seeking [52]. Recently, Moorman et al. [32] observed differential OX activation in different aspects of ethanol seeking. While all alcohol-related behaviors induced activation of OX neurons, the OX neuronal activation in the LH was positively correlated with context-induced reinstatement and home-cage preference, while the DMH and PEF OX neuronal activation was positively correlated with context-induced reinstatement, home-cage consumption, and ethanol preference. Interestingly, although there was increased activation of OX neurons, no correlation was seen with cue-induced reinstatement [32]. Moreover Hamlin et al. [56] and Dayas et al. [52] examined the distinctive aspects of alcohol seeking (context- vs. cue-induced reinstatement), which may contribute to the different results.

12 Future Avenues of Research

In the past 10 years since the OX system was implicated in alcohol consumption and seeking behaviors, many studies have examined the aspects of this relationship and expanded our knowledge; however, many questions remain. Although the dissection of the role of OX at specific loci has begun, the precise actions of OX_1 versus OX_2 within specific brain regions and their involvement in different aspects of alcohol-related behaviors is complex and requires further research. Furthermore, the specificity for alcohol versus natural reward and arousal processes is of importance. Determining these discrepancies may allow for targeting of specific alcohol-related behaviors, while leaving other behaviors such as natural reward seeking, and arousal intact, and lead to therapeutics with limited off target adverse effects.

One aspect of alcohol seeking that remains insufficiently investigated is the impact of OX on stress-induced reinstatement. The few studies that have examined the effect of OX on stress-induced relapse of alcohol seeking did so using the pharmacological stressor yohimbine, with no published studies to date examining footshock-induced relapse of alcohol seeking. With recent uncertainty of the effects of yohimbine on cue reactivity [69], the overlapping and distinct mechanisms between these two stress models and the involvement of the OX system require elucidation. A possible target for OX in stress-induced reinstatement may be the extended amygdala, including the CeA and the BNST, where OX neurons send afferent projections [16, 100]. Moreover, the NI, a hindbrain region, implicated in stress-induced reinstatement of alcohol seeking [66, 67] may be of interest as it receives an OX afferent input from the hypothalamus [65]. Furthermore, the recent study by Moorman et al. [32] examined the topographic activation of hypothalamic OX neuronal populations following ethanol self-administration, context- and cue-induced reinstatement of alcohol seeking [32]. However the activation of OX neurons following stress-induced reinstatement is not yet elucidated. Further dissociation of OX activation following stress-induced reinstatement of ethanol seeking will allow further comparison and clarification of the dichotomy of OX neurons in specific alcohol-seeking behaviors.

The development and proven preclinical utility of optogenetic and chemogenetic techniques that allow reversible inactivation of specific pathways may prove useful in elucidating any topographical dichotomy of OX neurons. These techniques allow for spatial, temporal, and cell type-selective modulation of neuronal circuits and have been utilized in the dissection of many neuronal circuits, including those relevant to addiction. Currently these techniques have not been employed to determine the role of OX pathways in ethanol seeking even though they are widely available and relatively cheap. However, the possible impact of collateral projections must be considered when planning such studies. For example, a recent study showed a collateral OX projection to the NAcSh and PVT from the LH. This arose mostly from the central region of the LH, which included part of the PEF and part of the LH (amounting to 1.6% of total OX projections). The OX fiber density was much less in the NAcSh compared to the PVT, suggesting dual OX modulation of the NAcSh and PVT [110]. Therefore, anatomically discrete antagonism of OX receptors may not encompass the full story; other compensatory/simultaneous mechanisms could occur in unison to drive a final behavioral readout.

While the past studies focused on direct impacts of OX on alcohol consumption and seeking, more recently this has been expanded to examine the interactions between other peptide systems and their impact upon OX system regulation. Evidence suggests that hypothalamic peptides including OX interact in multiple and complex ways in the modulation of alcohol drinking. Research by the Leibowitz laboratory has focused on positive and negative feedback loops where hypothalamic peptides interact to regulate alcohol intake in rodent models [111]. These peptides include MCH, NPY, SP, galanin, enkephalin, dynorphin (DYN), melanocortins, and CRF, which have all been implicated in aspects of alcohol seeking. The contribution of such complex network-level interactions requires elucidation. Furthermore, OX is co-expressed with other neuropeptides within the same neuron. For example, DYN is co-expressed and likely co-released with OX, although these peptides have largely opposing behavioral actions [112].

The OX system has an established role in alcohol use and abuse and is a candidate target for future pharmacotherapies for AUD. Further distinction between DORA and SORA potential advantages may help in this regard [6, 113]. Recently, the US FDA approved suvorexant, a DORA registered for the treatment of insomnia. Selective OX_2 receptor antagonists are beginning to enter into clinical investigation for sleep disorders; however, OX_1 receptor antagonists currently remain in preclinical studies [6].

References

1. Rehm J, Mathers C, Popova S et al (2009) Global burden of disease and injury and economic cost attributable to alcohol use and alcohol-use disorders. Lancet 373:2223–2233. doi:10.1016/S0140-6736(09)60746-7
2. Wagner F (2002) From first drug use to drug dependence; developmental periods of risk for dependence upon marijuana, cocaine, and alcohol. Neuropsychopharmacology 26:479–488. doi:10.1016/S0893-133X(01)00367-0
3. Anton RF (2001) Carbohydrate-deficient transferrin for detection and monitoring of sustained heavy drinking. Alcohol 25:185–188. doi:10.1016/S0741-8329(01)00165-3
4. Jupp B, Lawrence AJ (2010) New horizons for therapeutics in drug and alcohol abuse. Pharmacol Ther 125:138–168. doi:10.1016/j.pharmthera.2009.11.002
5. Cowen MS, Chen F, Lawrence AJ (2004) Neuropeptides: implications for alcoholism. J Neurochem 89:273–285. doi:10.1111/j.1471-4159.2004.02394.x
6. Boss C, Roch C (2015) Recent trends in orexin research – 2010 to 2015. Bioorg Med Chem Lett 25:2875–2887. doi:10.1016/j.bmcl.2015.05.012
7. Wayner MJ, Greenberg I, Carey RJ, Nolley D (1971) Ethanol drinking elicited during electrical stimulation of the lateral hypothalamus. Physiol Behav 7:793–795. doi:10.1016/0031-9384(71)90152-1
8. Lestang I, Cardo B, Roy MT, Velley L (1985) Electrical self-stimulation deficits in the anterior and posterior parts of the medial forebrain bundle after ibotenic acid lesion of the middle lateral hypothalamus. Neuroscience 15:379–388. doi:10.1016/0306-4522(85)90220-9
9. de Lecea L, Kilduff TS, Peyron C et al (1998) The hypocretins: hypothalamus-specific peptides with neuroexcitatory activity. Proc Natl Acad Sci 95:322–327. doi:10.1073/pnas.95.1.322
10. Sakurai T, Amemiya A, Ishii M et al (1998) Orexins and orexin receptors: a family of hypothalamic neuropeptides and G protein-coupled receptors that regulate feeding behavior. Cell 92:573–585. doi:10.1016/S0092-8674(00)80949-6
11. Chemelli RM, Willie JT, Sinton CM et al (1999) Narcolepsy in orexin knockout mice. Cell 98:437–451. doi:10.1016/S0092-8674(00)81973-X
12. Lin L, Faraco J, Li R et al (1999) The sleep disorder canine narcolepsy is caused by a mutation in the hypocretin (orexin) receptor 2 gene. Cell 98:365–376. doi:10.1016/S0092-8674(00)81965-0
13. Nishino S, Ripley B, Overeem S et al (2000) Hypocretin (orexin) deficiency in human narcolepsy. Lancet 355:39–40. doi:10.1016/S0140-6736(99)05582-8
14. Nishino S (2007) The hypocretin/orexin receptor: therapeutic prospective in sleep disorders. Expert Opin Investig Drugs 16(11):1785–1797
15. Sakurai T (2007) The neural circuit of orexin (hypocretin): maintaining sleep and wakefulness. Nat Rev Neurosci 8:171–181. doi:10.1038/nrn2092
16. Peyron C, Tighe DK, van den Pol AN et al (1998) Neurons containing hypocretin (orexin) project to multiple neuronal systems. J Neurosci 18:9996–10015

17. Fadel J, Deutch A (2002) Anatomical substrates of orexin–dopamine interactions: lateral hypothalamic projections to the ventral tegmental area. Neuroscience 111(2):379–387
18. Winsky-Sommerer R, Boutrel B, De Lecea L (2003) The role of the hypocretinergic system in the integration of networks that dictate the states of arousal. Drug News Perspect 16:504–512
19. Harris GC, Wimmer M, Aston-Jones G (2005) A role for lateral hypothalamic orexin neurons in reward seeking. Nature 437:556–559. doi:10.1038/nature04071
20. Boutrel B, Kenny PJ, Specio SE et al (2005) Role for hypocretin in mediating stress-induced reinstatement of cocaine-seeking behavior. Proc Natl Acad Sci 102:19168–19173. doi:10.1073/pnas.0507480102
21. Lawrence AJ, Cowen MS, Yang H-J et al (2006) The orexin system regulates alcohol-seeking in rats. Br J Pharmacol 148:752–759. doi:10.1038/sj.bjp.0706789
22. Barson JR, Ho HT, Leibowitz SF (2015) Anterior thalamic paraventricular nucleus is involved in intermittent access ethanol drinking: role of orexin receptor 2. Addict Biol 20:469–481
23. Sterling ME, Karatayev O, Chang G-Q et al (2015) Model of voluntary ethanol intake in zebrafish: effect on behavior and hypothalamic orexigenic peptides. Behav Brain Res 278:29–39. doi:10.1016/j.bbr.2014.09.024
24. Morganstern I, Chang G-Q, Barson JR et al (2010) Differential effects of acute and chronic ethanol exposure on orexin expression in the perifornical lateral hypothalamus. Alcohol Clin Exp Res 34:886–896. doi:10.1111/j.1530-0277.2010.01161.x
25. Olney JJ, Navarro M, Thiele TE (2015) Binge-like consumption of ethanol and other salient reinforcers is blocked by orexin-1 receptor inhibition and leads to a reduction of hypothalamic orexin immunoreactivity. Alcohol Clin Exp Res 39:21–29. doi:10.1111/acer.12591
26. Holtz NA, Zlebnik NE, Carroll ME (2012) Differential orexin/hypocretin expression in addiction-prone and -resistant rats selectively bred for high (HiS) and low (LoS) saccharin intake. Neurosci Lett 522:12–15. doi:10.1016/j.neulet.2012.05.066
27. Dess NK, Badia-Elder NE, Thiele TE et al (1998) Ethanol consumption in rats selectively bred for differential saccharin intake. Alcohol 16:275–278
28. Perry JL, Morgan AD, Anker JJ et al (2006) Escalation of i.v. cocaine self-administration and reinstatement of cocaine-seeking behavior in rats bred for high and low saccharin intake. Psychopharmacology (Berl) 186:235–245. doi:10.1007/s00213-006-0371-x
29. Dess NK, Chapman CD, Monroe D (2009) Consumption of SC45647 and sucralose by rats selectively bred for high and low saccharin intake. Chem Senses 34:211–220. doi:10.1093/chemse/bjn078
30. Barson JR, Karatayev O, Gaysinskaya V et al (2012) Effect of dietary fatty acid composition on food intake, triglycerides, and hypothalamic peptides. Regul Pept 173:13–20. doi:10.1016/j.regpep.2011.08.012
31. Carvajal F, Alcaraz-Iborra M, Lerma-Cabrera JM et al (2015) Orexin receptor 1 signaling contributes to ethanol binge-like drinking: pharmacological and molecular evidence. Behav Brain Res 287:230–237. doi:10.1016/j.bbr.2015.03.046
32. Moorman DE, James MH, Kilroy EA, Aston-Jones G (2016) Orexin/hypocretin neuron activation is correlated with alcohol seeking and preference in a topographically specific manner. Eur J Neurosci. doi:10.1111/ejn.13170
33. Richards JK, Simms JA, Steensland P et al (2008) Inhibition of orexin-1/hypocretin-1 receptors inhibits yohimbine-induced reinstatement of ethanol and sucrose seeking in Long-Evans rats. Psychopharmacology (Berl) 199:109–117. doi:10.1007/s00213-008-1136-5
34. Shoblock JR, Welty N, Aluisio L et al (2011) Selective blockade of the orexin-2 receptor attenuates ethanol self-administration, place preference, and reinstatement. Psychopharmacology (Berl) 215:191–203. doi:10.1007/s00213-010-2127-x
35. Moorman DE, Aston-Jones G (2009) Orexin-1 receptor antagonism decreases ethanol consumption and preference selectively in high-ethanol–preferring Sprague–Dawley rats. Alcohol 43:379–386. doi:10.1016/j.alcohol.2009.07.002

36. Srinivasan S, Simms JA, Nielsen CK et al (2012) The dual orexin/hypocretin receptor antagonist, almorexant, in the ventral tegmental area attenuates ethanol self-administration. PLoS One 7:e44726. doi:10.1371/journal.pone.0044726
37. Anderson RI, Becker HC, Adams BL et al (2014) Orexin-1 and orexin-2 receptor antagonists reduce ethanol self-administration in high-drinking rodent models. Front Neurosci 8:33. doi:10.3389/fnins.2014.00033
38. Cason AM, Smith RJ, Tahsili-Fahadan P et al (2010) Role of orexin/hypocretin in reward-seeking and addiction: implications for obesity. Physiol Behav 100:419–428. doi:10.1016/j.physbeh.2010.03.009
39. Jupp B, Krstew E, Dezsi G, Lawrence AJ (2011) Discrete cue-conditioned alcohol-seeking after protracted abstinence: pattern of neural activation and involvement of orexin1 receptors. Br J Pharmacol 162:880–889. doi:10.1111/j.1476-5381.2010.01088.x
40. Thiele TE, Navarro M (2014) "Drinking in the dark" (DID) procedures: a model of binge-like ethanol drinking in non-dependent mice. Alcohol 48:235–241
41. McElhinny CJ, Lewin AH, Mascarella SW et al (2012) Hydrolytic instability of the important orexin 1 receptor antagonist SB-334867: possible confounding effects on in vivo and in vitro studies. Bioorg Med Chem Lett 22:6661–6664. doi:10.1016/j.bmcl.2012.08.109
42. Brown R, Khoo S, Lawrence A (2013) Central orexin (hypocretin) 2 receptor antagonism reduces ethanol self-administration, but not cue-conditioned ethanol-seeking, in ethanol-preferring rats. Int J Neuropsychopharmacol 16:2067–2079. doi:10.1017/S1461145713000333
43. Sterling ME, Chang G-Q, Karatayev O et al (2016) Effects of embryonic ethanol exposure at low doses on neuronal development, voluntary ethanol consumption and related behaviors in larval and adult zebrafish: role of hypothalamic orexigenic peptides. Behav Brain Res. doi:10.1016/j.bbr.2016.01.013
44. Steiner MA, Lecourt H, Strasser DS et al (2011) Differential effects of the dual orexin receptor antagonist almorexant and the GABA(A)-α1 receptor modulator zolpidem, alone or combined with ethanol, on motor performance in the rat. Neuropsychopharmacology 36:848–856. doi:10.1038/npp.2010.224
45. Brisbare-Roch C, Dingemanse J, Koberstein R et al (2007) Promotion of sleep by targeting the orexin system in rats, dogs and humans. Nat Med 13:150–155. doi:10.1038/nm1544
46. Schneider ER, Rada P, Darby RD et al (2007) Orexigenic peptides and alcohol intake: differential effects of orexin, galanin, and ghrelin. Alcohol Clin Exp Res 31:1858–1865. doi:10.1111/j.1530-0277.2007.00510.x
47. Chen Y-W, Barson JR, Chen A, Hoebel BG, Leibowitz SF (2014) Hypothalamic peptides controlling alcohol intake: differential effects on microstructure of drinking bouts. Alcohol 48:657–664
48. Martin-Fardon R, Weiss F (2014) N-(2-methyl-6-benzoxazolyl)-N′-1,5-naphthyridin-4-yl urea (SB334867), a hypocretin receptor-1 antagonist, preferentially prevents ethanol seeking: comparison with natural reward seeking. Addict Biol 19:233–236. doi:10.1111/j.1369-1600.2012.00480.x
49. Mahler SV, Aston-Jones GS (2012) Fos activation of selective afferents to ventral tegmental area during cue-induced reinstatement of cocaine seeking in rats. J Neurosci 32:13309–13326. doi:10.1523/JNEUROSCI.2277-12.2012
50. Mayannavar S, Rashmi KS, Rao YD, Yadav S, Ganaraja B (2016) Effect of Orexin A antagonist (SB-334867) infusion into the nucleus accumbens on consummatory behavior and alcohol preference in Wistar rats. Indian J Pharm 48:53
51. Dhaher R, Hauser SR, Getachew B et al (2010) The orexin-1 receptor antagonist SB-334867 reduces alcohol relapse drinking, but not alcohol-seeking, in alcohol-preferring (P) rats. J Addict Med 4:153–159. doi:10.1097/ADM.0b013e3181bd893f
52. Dayas CV, McGranahan TM, Martin-Fardon R, Weiss F (2008) Stimuli linked to ethanol availability activate hypothalamic CART and orexin neurons in a reinstatement model of relapse. Biol Psychiatry 63:152–157. doi:10.1016/j.biopsych.2007.02.002

53. Brown RM, Kim AK, Khoo SY-S et al (2015) Orexin-1 receptor signalling in the prelimbic cortex and ventral tegmental area regulates cue-induced reinstatement of ethanol-seeking in iP rats. Addict Biol. doi:10.1111/adb.12251
54. Millan EZ, Furlong TM, McNally GP (2010) Accumbens shell-hypothalamus interactions mediate extinction of alcohol seeking. J Neurosci 30:4626–4635. doi:10.1523/JNEUROSCI.4933-09.2010
55. Schacht JP, Anton RF, Myrick H (2013) Functional neuroimaging studies of alcohol cue reactivity: a quantitative meta-analysis and systematic review. Addict Biol 18:121–133. doi:10.1111/j.1369-1600.2012.00464.x
56. Hamlin AS, Newby J, McNally GP (2007) The neural correlates and role of D1 dopamine receptors in renewal of extinguished alcohol-seeking. Neuroscience 146:525–536. doi:10.1016/j.neuroscience.2007.01.063
57. Prasad AA, McNally GP (2014) Effects of vivo morpholino knockdown of lateral hypothalamus orexin/hypocretin on renewal of alcohol seeking. PLoS One 9:e110385. doi:10.1371/journal.pone.0110385
58. Marchant NJ, Hamlin AS, McNally GP (2009) Lateral hypothalamus is required for context-induced reinstatement of extinguished reward seeking. J Neurosci 29:1331–1342. doi:10.1523/JNEUROSCI.5194-08.2009
59. Sinha R (2001) How does stress increase risk of drug abuse and relapse? Psychopharmacology (Berl) 158:343–359. doi:10.1007/s002130100917
60. Samson WK, Taylor MM, Follwell M, Ferguson AV (2002) Orexin actions in hypothalamic paraventricular nucleus: physiological consequences and cellular correlates. Regul Pept 104:97–103. doi:10.1016/S0167-0115(01)00353-6
61. Kuru M, Ueta Y, Serino R et al (2000) Centrally administered orexin/hypocretin activates HPA axis in rats. Neuroreport 11:1977–1980. doi:10.1097/00001756-200006260-00034
62. Winsky-Sommerer R, Yamanaka A, Diano S et al (2004) Interaction between the corticotropin-releasing factor system and hypocretins (orexins): a novel circuit mediating stress response. J Neurosci 24:11439–11448. doi:10.1523/JNEUROSCI.3459-04.2004
63. Wang B, You Z-B, Wise RA (2009) Reinstatement of cocaine seeking by hypocretin (orexin) in the ventral tegmental area: independence from the local corticotropin-releasing factor network. Biol Psychiatry 65:857–862. doi:10.1016/j.biopsych.2009.01.018
64. Plaza-Zabala A, Martín-García E, de Lecea L et al (2010) Hypocretins regulate the anxiogenic-like effects of nicotine and induce reinstatement of nicotine-seeking behavior. J Neurosci 30:2300–2310. doi:10.1523/JNEUROSCI.5724-09.2010
65. Blasiak A, Siwiec M, Grabowiecka A et al (2015) Excitatory orexinergic innervation of rat nucleus incertus – implications for ascending arousal, motivation and feeding control. Neuropharmacology 99:432–447. doi:10.1016/j.neuropharm.2015.08.014
66. Kastman HE, Blasiak A, Walker L, Siwiec M, Krstew EV, Gundlach AL, Lawrence AJ (2016) Nucleus incertus Orexin2 receptors mediate alcohol seeking in rats. Neuropharmacology 110:82–91
67. Ryan PJ, Kastman HE, Krstew EV et al (2013) Relaxin-3/RXFP3 system regulates alcohol-seeking. Proc Natl Acad Sci U S A 110:20789–20794. doi:10.1073/pnas.1317807110
68. Walker AW, Smith CM, Chua BE et al (2015) Relaxin-3 receptor (RXFP3) signalling mediates stress-related alcohol preference in mice. PLoS One 10:e0122504. doi:10.1371/journal.pone.0122504
69. Chen Y-W, Fiscella KA, Bacharach SZ et al (2015) Effect of yohimbine on reinstatement of operant responding in rats is dependent on cue contingency but not food reward history. Addict Biol 20:690–700. doi:10.1111/adb.12164
70. McDougle CJ, Price LH, Heninger GR et al (1995) Noradrenergic response to acute ethanol administration in heathly subjects: comparison with intravenous yohimbine. Psychopharmacology (Berl) 118:127–135. doi:10.1007/BF02245830

71. Umhau JC, Schwandt ML, Usala J et al (2011) Pharmacologically induced alcohol craving in treatment seeking alcoholics correlates with alcoholism severity, but is insensitive to acamprosate. Neuropsychopharmacology 36:1178–1186. doi:10.1038/npp.2010.253

72. See RE, Waters RP (2010) Pharmacologically-induced stress: a cross-species probe for translational research in drug addiction and relapse. Am J Transl Res 3:81–89

73. Mantsch JR, Baker DA, Funk D et al (2016) Stress-induced reinstatement of drug seeking: 20 years of progress. Neuropsychopharmacology 41:335–356. doi:10.1038/npp.2015.142

74. Lê AD, Harding S, Juzytsch W et al (2000) The role of corticotrophin-releasing factor in stress-induced relapse to alcohol-seeking behavior in rats. Psychopharmacology (Berl) 150:317–324. doi:10.1007/s002130000411

75. Funk D, Li Z, Lê AD (2006) Effects of environmental and pharmacological stressors on c-fos and corticotropin-releasing factor mRNA in rat brain: relationship to the reinstatement of alcohol seeking. Neuroscience 138:235–243. doi:10.1016/j.neuroscience.2005.10.062

76. Marinelli PW, Funk D, Juzytsch W et al (2007) The CRF1 receptor antagonist antalarmin attenuates yohimbine-induced increases in operant alcohol self-administration and reinstatement of alcohol seeking in rats. Psychopharmacology (Berl) 195:345–355. doi:10.1007/s00213-007-0905-x

77. Zhou L, Smith RJ, Do PH et al (2012) Repeated orexin 1 receptor antagonism effects on cocaine seeking in rats. Neuropharmacology 63:1201–1207. doi:10.1016/j.neuropharm.2012.07.044

78. Ziółkowski M, Czarnecki D, Budzyński J et al (2015) Orexin in patients with alcohol dependence treated for relapse prevention: a pilot study. Alcohol Alcohol. doi:10.1093/alcalc/agv129

79. Voorhees CM, Cunningham CL (2011) Involvement of the orexin/hypocretin system in ethanol conditioned place preference. Psychopharmacology (Berl) 214:805–818. doi:10.1007/s00213-010-2082-6

80. Becker H (1999) Alcohol withdrawal: neuroadaptation and sensitization. CNS Spectr 4:38–40, 57–65

81. Littleton J (1999) Neurochemical mechanisms underlying alcohol withdrawal. Alcohol Res Health 22:13

82. Georgescu D, Zachariou V, Barrot M et al (2003) Involvement of the lateral hypothalamic peptide orexin in morphine dependence and withdrawal. J Neurosci 23:3106–3111

83. Sharf R, Sarhan M, Dileone RJ (2008) Orexin mediates the expression of precipitated morphine withdrawal and concurrent activation of the nucleus accumbens shell. Biol Psychiatry 64:175–183. doi:10.1016/j.biopsych.2008.03.006

84. von der Goltz C, Koopmann A, Dinter C et al (2011) Involvement of orexin in the regulation of stress, depression and reward in alcohol dependence. Horm Behav 60:644–650. doi:10.1016/j.yhbeh.2011.08.017

85. Bayerlein K, Kraus T, Leinonen I et al (2011) Orexin A expression and promoter methylation in patients with alcohol dependence comparing acute and protracted withdrawal. Alcohol 45:541–547. doi:10.1016/j.alcohol.2011.02.306

86. Narita M, Nagumo Y, Hashimoto S et al (2006) Direct involvement of orexinergic systems in the activation of the mesolimbic dopamine pathway and related behaviors induced by morphine. J Neurosci 26:398–405. doi:10.1523/JNEUROSCI.2761-05.2006

87. Mang GM, Dürst T, Bürki H et al (2012) The dual orexin receptor antagonist almorexant induces sleep and decreases orexin-induced locomotion by blocking orexin 2 receptors. Sleep 35:1625–1635. doi:10.5665/sleep.2232

88. Rodgers RJ, Halford JCG, Nunes de Souza RL et al (2001) SB-334867, a selective orexin-1 receptor antagonist, enhances behavioural satiety and blocks the hyperphagic effect of orexin-A in rats. Eur J Neurosci 13:1444–1452. doi:10.1046/j.0953-816x.2001.01518.x

89. Hoch M, Hay JL, Hoever P et al (2013) Dual orexin receptor antagonism by almorexant does not potentiate impairing effects of alcohol in humans. Eur Neuropsychopharmacol 23:107–117. doi:10.1016/j.euroneuro.2012.04.012

90. Robinson TE, Berridge KC (2008) Review. The incentive sensitization theory of addiction: some current issues. Philos Trans R Soc Lond B Biol Sci 363:3137–3146. doi:10.1098/rstb. 2008.0093
91. Macedo GC, Kawakami SE, Vignoli T et al (2013) The influence of orexins on ethanol-induced behavioral sensitization in male mice. Neurosci Lett 551:84–88. doi:10.1016/j.neulet. 2013.07.010
92. Winrow CJ, Tanis KQ, Reiss DR et al (2010) Orexin receptor antagonism prevents transcriptional and behavioral plasticity resulting from stimulant exposure. Neuropharmacology 58:185–194. doi:10.1016/j.neuropharm.2009.07.008
93. Borgland SL, Taha SA, Sarti F et al (2006) Orexin A in the VTA is critical for the induction of synaptic plasticity and behavioral sensitization to cocaine. Neuron 49:589–601. doi:10.1016/j.neuron.2006.01.016
94. Hutcheson DM, Quarta D, Halbout B et al (2011) Orexin-1 receptor antagonist SB-334867 reduces the acquisition and expression of cocaine-conditioned reinforcement and the expression of amphetamine-conditioned reward. Behav Pharmacol 22:173–181. doi:10.1097/FBP. 0b013e328343d761
95. Thompson JL, Borgland SL (2011) A role for hypocretin/orexin in motivation. Behav Brain Res 217:446–453. doi:10.1016/j.bbr.2010.09.028
96. Stettner GM, Kubin L, Volgin DV (2011) Antagonism of orexin 1 receptors eliminates motor hyperactivity and improves homing response acquisition in juvenile rats exposed to alcohol during early postnatal period. Behav Brain Res 221:324–328. doi:10.1016/j.bbr.2011.03.028
97. Chang G-Q, Karatayev O, Liang SC et al (2012) Prenatal ethanol exposure stimulates neurogenesis in hypothalamic and limbic peptide systems: possible mechanism for offspring ethanol overconsumption. Neuroscience 222:417–428. doi:10.1016/j.neuroscience.2012.05.066
98. Korotkova TM, Eriksson KS, Haas HL, Brown RE (2002) Selective excitation of GABAergic neurons in the substantia nigra of the rat by orexin/hypocretin in vitro. Regul Pept 104:83–89. doi:10.1016/S0167-0115(01)00323-8
99. Martin G, Fabre V, Siggins GR, de Lecea L (2002) Interaction of the hypocretins with neurotransmitters in the nucleus accumbens. Regul Pept 104:111–117. doi:10.1016/S0167-0115(01)00354-8
100. Schmitt O, Usunoff KG, Lazarov NE et al (2012) Orexinergic innervation of the extended amygdala and basal ganglia in the rat. Brain Struct Funct 217:233–256. doi:10.1007/s00429-011-0343-8
101. Chen Y-W, Barson JR, Chen A et al (2013) Glutamatergic input to the lateral hypothalamus stimulates ethanol intake: role of orexin and melanin-concentrating hormone. Alcohol Clin Exp Res 37:123–131. doi:10.1111/j.1530-0277.2012.01854.x
102. Cannella N, Economidou D, Kallupi M et al (2009) Persistent increase of alcohol-seeking evoked by neuropeptide S: an effect mediated by the hypothalamic hypocretin system. Neuropsychopharmacology 34:2125–2134. doi:10.1038/npp.2009.37
103. Ubaldi M, Giordano A, Severi I et al (2015) Activation of hypocretin-1/orexin-A neurons projecting to the bed nucleus of the stria terminalis and paraventricular nucleus is critical for reinstatement of alcohol seeking by neuropeptide S. Biol Psychiatry. doi:10.1016/j.biopsych. 2015.04.021
104. Barson JR, Poon K, Ho HT, Alam MI, Sanzalone L, Leibowitz SF (2015) Substance P in the anterior thalamic paraventricular nucleus: promotion of ethanol drinking in response to orexin from the hypothalamus. Addict Biol. doi:10.1111/adb.12288
105. Marcus JN, Aschkenasi CJ, Lee CE et al (2001) Differential expression of orexin receptors 1 and 2 in the rat brain. J Comp Neurol 435:6–25. doi:10.1002/cne.1190
106. Zhu Y, Miwa Y, Yamanaka A et al (2003) Orexin receptor type-1 couples exclusively to pertussis toxin-insensitive G-proteins, while orexin receptor type-2 couples to both pertussis toxin-sensitive and -insensitive G-proteins. J Pharmacol Sci 92:259–266. doi:10.1254/jphs.92.259
107. Harris GC, Aston-Jones G (2006) Arousal and reward: a dichotomy in orexin function. Trends Neurosci 29:571–577

108. Baldo BA, Gual-Bonilla L, Sijapati K et al (2004) Activation of a subpopulation of orexin/hypocretin-containing hypothalamic neurons by GABAA receptor-mediated inhibition of the nucleus accumbens shell, but not by exposure to a novel environment. Eur J Neurosci 19:376–386. doi:10.1111/j.1460-9568.2004.03093.x

109. Harris GC, Wimmer M, Randall-Thompson JF, Aston-Jones G (2007) Lateral hypothalamic orexin neurons are critically involved in learning to associate an environment with morphine reward. Behav Brain Res 183:43–51. doi:10.1016/j.bbr.2007.05.025

110. Lee EY, Lee HS (2016) Dual projections of single orexin- or CART-immunoreactive, lateral hypothalamic neurons to the paraventricular thalamic nucleus and nucleus accumbens shell in the rat: light microscopic study. Brain Res. doi:10.1016/j.brainres.2015.12.062

111. Barson JR, Leibowitz SF (2016) Hypothalamic neuropeptide signaling in alcohol addiction. Prog Neuropsychopharmacology Biol Psychiatry 65:321–329

112. Muschamp JW, Hollander JA, Thompson JL et al (2014) Hypocretin (orexin) facilitates reward by attenuating the antireward effects of its cotransmitter dynorphin in ventral tegmental area. Proc Natl Acad Sci U S A 111:E1648–E1655. doi:10.1073/pnas.1315542111

113. Khoo SY-S, Brown RM (2014) Orexin/hypocretin based pharmacotherapies for the treatment of addiction: DORA or SORA? CNS Drugs 28:713–730

A Decade of Orexin/Hypocretin and Addiction: Where Are We Now?

Morgan H. James, Stephen V. Mahler, David E. Moorman, and Gary Aston-Jones

Abstract One decade ago, our laboratory provided the first direct evidence linking orexin/hypocretin signaling with drug seeking by showing that activation of these neurons promotes conditioned morphine-seeking behavior. In the years since, contributions from many investigators have revealed roles for orexins in addiction for all drugs of abuse tested, but only under select circumstances. We recently proposed that orexins play a fundamentally unified role in coordinating "motivational activation" under numerous behavioral conditions, and here we unpack this hypothesis as it applies to drug addiction. We describe evidence collected over the past 10 years that elaborates the role of orexin in drug seeking under circumstances where high levels of effort are required to obtain the drug, or when motivation for drug reward is augmented by the presence of external stimuli like drug-associated cues/contexts or stressors. Evidence from studies using traditional self-administration and reinstatement models, as well as behavioral economic analyses of drug demand elasticity,

M.H. James
Brain Health Institute, Rutgers University/Rutgers Biomedical and Health Sciences, Piscataway, NJ 08854, USA
Florey Institute of Neuroscience and Mental Health, University of Melbourne, Parkville, VIC 2337, Australia

S.V. Mahler
Department of Neurobiology and Behavior, University of California, Irvine, Irvine, CA 92967, USA

D.E. Moorman
Department of Psychological and Brain Sciences & Neuroscience and Behavior Graduate Program, University of Massachusetts Amherst, Amherst, MA 01003, USA

G. Aston-Jones (✉)
Brain Health Institute, Rutgers University/Rutgers Biomedical and Health Sciences, Piscataway, NJ 08854, USA
e-mail: gsa35@ca.rutgers.edu

© Springer International Publishing AG 2016
Curr Topics Behav Neurosci (2017) 33: 247–282
DOI 10.1007/7854_2016_57
Published Online: 18 December 2016

247

clearly delineates a role for orexin in modulating motivational, rather than the primary reinforcing aspects of drug reward. We also discuss the anatomical interconnectedness of the orexin system with wider motivation and reward circuits, with a particular focus on how orexin modulates prefrontal and other glutamatergic inputs onto ventral tegmental area dopamine neurons. Last, we look ahead to the next decade of the research in this area, highlighting the recent FDA approval of the dual orexin receptor antagonist suvorexant (Belsomra®) for the treatment of insomnia as a promising sign of the potential clinical utility of orexin-based therapies for the treatment of addiction.

Keywords Addiction • Alcohol • Behavioral economics • Cocaine • Dopamine • Drugs of abuse • Glutamate • Heroin • Hypocretin • Motivation • Orexin • Reward • VTA

Contents

1   Introduction

Orexin peptides A and B (also called hypocretin 1 and 2) are produced by a limited number of neurons within the hypothalamus, ranging mediolaterally from the dorsomedial hypothalamic nucleus (DMH) through the perifornical area (PFA) into the lateral hypothalamus (LH) proper [1, 2]. Orexins signal via two G-protein-coupled receptors (orexin 1 and 2 receptors; Ox1R and Ox2R) that are distributed throughout the central nervous system [1, 2]. Orexin A has equal affinity for Ox1R and Ox2R, whereas orexin-B preferentially binds Ox2R [2]. Consistent with their anatomical position within the lateral hypothalamus, orexin neurons were originally identified as playing a key role in feeding behavior, with exogenous application of orexins shown to increase food intake [2, 3], and this effect being blocked by antagonism of orexin receptors [4]. This role for orexins in appetitive behavior has since been extended to a range of reward-related processes, including drug abuse and addiction [5–12].

Indirect evidence for a role for orexins in addiction came from a series of interesting clinical observations of patients suffering from narcolepsy with cataplexy, a neurological disorder associated with frequent sleep/wake transitions resulting from an almost complete loss of neurons producing orexin [13, 14]. A number of clinical case studies reported that narcoleptic patients rarely develop stimulant abuse, despite long-term treatment with stimulant medications including amphetamines [15, 16]. Consistent with this, Georgescu et al. [17] reported that orexin cells are activated (as seen with the immediate early gene product, Fos) after chronic morphine and withdrawal, and orexin knockout mice develop attenuated physical signs of morphine dependence compared to wild type controls.

In 2005, our laboratory directly implicated orexin signaling in drug-seeking behavior by showing that activation of LH orexin neurons is strongly correlated with conditioned preference for environmental contexts associated with cocaine or morphine [18] (Fig. 1). This study also showed that local chemical stimulation of LH orexin neurons reinstates extinguished drug seeking, and that this effect is blocked by systemic administration of an Ox1R antagonist (Fig. 1). Soon after, Boutrel and colleagues reported that central infusions of orexin peptide reinstate extinguished cocaine seeking and systemic injections of the Ox1R antagonist block cocaine seeking elicited by stress [19]. Together, these papers prompted a surge in preclinical studies examining the involvement of the orexins in various animal models of addiction, with >200 journal articles on orexin/hypocretin and addiction now published. As this number indicates, significant progress has been made in the last decade in understanding the specific roles for orexins in drug seeking and how they interact with the brain reward circuitry to drive these processes.

We recently articulated a hypothesis that the orexin system is preferentially engaged by situations of high motivational relevance, such as during physiological need states, exposure to threats, or reward opportunities [11]. We suggest that orexin neurons translate this "motivational activation" into organized suites of psychological and physiological processes that support adaptive behaviors (Fig. 2). With respect to addiction behaviors, we and others have argued that orexin neurons are not involved in modulating the primary reinforcing effects of drugs of abuse per se, but are critical to regulating motivated responding for these drugs [7, 11, 20]. Key to this is the observation that motivation for drug reward can be enhanced by various external stimuli, including drug-associated stimuli and stress, and that this occurs in an orexin-dependent manner [21].

Here, we seek to synthesize the findings of the past 10 years that support the "motivational activation" hypothesis as it relates to drug addiction. We present evidence from studies using a range of behavioral models of addiction that highlight a motivational role for orexin in drug seeking. Next, we provide an overview of how

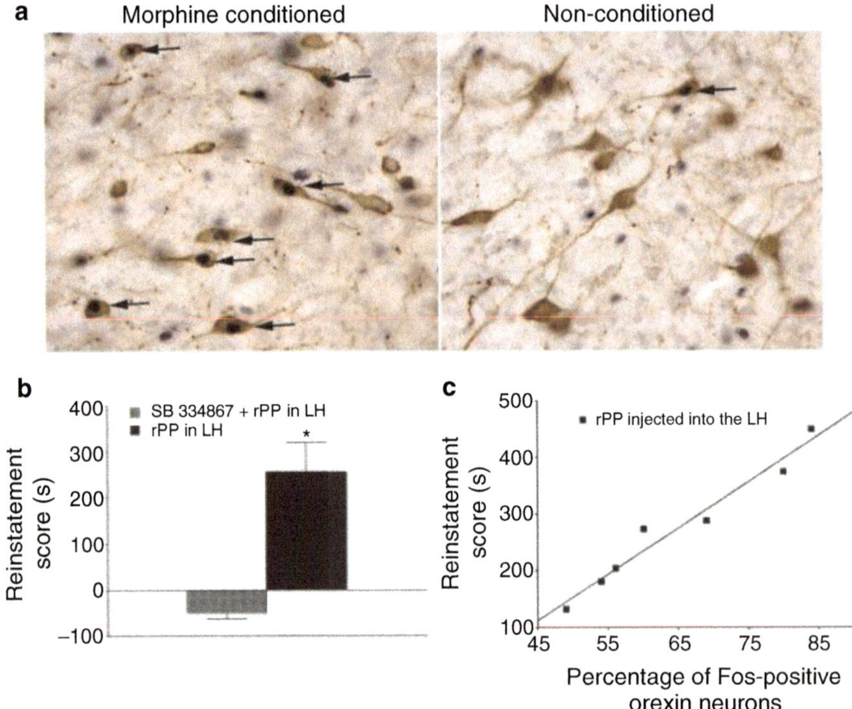

Fig. 1 First direct evidence that orexin signaling is involved in drug seeking behavior. (**a**) Using a two-chamber, non-biased, conditioned-place preference paradigm, animals were trained to associate one chamber with drug (morphine or cocaine) reward, whereas the other chamber was associated with no reward. Conditioned animals displayed reward-seeking by spending significantly more time in the reward-paired chamber. To determine if orexin neurons were stimulated during the expression of this preference, tissue was dual-labeled for both orexin and the immediate early gene protein Fos (a marker of neuronal stimulation). Only conditioned animals that exhibited a preference for the reward-paired chamber showed increased Fos expression in LH orexin cells. Similar percentages or orexin neurons in LH were activated following preference testing for morphine or cocaine. (**b**) Additional experiments sought to demonstrate that activation of orexin neurons could reinstate extinguished drug seeking. Animals were submitted to morphine conditioned place preference conditioning and then this behavior was extinguished by repeatedly exposing the rats to the chambers without morphine administration. To activate orexin neurons, animals received intra-LH infusions of the Y4 receptor agonist rPP (rat pancreatic polypeptide), as orexin neurons express the Y4 receptor and activation of this receptor by rPP induces Fos in these neurons. Infusions of rPP into LH robustly reinstated an extinguished morphine place preference, and this effect was blocked by pretreatment with the selective orexin 1 receptor antagonist SB-334867. (**c**) rPP-induced Fos expression in orexin neurons was strongly correlated with reinstatement score. Figure adapted from Harris et al. [18]

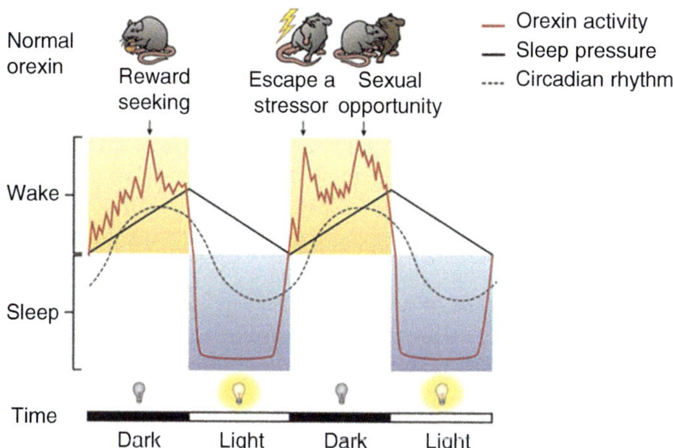

Fig. 2 "Motivational activation": a unifying hypothesis of orexin/hypocretin function. We propose that orexin neurons have a fundamentally unified role that spans a variety of processes, including arousal, reward seeking, homeostatic regulation, and stress, which we term "motivational activation." Orexin neuron activity (*red line*) has a circadian-like pattern but also exhibits phasic bursts during waking as a function of motivational state and adaptive behavior, such as reward seeking. We hypothesize that both suprachiasmatic nucleus-dependent tonic excitation and phasic orexin neuron burst firing help to produce and maintain wakefulness – the latter as a consequence of increased motivation to engage adaptive challenges or opportunities. Adapted from Mahler et al. [11]

orexin neurons interact with motivational and reward circuits to drive these processes, with a particular emphasis on how orexin neurons modulate the efficacy of glutamatergic inputs onto VTA neurons. We also discuss heterogeneity in orexin cell subpopulations, and the anatomical and functional characteristics that may contribute to the diversity of orexin neuron function. Finally, we seek to identify key gaps in our knowledge regarding orexin function in addiction that will be important to address in the coming decade of research in this field. Throughout this article, we focus primarily on psychostimulant drugs, as the role of orexin in alcohol use and abuse is covered elsewhere in this volume [22].

2 Orexin Signaling and Self-Administration Behavior

2.1 Traditional Self-Administration Models

Studies investigating the role of orexin in drug reward have frequently utilized the self-administration model, which allows investigators to examine responding for drug under schedules of reinforcement that differ in terms of the amount of effort required to obtain the drug. Under low fixed-ratio (FR) schedules of reinforcement (e.g., FR1, FR3), animals can easily maintain a preferred blood level of drug [23–25] and responding under these schedules allows insight into the primary reinforcing properties of the drug. Systemic administration of the selective Ox1R antagonist SB-334867 has no effect on cocaine or amphetamine self-administration under FR1 or FR3 schedules of reinforcement [19, 26–28], suggesting limited involvement of the orexin system in modulating the hedonic properties of psychostimulants.

Instead, orexin signaling is required for psychostimulant self-administration when higher levels of effort are required to obtain the drug. Systemic administration of SB-334867, as well as genetic deletion of Ox1R receptors, greatly reduces responding for cocaine on an FR5 schedule of reinforcement [29]. Similarly, under a progressive ratio (PR) schedule of reinforcement, where animals must expend exponentially increasing amounts of effort to obtain each subsequent drug infusion, systemic or intra-VTA injections of SB-334867 reduce the maximal effort the animal is willing the exert (as indicated by reduced breakpoints) [20]. Further, when motivation for cocaine is increased by intermittently restricting access to the drug, intra-VTA infusions of the orexin-A peptide increase, whereas intra-VTA infusions SB-334867 decrease, cocaine self-administration [26, 30]. These findings are consistent with our recent results using behavioral economic procedures (described in Section 4) [21]. Importantly, these effects appear to be mediated primarily by the Ox1R, as selective Ox2R antagonists have no effect on highly motivated responding for cocaine [31].

2.2 *Intracranial Self-Stimulation Models*

Findings from self-administration studies are closely mirrored in studies that utilize the intracranial self-stimulation (ICSS) paradigm. In this paradigm, instead of responding for a drug reward, animals are required to make an operant response (generally turning a wheel) to deliver a rewarding electrical current into LH. When animals are trained to respond for LH stimulation under a continuous reinforcement schedule that requires low levels of effort (FR1) to obtain rewarding stimulation, SB-334867 has no effect on stimulation reward threshold, nor does it affect the ability of cocaine to reduce the minimum level of electrical current necessary to support self-stimulation [32]. In contrast however, the effects of cocaine on ICSS threshold are blocked by systemic SB-334867 when animals are trained to respond for ICSS under a discrete trial current-threshold procedure, which requires greater degrees of effort and motivation to obtain ICSS [29].

2.3 *Conclusions*

In summary, data from self-administration and ICSS studies point to a unique role for orexin in driving responding for psychostimulant reward under conditions of enhanced motivation. Orexin is required for psychostimulant self-administration behavior or

ICSS responding only under schedules that require high amounts of effort to earn drug reward (FR5, PR). This is further supported by studies investigating orexin signaling in economic demand for psychostimulants (discussed in Section 4). In contrast, orexin is not necessary for the primary reinforcing effects of cocaine or amphetamine, or ICSS. Consistent with a more global role for orexin in modulating motivated behavior, SB-334867 also attenuates high-effort (FR5, PR) responding for other drugs of abuse, including alcohol [33–35], nicotine [36], and heroin [37], as well as high fat chocolate food (but not regular food or sucrose; [20, 33]. Unlike cocaine however, SB-334867 also affects low-effort (FR1 and FR3) responding for alcohol, nicotine, and heroin, indicating that there may be a role for orexin in mediating hedonic properties of some drugs [33, 36–38].

3 Orexin and Drug Seeking Elicited by External Stimuli

Environmental cues that are associated with drug use evoke motivational states that drive drug-seeking behavior, even in the absence of the drug. These cues engage the orexin system, which in turn acts to translate cue-induced motivation into appetitive actions. Studies investigating these processes utilize a variety of behavioral models, including conditioned place preference (CPP), operant reinstatement, and conditioned reinforcement paradigms. Exposure to stress is another potent stimulus known to drive drug seeking. Below, we discuss evidence from each of these paradigms that point to a central role for the orexin system in driving externally motivated drug-seeking behavior.

3.1 Orexin and Conditioned Place Preference

Harris et al. showed that activation of lateral (but not medial) hypothalamic orexin neurons is strongly correlated with CPP for cocaine and morphine [18] (Fig. 1). This study also reported that local chemical stimulation of LH orexin neurons reinstates extinguished CPP behavior, and that this effect was blocked by systemic administration of the Ox1R antagonist, SB-334867. Subsequent studies showed that systemic SB-334867 blocks the expression of CPP for amphetamine [27], and the dual OxR1 + OxR2 antagonist almorexant prevents the expression of cocaine CPP [39]. Expression of cocaine CPP depends on inputs to orexin neurons from the rostral lateral septum (LSr): Fos expression in LH-projecting LSr neurons is positively correlated with CPP, and inhibiting LSr cells prevents expression of CPP, and of Fos in LH orexin neurons [40]. Together, these findings are consistent with the notion that orexin neurons are engaged by stimuli (or environments) of high motivational relevance, and that this motivational activation is critical for the expression of reward-seeking behavior [11, 41].

3.2 *Reinstatement Behavior*

The orexin system is critically involved in the process of translating cue-induced motivation into appetitive actions. These processes are often probed experimentally using the reinstatement model of drug seeking whereby environmental cues paired with drug during self-administration are used to reinstate drug seeking following a period of abstinence or extinction training [12, 42]. Drug-associated cues are typically classified as either: (1) discrete; such as light or tone cues that are paired with cocaine infusions; (2) discriminative; whereby drug delivery is response-contingent but stimuli (lights, tones) predict drug availability (i.e., are not-response contingent); or (3) contextual; where animals are trained to self-administer and extinguish in two separate contexts with distinguishable compound cues (light, sound, olfactory or tactile cues), and reinstatement behavior is assessed by returning the animal to the original context where they learned to self-administer.

Systemic administration of SB-334867 dose-dependently blocks reinstatement of extinguished cocaine seeking elicited by discrete [21, 28, 43], discriminative [44], or contextual cues in rats [45] (Fig. 3). Similarly, systemic SB-334867 attenuates reinstatement of cocaine seeking when animals are returned to the self-administration chamber after either 1 day or 2 weeks of abstinence [45] (Fig. 3). These effects are driven by signaling at Ox1R, as pretreatment with the Ox2R antagonist 4PT had no effect on cue-induced cocaine seeking [27]. SB-334867 is also effective at blocking reinstatement of alcohol [34, 35, 46, 47], nicotine [48], and remifentanil [49] seeking, and there is evidence that Ox2R antagonists block reinstatement of nicotine seeking [50], but not cue-induced alcohol seeking [51].

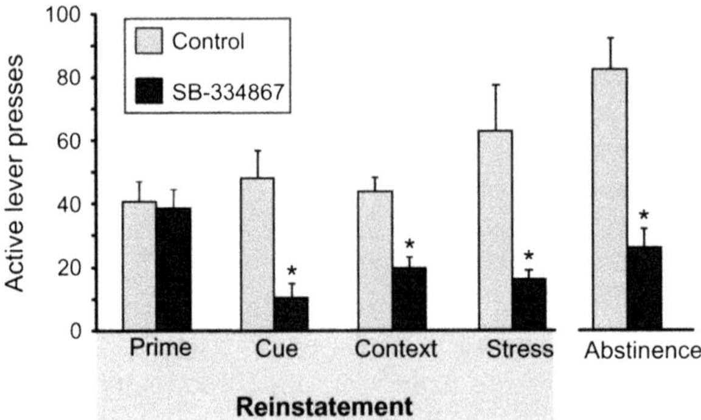

Fig. 3 Effects of SB-334867 treatment on reinstatement of cocaine seeking behavior. Systemic treatment with the orexin 1 receptor antagonist SB-334867 (30 mg/kg, i.p.) blocks reinstatement of cocaine seeking elicited by discrete cues, contextual cues, and stress. In contrast, SB-334867 has no effect on cocaine seeking elicited by a priming injection of cocaine. Figure adapted from Mahler et al. [12]

3.3 Conditioned Reinforcement

A model related to cue-induced drug seeking is the conditioned reinforcement (CR) paradigm, whereby animals learn to perform a novel response (e.g., nose poke) to receive a cue that was previously associated with drug delivery via another response (e.g., lever press) [52, 53]. Rats that were treated systemically with SB-334867 over 5 consecutive days of FR1 cocaine self-administration training showed attenuated responding for the cocaine-associated cue in a subsequent CR test [27]. This finding points to a role for orexin in ascribing motivational significance to discrete cues during self-administration training. In contrast however, Smith et al. [28] reported that SB-334867 administered systemically prior to a single cocaine-cue conditioning session had no effect on the ability of those cues to elicit reinstatement of cocaine seeking following extinction training. The reasons for the discrepancy in these two sets of findings are unclear, but may be related to differences in motivational significance acquired by the cue over a single versus repeated conditioning trials, or to different roles for orexin in operant vs Pavlovian conditioning. Importantly, treatment with SB-334867 prior to the CR test also attenuates responding [27], which is consistent with findings outlined above relating to cue-induced reinstatement, and supports a role for orexin in cue-driven motivated responding.

3.4 Stress

Orexin neurons are well positioned to modulate the behavioral and physiological responses to stress via bidirectional connections with various stress-relevant regions including bed nucleus of the stria terminalis (BNST), central amygdala (CeA), paraventricular nucleus of the hypothalamus (PVN), and medial prefrontal cortex (mPFC; for an overview, see James et al. [54]). Indeed, orexin neurons are activated by various stressors [8, 55–58] and optogenetic stimulation of orexin neurons reduces time spent in the interaction zone in a social interaction test and increases freezing in an open field test [59, 60]. Further, orexin receptor antagonists reduce anxiety behavior, and attenuate neuroendocrine responses to stress [48, 55, 61, 62]. These findings are consistent with a global role for orexin involvement in situations of high motivational relevance and translating this into adaptive behavior [7, 11, 20].

Drug-seeking behavior is strongly influenced by stress. Extinguished drug seeking can be reinstated by exposure to a range of stressors, including brief (15 min) intermittent, unpredictable electrical footshock [63, 64], food deprivation [65], or a range of pharmacological stressors, including the anxiogenic drug yohimbine [66, 67]. Systemic administration of SB-334867 reduces reinstatement of cocaine seeking elicited by yohimbine in both male and female rats [43]. Further, reinstatement of cocaine seeking elicited by intracerebroventricular (ICV) infusions of orexin-A is blocked by systemic administration of a non-selective CRF receptor antagonist [19]. The BNST appears to be an important brain region for these processes, as SB-334867 blocks yohimbine-induced changes in excitatory transmission in dorsal anterolateral BNST that are associated with the reinstatement of cocaine seeking, and these changes are absent in prepro-orexin knockout mice [68]. Orexin may also act at the nucleus accumbens shell (NAcSh) to mediate stress-induced drug seeking, as local infusions of either SB-334867 or TCS-OX2-29 block footshock-induced reinstatement of morphine seeking [69]. In contrast, the VTA and dentate gyrus of the hippocampus are unlikely sites of action, as local infusions of SB-334867 into these sites do not affect stress-induced seeking of cocaine or morphine, respectively [70, 71].

3.5 Conclusions

In summary, orexin signaling is necessary for motivated psychostimulant seeking elicited by external triggers like discrete drug-associated cues, contexts associated with prior drug use, or stress. Such stimuli are known to elicit craving and relapse behavior in human addicts [72–77]. This may reflect a role for orexin in conditioned responding for highly salient natural rewards, including certain foods and sex [20, 78, 79]. The orexin system may therefore offer a potential target for pharmacotherapies designed to reduce the risk of relapse during abstinence [10, 80, 81]. We propose that by decreasing OX1R signaling, and thereby attenuating responses of VTA dopamine neurons to cues associated with these rewards (discussed in Section 5.2.1), it may be possible to preferentially attenuate exaggerated craving responses without greatly affecting baseline reward processing. Indeed, many preclinical studies now suggest that natural reward seeking is unaffected following treatment with orexin receptor antagonists at doses that block drug-seeking behavior [9, 33, 44, 46].

4 Orexin and Behavioral Economics

4.1 Behavioral Economics Models

Behavioral-economic (BE) analyses of demand elasticity for drug and other rewards offer a sophisticated means by which to examine separately both reinforcing properties of a drug, and motivation to obtain and consume it. In a typical BE experiment, consumption of drug is plotted as a function of drug price, and both humans and animals are similarly responsive to increased drug price (in humans, monetary cost; in rats, lever presses). In rats, we employ a within-session BE paradigm [82, 83], and apply an exponential demand equation to generate estimates of preferred cocaine intake at zero cost (Q_0, quantified by where computed demand curves intercept the ordinate) [84]. As such, Q_0 can be considered a measure of hedonic set point for a drug, or the consumption of a drug under 'free' (no cost) conditions. In addition, the slope of the demand curve conveys demand elasticity, or the sensitivity of drug seeking to increases in price. This elasticity parameter, termed alpha, is therefore an index of the animal's motivation to obtain drug, or to defend preferred blood levels of drug. Importantly, Q_0 and alpha are entirely orthogonal in cocaine demand tests, meaning that each parameter can be examined concurrently and independently [82, 85].

As described above, orexin is specifically involved in modulating motivated responding, as well as drug seeking elicited by external stimuli. A recent study in our laboratory used BE demand curve analysis to investigate these processes directly [21]. Different groups of rats were trained to self-administer cocaine with and without drug-paired discrete stimuli (light + tone), and demand curves for both groups of rats were calculated after pretreatment with SB-334867 or vehicle. Motivation (alpha) for cocaine was significantly higher when cocaine-associated cues were present and systemic administration of SB-334867 decreased motivation (increased alpha values) only in the presence of these cocaine-associated cues (Fig. 4). Further, SB-334867 had no effect on the hedonic set point (Q_0) for cocaine, regardless of whether cocaine-associated cues were present. These findings are highly consistent with a previous study using a similar BE thresholding procedure to demonstrate that SB-334867 does not affect cocaine consumption, but reduces the maximal number of responses that rats were willing to make to maintain self-administration (P_{Max}) [26].

The observation that cocaine-paired discrete stimuli increased motivation for cocaine reward is consistent with previous findings that drug-paired stimuli take on motivational properties that drive responding independently of the reinforcing effects of the cocaine itself [86]. Interestingly, this cue-driven enhancement of motivation for cocaine was blocked by systemic SB-334867, indicating that signaling at Ox1Rs increases the reinforcing efficacy of cocaine-associated cues. These findings complement those outlined above, showing an important role for orexin signaling in stimulus-

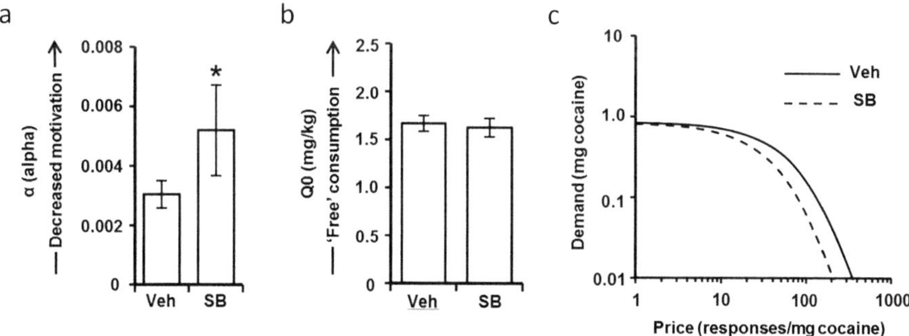

Fig. 4 Effects of SB-334867 treatment on economic demand for cocaine. (**a**) Pretreatment with SB-334867 increased cocaine demand elasticity (α), reflecting a decrease in motivation for cocaine. (**b**) SB-334867 had no effect on free cocaine consumption (Q_0), consistent with previous studies showing that the orexin system is not involved in mediating the hedonic properties of cocaine. (**c**) Demand curves representing increased sensitivity of demand to price following SB-334867 treatment. Figure adapted from Bentzley et al. [21]

driven drug seeking. In contrast, blockade of orexin signaling had no effect on "free" consumption of cocaine (Q_0). Together, these findings are strongly consistent with the "motivational activation" hypothesis of orexin function [11].

5 Interaction Between Orexin Neurons and Brain Reward Circuitry

Orexin signaling is crucial for highly motivated and conditioned drug seeking, but how do orexin neurons interact with wider reward circuits in the brain to modulate motivation? Examination of the afferent and efferent projection patterns of orexin neurons reveals that these cells are highly interconnected with key motivational and reward regions. The topography of the orexin system is described in detail elsewhere [87, 88], as well as in other chapters of this volume. Here, we summarize studies that have provided anatomical or functional evidence of orexin-related circuits in psychostimulant-seeking behavior. In particular, we focus on orexin projections to the midbrain dopamine (DA) system, and how orexin modulates glutamatergic inputs to VTA, including those that arise from the prefrontal cortex.

5.1 Orexin Afferents

The first comprehensive study of inputs to orexin neurons was carried out by Sakurai and colleagues [89], using the retrograde tracer, tetanus toxin, that was genetically encoded to specifically reveal inputs to orexin neurons. Using this approach, the authors identified strong projections from various reward-related regions, including medial prefrontal cortex (mPFC), BNST, basal forebrain, basolateral and medial amygdala, arcuate nucleus, PVN, and median raphe nucleus (MRN). A subsequent study by Yoshida et al. [90] used the retrograde tracer cholera toxin B subunit (CTb) to identify afferents to LH and DMH orexin fields, and confirmed retrogradely labeled inputs using anterograde tracers and identification of appositions with orexin neurons. Using this approach, these authors confirmed many of the inputs identified by Sakurai et al., as well as strong projections from NAc shell, dorsolateral septum, ventral pallidum, central amygdala, VTA, and ventromedial and anterior hypothalamus. For a summary of afferent projections to orexin neurons from reward-related structures, see Fig. 5.

To identify inputs to orexin neurons that are important for reward seeking, our laboratory examined Fos expression in various forebrain regions during the expression of cocaine CPP in animals that had previously received injections of the retrograde tracer CTb into the lateral or medial orexin fields [40]. This approach revealed that neurons in the rostral lateral septum and ventral BNST that project to the lateral orexin field are uniquely activated during cocaine CPP, and the magnitude of this activation

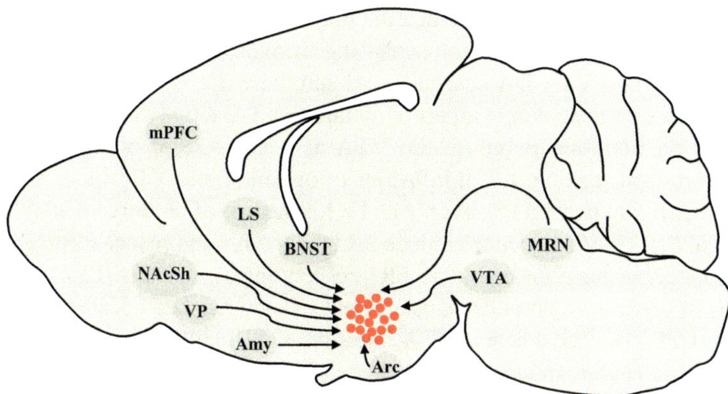

Fig. 5 Afferent projections to orexin neurons from reward-related structures. A mid-sagittal section of a rat brain illustrating loci involved in drug seeking that provide input onto orexin neurons. *Amy* amygdala (central, medial, and basolateral amygdala provide input to orexin neurons), *Arc* arcuate nucleus of the hypothalamus, *BNST* bed nucleus of the stria terminalis, *LS* lateral septum, *mPFC* medial prefrontal cortex (both prelimbic and infralimbic structures), *MRN* medial raphe nucleus, *NAcSh* nucleus accumbens shell, *VP* ventral pallidum, *VTA* ventral tegmental area

was correlated with the degree of CPP expression. In a follow-up experiment, bilateral disconnection of the septum-LH orexin pathway attenuated the expression of cocaine CPP, confirming a functional role for this input in the expression of reward seeking [40].

Very few studies have otherwise been carried out examining the functional role of orexin afferents in reward seeking. One region of potential interest is the NAc shell, as this region projects strongly to the LH orexin area [90–92] and inactivation of this region induces Fos expression in the LH orexin field and promotes alcoholic beer seeking [93] and feeding behavior [94]. Another study, while not identifying the source of inputs, showed that excitatory drive from glutamatergic neurons onto orexin neurons is increased following cocaine exposure [95]. Given the apparent role for the orexin system in driving motivational behavior, future studies may benefit from focusing on inputs from regions involved in motivational processes, including prefrontal cortex and striatopallidal structures.

5.2 Orexin Efferents

5.2.1 Ventral Tegmental Area

Neurons in the VTA, along with the neighboring substantia nigra pars compacta, are major sources of DA in the brain [96]. VTA DA neurons mediate the effects of drugs of abuse, including psychostimulants, and play a key role in modulating motivational processes of drug-seeking behavior [6, 97–99]. VTA neurons express both Ox1R and Ox2R, although there are a limited number of orexin-containing synapses in this region, suggesting that the majority of orexin input to VTA may be nonsynaptic *en passant* fibers or non-synapsing terminals [100]. Regardless, orexin-A directly depolarizes VTA DA neurons [101] and there is considerable evidence that orexin facilitates glutamatergic inputs to modulate VTA activity and reward seeking.

A growing literature points to the VTA as a key site of orexin signaling in motivated reward seeking. Local infusions of orexin A into VTA increase cocaine self-administration under a PR, but not an FR1, schedule of reinforcement [30]. Similarly, local SB-334867 injections reduce PR breakpoints and reduce effort expended in a BE task, but have no effect on FR1 responding for cocaine [26]. Intra-VTA infusions of orexin-A also elicit reinstatement of extinguished cocaine seeking and morphine CPP [18, 71], whereas VTA injection of SB-334867 blocks both discrete and discriminative cue-induced reinstatement of cocaine seeking [102, 103]. Similar findings have been reported for cue-induced reinstatement of ethanol seeking [104].

Orexin signaling in VTA increases DA release in forebrain structures like NAc, which is known to be important for reward seeking behavior [105–112]. Infusions of orexin-A into VTA increase DA release in prefrontal cortex and ventral striatum [30, 113], whereas infusions of SB-334867 reduce cocaine-induced DA in NAc [71], showing clear modulation of VTA DA neurons by orexin inputs.

VTA DA neuron firing is regulated by glutamate inputs [114–117], and orexin plays an important role in modulating these. Borgland et al. first demonstrated that orexin modulates glutamatergic signaling in VTA DA neurons [118]. Bath application of orexin-A facilitates glutamatergic transmission in VTA DA neurons, initially via facilitation of NMDA signaling, and later, via facilitation of AMPA signaling – both via postsynaptic mechanisms. VTA Ox1R antagonism also blocks cocaine-induced LTP of excitatory DA afferents in vitro, and blocks locomotor sensitization to cocaine in vivo. Orexin-B also enhances glutamatergic inputs to VTA DA neurons, but does so via presynaptic as well as postsynaptic mechanisms [119].

We followed up these findings by demonstrating a functional interaction between glutamate and orexin in VTA mediation of conditioned cocaine seeking [103]. First, we showed that intra-VTA SB-334867 blocked cue-induced reinstatement of cocaine seeking, as did intra-VTA blockade of AMPA receptors with CNQX. In addition, blockade of VTA orexin in one cerebral hemisphere, with simultaneous blockade of AMPA signaling in the contralateral hemisphere, similarly blocked cue reinstatement, indicating necessity of concurrent orexin and AMPA signaling for cues to elicit cocaine seeking. Moreover, we showed that reductions in cocaine seeking by intra-VTA SB-334867 were reversed when we concurrently administered the AMPA allosteric modulator PEPA into VTA, to potentiate AMPA signaling. Together, these results show that by facilitating AMPA signaling in DA neurons, orexin in VTA potentiates both DA neuron activity and motivated behavioral responses to cocaine cues. Although orexin potentiates NMDA signaling in VTA neurons as well [118, 119], we did not see reductions in cue-induced cocaine seeking following intra-VTA blockade of NMDA receptors with AP-5 microinjections. VTA NMDA transmission is most important for acquisition of reward learning (presumably via induction of LTP; [117]), whereas AMPA may instead mediate expression of cocaine seeking elicited by previously learned cocaine cues [12, 103]. Therefore, orexin is likely capable of facilitating both reward learning, and expression of motivation elicited by prior learning, via facilitation of glutamate transmission in VTA.

The sources of glutamate inputs to VTA DA neurons that might be facilitated by orexin include hypothalamic, mid- and hind-brain nuclei, and/or the large projection from prelimbic mPFC [120, 121]. Although mPFC electrical stimulation usually elicits only weak enhancements of DA neuron firing in anesthetized animals [122, 123], we asked whether orexin application to VTA would increase the efficacy of mPFC stimulation-induced DA neuron activation [124]. Indeed, application of a low concentration of orexin-A that does not itself affect DA neuron firing robustly facilitated DA neuron firing elicited by prelimbic stimulation. This demonstrates that mPFC inputs to VTA DA neurons are facilitated by orexin-A in vivo. This said, we note that we did not see activation (as measured with Fos) in VTA afferent neurons from either mPFC or hypothalamic orexin neurons themselves during cue-induced cocaine seeking (compared to various control behaviors; [125]). Intriguingly, non-orexin-A expressing-, VTA-projecting neurons in the LH orexin field were instead activated during cued cocaine seeking.

5.2.2 Other Efferent Regions

Beyond the VTA, only a handful of regions have been investigated as potential targets of orexin signaling in drug seeking, and the large majority of these studies have been carried out in alcohol-seeking models. Perhaps the next most comprehensively studied orexin efferent region is the paraventricular thalamus (PVT), a midline thalamic structure that is more densely innervated by orexin-positive fibers than any other forebrain structure [88]. The PVT expresses both Ox1R and Ox2R in high abundance [87], and bath application of orexin peptide potently depolarizes PVT neurons [126]. Recent evidence has implicated the PVT as a key structure in brain reward circuitry, and a clear role has been shown for this structure in a range of drug-seeking paradigms [9, 81, 127–133]. Dayas and colleagues provided the first evidence of an LH orexin-PVT circuit in drug seeking, by showing that PVT neurons that are activated by drug-associated cues are closely apposed by orexin-positive fibers [129]. More direct evidence has come from studies showing that infusions of orexin-A and -B increase ethanol drinking and reinstate extinguished cocaine-seeking behavior. These effects appear to be mediated primarily via Ox2Rs, as Ox2R (but not Ox1R) levels in PVT are increased following ethanol drinking, and local infusions of the Ox2R antagonist TCS-OX2-29, but not the OxR1 antagonist SB-334867, reduces alcohol drinking [128]. A recent study suggests that signaling at the OxR2 in the PVT is also important for psychostimulant seeking, as infusions of orexin A into the PVT reinstated cocaine seeking and this was blocked by co-administration of TCS-OX2-29 but not SB-334867 [134]. These findings are consistent with an earlier demonstration that infusions of SB-334867 into the PVT have no effect on reinsatement of cocaine seeking elicited by discriminative cues [102].

The NAc is another potentially important site of orexin action in drug seeking, as this region expresses moderate to high levels of Ox2R [87, 135] and both orexin-A and B alter activity of NAc neurons [136, 137]. The effects of intra-NAc infusions of orexin receptor agonists or antagonists have not been examined for psychostimulants, but local infusions of SB-334867 or TCS-29-029 in NAcSh prevent stress-induced reinstatement of morphine conditioned place preference [69], and injections of TCS-OX2-029 in NAcC reduced responding for ethanol [51]. Also, injections of orexin-A into the rostral half of medial NAcSh increase the hedonic "liking" for sucrose taste [138], as do orexin-A injections into the posterior half of the ventral pallidum [139].

The BNST may be another target for orexin signaling in drug reward, as infusions of orexin-A into this region enhance alcohol seeking [140]. Also, as described above, OxR1 signaling in BNST appears to play an important role in stress-induced drug seeking. More recently, stress-induced reinstatement of alcohol seeking was shown to be dependent on signaling at the OX_2R in the nucleus incertus, a small pontine region [141]. In the cortex, infusions of SB-334867 into the prelimbic cortex attenuate cue-induced reinstatement of ethanol seeking [104], and local infusions of SB-334867 in the insular cortex reduce nicotine self-administration behavior [36]. These findings highlight potential targets for orexin signaling beyond the mesocorticolimbic DA system that may be important in modulating psychostimulant-seeking behavior. For a summary of regions shown to be important sites of orexin signaling in drug seeking, see Fig. 6.

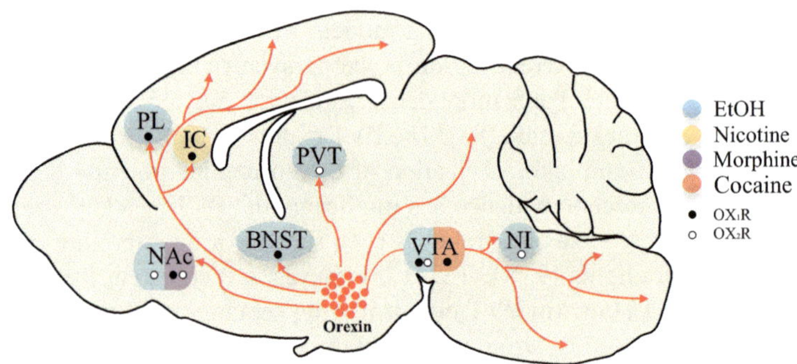

Fig. 6 Sites of orexin signaling in drug-seeking behavior. Orexin neurons in the lateral hypothalamus project to a number of reward structures involved in drug-seeking behavior. The ventral tegmental area (VTA) has been identified as an important site of orexin signaling in cocaine seeking behavior (*red*). Intra-VTA infusions of the orexin receptor 1 antagonist SB-334867 attenuate reinstatement of cocaine seeking elicited by discriminative stimuli, discrete stimuli, and stress [71, 102, 103]. Similarly, intra-VTA infusions of SB-334867 block cue-induced reinstatement of ethanol seeking [104], and infusions of the dual orexin receptor antagonist almorexant into the VTA reduce alcohol drinking [142]. Ethanol-seeking behaviors have been shown to be dependent on orexin signaling at other loci, including the prelimbic cortex (PL; [104]), nucleus accumbens (NAc; [51]), the anterior paraventricular thalamus (PVT; [128]), the bed nucleus of the stria terminalis [140], and the nucleus incertus (NI; [141]). Signaling at both the OX1R and OX2 in the NAc is important for stress-induced reinstatement of morphine conditioned place preference [69], and infusions of SB-334867 into the insular cortex (IC) reduce nicotine seeking behavior [36]. (PNAS)

6   Functional Heterogeneity of Orexin Neurons: A Medial/ Lateral Dichotomy in Orexin Cell Function?

Orexin neurons participate in a wide variety of functions including, but not limited to arousal, stress, and appetitive motivation [11, 143, 144]. A question arises as to how the orexin system participates in such diverse, and even apparently contradictory functions (e.g., stress and reward). One possibility is that there are heterogeneous subpopulations of orexin neurons that selectively participate in specific behavioral or physiological functions. There are multiple potential sources of heterogeneity in the orexin system. One of the more prominent factors appears to be differences in circuit functions of orexin neurons located in different hypothalamic subregions. Our laboratory proposed that orexin neurons in LH preferentially encode reward motivation whereas orexin neurons in DMH and PFA process arousal and stress [8]. This proposal was initially based on our demonstration that LH orexin neurons were Fos-activated by cues associated with food, morphine, and cocaine but not footshock, whereas PFA-DMH orexin neurons were preferentially activated by arousal/anxiety-inducing footshock stimulation [8]. Furthermore, chemical activation of LH, but not PFA-DMH, orexin neurons reinstated preference for morphine [18]. We also noted studies in which activation of LH, and not DMH, orexin neurons was associated with antipsychotic-induced weight gain [145] as well as increased feeding associated with NAcSh inhibition [94]. Conversely, PFA-DMH orexin neuron Fos activation was preferentially associated with wakefulness and arousal [145–147] as well as footshock, restraint, and cold stress [148, 149].

Since the publication of Harris and Aston-Jones [8], a number of studies have been performed investigating the activation of orexin neurons in a variety of appetitive and aversive contexts. In many cases, the distinction between lateral neurons encoding appetitive motivated behaviors and medial neurons encoding arousal or aversive motivated behaviors has been replicated. Work from our lab has shown preferential encoding of drug preference in LH orexin neurons [40, 150–152] and preferential signaling of arousal by PFA/DMH orexin neurons [153]. A number of other groups have also found preferential encoding of reward seeking by lateral orexin neurons. Context-induced reinstatement (ABA renewal) of beer seeking induced Fos activation of PFA and LH, but not DMH, orexin neurons, and the activity of LH orexin neurons (and not PFA/DMH) was correlated with strength of renewal [154]. Increased Fos activation of DMH and PFA, but not LH orexin neurons was observed following ABA renewal of cocaine seeking, though this increase was also seen in ABB controls in which animals were tested in their extinguished environment (B) instead of the conditioned environment (A) [91]. This is similar to the increased Fos observed in PFA and DMH (but not LH) orexin neurons during extinction from sucrose self-administration [78].

The increased reinstatement for cocaine seeking following ICV infusions of neuropeptide S depended primarily on LH orexin neurons [155]. Alcohol seeking induced by baclofen/muscimol infusions in the NAc induced Fos activation of PFA and LH, but not DMH, orexin neurons [93]. Morphine CPP preferentially activated LH, and not PFH or DMH, orexin neurons [156]. In a recent report, LH orexin neurons exhibited stronger firing in response to reward- vs. punishment-predicting cues [157], though the authors noted that LH orexin neurons also exhibited activity correlations with arousal in their previous studies [158].

Also supporting lateral vs. medial differences in orexin neuron function is the finding that medial orexin neurons preferentially encode arousing or aversive stimuli, events, and outcomes. Unpredictable chronic mild stress, a rodent model of depression, increased Fos expression in PFA/DMH orexin neurons, but not LH [159]. Naloxone-induced morphine withdrawal activates DMH and PFA, but not LH orexin neurons [160]. Anxiogenic doses of acute nicotine increased Fos in PFA/DMH but not LH, whereas cue-induced reinstatement of nicotine seeking increased Fos in PFA and LH but not DMH orexin neurons [48, 161]. DMH and PFA, but not LH neurons are activated by panic induced by sodium lactate, CO_2, caffeine, or the anxiogenic drug FG-7142 [62, 162–164]. Inflammation using formalin injections into the hindpaw increased Fos in DMH and PFA, but not LH orexin neurons, and this activation was correlated with the extent of licking and grooming responses [165].

However, a medial/lateral dichotomy in orexin neuron function has not been observed in every study (see Tables 1 and 2). Stressful situations such as exposure to a brightly lit novel environment induced Fos activation of medial and lateral orexin neurons [172], as did a range of behaviors in mice, including progressive ratio responding for food reward [173]. In one study, restraint stress resulted in increased Fos in LH but not PFA/DMH orexin neurons [166]. Discriminative stimulus-driven reinstatement of alcohol seeking increased Fos activity in DMH, PFA, and LH orexin neurons [129]. Similarly, in a conditioned sucrose-seeking model, orexin neuron activation was observed across all orexin cell fields in food-restricted rats [78]. Cue-evoked feeding in sated animals produced preferential activation of PFA orexin neurons [174, 175], as did expectation of chocolate [176], while feeding induced by DAMGO infusions in the medial prefrontal cortex [177] or NAc [178] activated medial but not lateral orexin neurons. Finally, in a recent study from our laboratory, we found that the strength of context-induced reinstatement of ethanol seeking was correlated with the strength of Fos activation of LH and DMH (but not PFA) orexin neurons [151]. Together, these results indicate that the lateral vs. medial dichotomy in orexin neuron function may not always be as simple as encoding appetitive vs. aversive behaviors, respectively. The role of the PFA orexin neurons, which in some cases appears to associate more with appetitive and in other cases with aversive behaviors, needs to be investigated further.

As shown in Table 2, there may be a dichotomy whereby Pavlovian reward-associated behaviors (such as conditioned place preference) primarily engage LH and not PFA orexin neurons. In contrast, instrumental reward seeking, often performed in a reinstatement context in which the primary reinforcer is unavailable, recruits both LH and PFA neurons. One possible explanation for this observation is that instrumental reinstatement in the absence of a reward produces both motivation to acquire reward along with frustration in not receiving reward, thus engaging both LH and PFA (and sometimes DMH) orexin neurons. Another possible explanation may arise from the fact that instrumental, and not Pavlovian, reward-related behaviors require increased arousal and/or motor activity, in addition to motivation for reward. The simultaneous increase in arousal/motor output as well as reward motivation thus activates both LH and PFA

Table 1 Summary of studies that have reported patterns of orexin neuron activation under conditions of stress or arousal

	Stimuli/paradigm	DMH orexin	PFA orexin	LH orexin	Reference
Stress	Acute amphetamine injection (1.5 mg/kg; i.p.)	↑ Fos[a]	↑ Fos[a]		Fadel et al. [145]
	Acute amphetamine injection (0.5 mg/kg; i.p.)	[b]	↑ Fos (correlated with wakefulness)	[b]	Estabrooke et al. [146]
	Acute caffeine injection (10, 30, 75 mg/kg; i.p.)	↑ Fos	↑ Fos	↑ Fos (10, 30 mg/kg dose)	Murphy et al. [147]
	Acute caffeine injection (50 mg/kg; i.p.)	↑ Fos[a]	↑ Fos[a]		Johnson et al. [163]
	Acute nicotine injection (0.8 mg/kg; s.c.)	↑ Fos[a]	↑ Fos[a]		Plaza-Zabala et al. [48]
	Footshock	↑ Fos	↑ Fos		Harris and Aston-Jones [8]
	Footshock	[b]	↑ Fos	[b]	Winsky-Sommerer et al. [149]
	Immobilization stress (30 min)	[b]	↑ Fos	[b]	Winsky-Sommerer et al. [149]
	Immobilization stress (20 min)	[c]	↑ Fos[c]	[c]	Sakamoto et al. [148]
	Immobilization stress (30 min; used to reinstate cocaine CPP)	[a]	[a]	↑ Fos	Tung et al. [166]
	Cold water stress	[b]	↑ Fos[b]	[b]	Sakamoto et al. [148]
	Carbon-dioxide stress	↑ Fos[a]	↑ Fos[a]		Sunanaga et al. [164] and Johnson et al. [167]
	Sodium lactate injections	↑ Fos	↑ Fos		Johnson et al. [62]
	FG-7142 injections (7.5 mg/kg)	↑ Fos[a]	↑ Fos[a]		Johnson et al. [163]
	Hindpaw formalin injections	↑ Fos (correlated with pain response)	↑ Fos (correlated with pain response)		Campbell et al. [165]

(continued)

Table 1 (continued)

	Stimuli/paradigm	DMH orexin	PFA orexin	LH orexin	Reference
	Naloxone-induced morphine withdrawal	↑ Fos	↑ Fos		Sharf et al. [160]
	Unpredictable chronic mild stress	↑ Fos[a]	↑ Fos[a]		Nollet et al. [159]
Arousal	Sleep deprivation	[b]	↑ Fos (correlated with wakefulness)	[b]	Estabrooke et al. [146]
	Active vs rest period	↑ Fos in active period	↑ Fos in active period (correlated with wakefulness)		Estabrooke et al. [146]
	Active vs rest period	↑ Fos in active period	↑ Fos in active period		Gompf and Aston-Jones [153]
	Running wheel confinement-induced resetting of circadian clock	↑ Fos	↑ Fos	↑ Fos	Webb et al. [168]
	Wakefulness induced by inhibition of basal forebrain	↑ Fos	↑ Fos	↑ Fos	Satoh et al. [111]
	Acute modafinil injection (150 mg/kg; i.p.)	[b]	↑ Fos	[b]	Chemelli et al. [169]

In cases where cell is blank, no effect of the experimental manipulation was observed in this region

[a]Indicates that PFA and DMH regions were not quantified separately

[b]Indicates that this region was not examined

[c]Indicates that in this paper, subregions were not quantified separately in the original paper – subregional effects are estimated here based on original figure depicting Orx+/Fos+ cells across entire orexin field

Table 2 Summary of studies that have reported patterns of orexin neuron activation in Pavlovian vs. instrumental reward seeking paradigms

		Stimuli/paradigm	DMH orexin	PFA orexin	LH orexin	Reference
Pavlovian	Opiates	Morphine CPP			↑ Fos (Fos correlated with CPP)	Harris and Aston-Jones [18], Harris et al. [150], and Lasheras et al. [156]
		Morphine CPP			↑ Fos in VTA-projecting neurons (Fos correlated with CPP)	Richardson and Aston-Jones [152]
	Cocaine	Cocaine CPP			↑ Fos (Fos correlated with CPP)	Harris and Aston-Jones [18]
		Cocaine CPP	a	a	↑ Fos in LS-projecting neurons (Fos in LH-projecting LS neurons correlated with CPP)	Sartor and Aston-Jones [40]
Instrumental	Cocaine	ABA renewal	↑ Fos (also seen in ABB)	↑ Fos (also been in ABB)		Hamlin et al. [91]
		Cued-reinstate-ment (with NPS infusions)		↑ Fos	↑ Fos	Kallupi et al. [155]
	Alcohol	Context-induced reinstatement	Fos corre-lated with seeking		Fos correlated with seeking	Moorman et al. [151]
		ABA renewal (beer)		↑ Fos	↑ Fos (Fos correlated with seeking)	Hamlin et al. [154]
		DS+ reinstatemet	↑ Fos	↑ Fos	↑ Fos	Dayas et al. [129]
		EtOH-seeking elicited by inhibi-tion of NAcSh		↑ Fos	↑ Fos	Millan et al. [93]

(continued)

Table 2 (continued)

		Stimuli/paradigm	DMH orexin	PFA orexin	LH orexin	Reference
		Homecage EtOH seeking		For correlated with seeking	Fos correlated with seeking	Moorman et al. [151]
	Nicotine	Cue-induced reinstatement		↑ Fos	↑ Fos	Plaza-Zabala [170]
	Methamphetamine	Methamphetamine self-administration		↑ Fos (not seen in meth-seeking rats)	↑ Fos (not seen in meth-seeking rats)	Cornish et al. [171]

In case where cell is blank, no effect of the experimental manipulation was observed in this region

[a]Indicates that PFA and DMH regions were not quantified separately

orexin neurons to facilitate an active reward-seeking behavior, hypotheses presented in previous reports [8, 173]. These proposals should be directly tested in behavioral paradigms designed to isolate subcomponents of reward seeking, arousal, motor behavior, etc., in order to identify specific relationships to orexin neuron activity.

In addition, it will be important to consider different behavioral categories beyond simply appetitive vs. aversive, which may further dissect selective contributions of orexin neuron populations. Future work will also benefit from characterizing real-time activation of orexin neurons (as shown, for example, in [157, 158, 179]), direct modulation of orexin neuron populations [59, 180–183], and correlating strength of activation with specific behaviors [18, 151, 154].

Differential anatomical connectivity of neurons within orexin subfields may provide a basis for functional differences between lateral vs medial orexin cell groups. In support of this possibility, a number of groups reported that different brain areas innervate lateral vs. medial orexin fields [90, 184]. Although somewhat speculative, an overview of the regions that preferentially project to the lateral vs. medial orexin fields reflects regions that could be broadly defined as appetitive/motivational vs. arousal/anxiety categories, respectively. For example, projections from dorsal mPFC in rats (prelimbic and cingulate), which are frequently associated with expression of conditioned fear [185], target medial orexin fields, whereas ventral medial prefrontal cortex (infralimbic) and orbitofrontal cortex, which are commonly associated with extinction of fear and reward seeking [185], preferentially target lateral regions of the orexin field [184]. As discussed above, studies in mice using genetic tools for identifying neuronal populations specifically projecting to orexin neurons revealed a range of brain areas including those related to sleep/wake (e.g., basal forebrain, median raphe, and ventrolateral preoptic area) as well as motivation and emotion (e.g., infralimbic medial prefrontal cortex, amygdala, NAc, BNST, and lateral septum) [89]. In these studies, however, lateral vs. medial differences were not considered indicating that the interaction between orexin neuron anatomical location and afferents may be a useful research path.

There is also evidence for differential physiological properties of orexin neurons based on projection targets; this provides another possible basis for functional heterogeneity among orexin cell groups. As outlined above, orexin neurons signal at a number of target structures to regulate drug seeking, including VTA [20, 30, 71, 102, 103], PVT [128, 131, 132], and other reward and motivation-related targets [12, 186]. Early work revealed that lateral (LH/PFA) orexin neurons preferentially projected to VTA in rats [187], although orexin neuron projections to VTA in mice appear to originate from both medial and lateral subdivisions [188]. Only a small number of studies to date have explored the function of orexin neurons based on their projection targets [150, 189]. Work from our lab demonstrated that VTA-projecting, but not locus coeruleus-projecting, orexin neurons were activated during morphine conditioned place preference [152]. Interestingly, the same study demonstrated that VTA-projecting LH, but not PFA/DMH orexin neurons exhibited Fos activation correlated with CPP scores. These and other results indicate that selective projections of specific orexin neuron populations may serve as a mechanism for differentiating functional ensembles.

Orexin neurons co-express other neurotransmitter molecules, and this provides another possible basis for functional differences in orexin neuron populations

[11]. Such co-expressed signaling molecules include classical neurotransmitters (eg. glutamate [190–192]) and neuropeptides (eg. dynorphin [193, 194]).

Finally, differences in physiological properties may also contribute to functional heterogeneity of orexin neuron subpopulations. Burdakov and colleagues have characterized two distinct populations of orexin neurons based on a combination of physiological (transient vs. persistent inhibitory responses to glucose administration in vitro) and anatomical (medial vs. lateral) location [195]. Additional studies may further delineate subpopulations of orexin neurons by endogenous physiological properties.

6.1 Summary

Different populations of orexin neurons may preferentially participate in specific functions within the greater context of motivational activation. Whether the primary differentiating factor is anatomical location within the medial/lateral axis of the hypothalamus, as our group proposed 10 years ago [8], has so far not been conclusively proven or dismissed. A likely future outcome will be the revelation that a confluence of factors – anatomical, hodological, physiological, and molecular – will differentiate orexin neurons into specific functional populations. Future integrative studies, using well-designed behavioral paradigms to isolate specific physiological or behavioral factors, will ultimately assist in understanding how the range of functionally distinct subpopulations of orexin neurons, and how these populations are segregated and how they interact, both with each other and with other hypothalamic neuronal populations.

7 Conclusions and Future Directions

The past decade has seen significant advancement of our understanding of the orexin system in reward and addiction. In drug-seeking paradigms, orexin neurons are preferentially engaged under circumstances where high levels of effort are required to obtain drug, or when motivation for drug is augmented by external stimuli such as drug-related cues or stress. These findings are consistent with what we believe constitutes a broader, fundamentally unified role for the orexin system in facilitating motivational states – both appetitive and aversive – and translating these states into behaviors that the organism uses to exploit a reward opportunity or respond to pressing threats – a process we termed "motivational activation" [11]. We suggest that this diverse, integrative role must be the result of heterogeneous functional connectivity with wider brain circuits that allow orexin neurons to modulate a range of behavioral and physiological outputs. Indeed, a priority of future studies must be to characterize this heterogeneity. As we discuss above, we believe that an important first step will be examining how different orexin subpopulations (for example, medial versus lateral) differ in terms of their precise afferent and efferent topography. Recent advances in cell-specific targeting approaches that allow for the selective regulation of orexin neurons and their afferents or efferents, using either optogenetics or chemogenetics (e.g., DREADDs) will allow these questions to be addressed at both an anatomical and functional level.

We note that the majority of studies of orexin in the context of drug abuse have been carried out in short-access models that may not fully recapitulate the behavioral features of addiction [196]. It is imperative that future studies examine how the orexin system is involved in the compulsive behavioral phenotypes that are precipitated by alternative models such long [196, 197] or intermittent [198] access models. Indeed, rats that are exposed to these paradigms exhibit behavioral characteristics reflective of increased motivational drive for drug, including escalated cocaine intake and compulsive drug use despite punishment. Currently, our laboratory is investigating how the orexin system may be involved in this motivational shift and whether this system can be targeted to reduce addiction-like behaviors.

In August 2014, the dual orexin receptor antagonist Suvorexant (Belsomra®) was approved by the US FDA for the treatment of insomnia. Clinical trials suggest that this drug is relatively well tolerated with few side effects reported at doses required to achieve sleep-promoting effects [199]. In addition, a number of other orexin receptor antagonist compounds are also currently at various stages of preclinical and clinical testing [10]. As such, there is significant interest in the potential use of orexin receptor antagonists for the treatment of other disorders associated with orexin dysfunction, including substance use disorders. Although we agree that these compounds may offer an exciting new approach to treating psychostimulant addiction – for which there are currently no effective pharmacotherapies – we also caution that our understanding of the orexin system in addiction is far from complete. This said, the rapid growth in our understanding of orexin in the last 10 years, including the introduction of the first orexin receptor antagonist for clinical use, bodes well for the next 10 years of research into this intriguing neuropeptide system.

Acknowledgments This work is supported by an NHMRC CJ Martin Fellowship (1072706) to M.H.J., and PHS grants R00 DA035251 to SVM, and R01 DA006214 to GAJ. Support for DEM: R21 DA041674.

References

1. de Lecea L, Kilduff TS, Peyron C, Gao X, Foye PE, Danielson PE, Fukuhara C, Battenberg EL, Gautvik VT, Bartlett FS 2nd, Frankel WN, van den Pol AN, Bloom FE, Gautvik KM, Sutcliffe JG (1998). The hypocretins: hypothalamus-specific peptides with neuroexcitatory activity. Proc Natl Acad Sci U S A 95(1):322–327
2. Sakurai T, Amemiya A, Ishii M, Matsuzaki I, Chemelli RM, Tanaka H, Williams SC, Richardson JA, Kozlowski GP, Wilson S, Arch JR, Buckingham RE, Haynes AC, Carr SA, Annan RS, McNulty DE, Liu WS, Terrett JA, Elshourbagy NA, Bergsma DJ, Yanagisawa M (1998) Orexins and orexin receptors: a family of hypothalamic neuropeptides and G protein-coupled receptors that regulate feeding behavior. Cell 92(4):573–585
3. Sweet DC, Levine AS, Billington CJ, Kotz CM (1999) Feeding response to central orexins. Brain Res 821(2):535–538

4. Haynes AC, Jackson B, Chapman H, Tadayyon M, Johns A, Porter RA, Arch JR (2000) A selective orexin-1 receptor antagonist reduces food consumption in male and female rats. Regul Pept 96(1–2):45–51

5. Aston-Jones G, Smith RJ, Sartor GC, Moorman DE, Massi L, Tahsili-Fahadan P, Richardson KA (2010) Lateral hypothalamic orexin/hypocretin neurons: a role in reward-seeking and addiction. Brain Res 1314:74–90

6. Baimel C, Bartlett SE, Chiou LC, Lawrence AJ, Muschamp JW, Patkar O, Tung LW, Borgland SL (2014) Orexin/hypocretin role in reward: implications for opioid and other addictions. Br J Pharmacol 172:334–348

7. Boutrel B, Cannella N, de Lecea L (2010) The role of hypocretin in driving arousal and goal-oriented behaviors. Brain Res 1314:103–111

8. Harris GC, Aston-Jones G (2006) Arousal and reward: a dichotomy in orexin function. Trends Neurosci 29(10):571–577

9. James MH, Yeoh JW, Graham BA, Dayas CV (2012) Insights for developing pharmacological treatments for psychostimulant relapse targeting hypothalamic peptide systems. J Addict Res Ther S4:008

10. Khoo SY, Brown RM (2014) Orexin/hypocretin based pharmacotherapies for the treatment of addiction: DORA or SORA? CNS Drugs 28(8):713–730

11. Mahler SV, Moorman DE, Smith RJ, James MH, Aston-Jones G (2014) Motivational activation: a unifying hypothesis of orexin/hypocretin function. Nat Neurosci 17 (10):1298–1303

12. Mahler SV, Smith RJ, Moorman DE, Sartor GC, Aston-Jones G (2012) Multiple roles for orexin/hypocretin in addiction. Prog Brain Res 198:79–121

13. Nishino S, Ripley B, Overeem S, Lammers GJ, Mignot E (2000) Hypocretin (orexin) deficiency in human narcolepsy. Lancet 355(9197):39–40

14. Peyron C, Faraco J, Rogers W, Ripley B, Overeem S, Charnay Y, Nevsimalova S, Aldrich M, Reynolds D, Albin R, Li R, Hungs M, Pedrazzoli M, Padigaru M, Kucherlapati M, Fan J, Maki R, Lammers GJ, Bouras C, Kucherlapati R, Nishino S, Mignot E (2000) A mutation in a case of early onset narcolepsy and a generalized absence of hypocretin peptides in human narcoleptic brains. Nat Med 6(9):991–997

15. Akimoto H, Honda Y, Takahashi Y (1960) Pharmacotherapy in narcolepsy. Dis Nerv Syst 21:704–706

16. Nishino S, Mignot E (1997) Pharmacological aspects of human and canine narcolepsy. Prog Neurobiol 52(1):27–78

17. Georgescu D, Zachariou V, Barrot M, Mieda M, Willie JT, Eisch AJ, Yanagisawa M, Nestler EJ, DiLeone RJ (2003) Involvement of the lateral hypothalamic peptide orexin in morphine dependence and withdrawal. J Neurosci 23(8):3106–3111

18. Harris GC, Wimmer M, Aston-Jones G (2005) A role for lateral hypothalamic orexin neurons in reward seeking. Nature 437(7058):556–559

19. Boutrel B, Kenny PJ, Specio SE, Martin-Fardon R, Markou A, Koob GF, de Lecea L (2005) Role for hypocretin in mediating stress-induced reinstatement of cocaine-seeking behavior. Proc Natl Acad Sci U S A 102(52):19168–19173

20. Borgland SL, Chang SJ, Bowers MS, Thompson JL, Vittoz N, Floresco SB, Chou J, Chen BT, Bonci A (2009) Orexin A/hypocretin-1 selectively promotes motivation for positive reinforcers. J Neurosci 29(36):11215–11225

21. Bentzley BS, Aston-Jones G (2015) Orexin-1 receptor signaling increases motivation for cocaine-associated cues. Eur J Neurosci 41(9):1149–1156

22. Walker LC, Lawrence AJ (2016) The role of orexins/hypocretins in alcohol use and abuse. Curr Top Behav Neurosci. doi:10.1007/7854_2016_55

23. Koob GF (1992) Neural mechanisms of drug reinforcement. Ann N Y Acad Sci 654:171–191

24. Wise RA (1987) Intravenous drug self-administration: a special case of positive reinforcement. In: Bozarth MA (ed) Methods of assessing the reinforcing properties of abused drugs. Springer, New York, pp 117–141

25. Wise RA (1997) Drug self-administration viewed as ingestive behaviour. Appetite 28(1):1–5
26. Espana RA, Oleson EB, Locke JL, Brookshire BR, Roberts DC, Jones SR (2010) The hypocretin-orexin system regulates cocaine self-administration via actions on the mesolimbic dopamine system. Eur J Neurosci 31(2):336–348
27. Hutcheson DM, Quarta D, Halbout B, Rigal A, Valerio E, Heidbreder C (2011) Orexin-1 receptor antagonist SB-334867 reduces the acquisition and expression of cocaine-conditioned reinforcement and the expression of amphetamine-conditioned reward. Behav Pharmacol 22(2):173–181
28. Smith RJ, See RE, Aston-Jones G (2009) Orexin/hypocretin signaling at the orexin 1 receptor regulates cue-elicited cocaine-seeking. Eur J Neurosci 30(3):493–503
29. Hollander JA, Pham D, Fowler CD, Kenny PJ (2012) Hypocretin-1 receptors regulate the reinforcing and reward-enhancing effects of cocaine: pharmacological and behavioral genetics evidence. Front Behav Neurosci 6:47
30. Espana RA, Melchior JR, Roberts DC, Jones SR (2011) Hypocretin 1/orexin A in the ventral tegmental area enhances dopamine responses to cocaine and promotes cocaine self-administration. Psychopharmacology (Berl) 214(2):415–426
31. Prince CD, Rau AR, Yorgason JT, España RA (2015) Hypocretin/orexin regulation of dopamine signaling and cocaine self-administration is mediated predominantly by hypocretin receptor 1. ACS Chem Nerosci 6(1):138–146
32. Riday TT, Fish EW, Robinson JE, Jarrett TM, McGuigan MM, Malanga CJ (2012) Orexin-1 receptor antagonism does not reduce the rewarding potency of cocaine in Swiss–Webster mice. Brain Res 1431:53–61
33. Jupp B, Krivdic B, Krstew E, Lawrence AJ (2011a) The orexin(1) receptor antagonist SB-334867 dissociates the motivational properties of alcohol and sucrose in rats. Brain Res 1391:54–59
34. Jupp B, Krstew E, Dezsi G, Lawrence AJ (2011b) Discrete cue-conditioned alcohol-seeking after protracted abstinence: pattern of neural activation and involvement of orexin(1) receptors. Br J Pharmacol 162(4):880–889
35. Lawrence AJ, Cowen MS, Yang HJ, Chen F, Oldfield B (2006) The orexin system regulates alcohol-seeking in rats. Br J Pharmacol 148(6):752–759
36. Hollander JA, Lu Q, Cameron MD, Kamenecka TM, Kenny PJ (2008) Insular hypocretin transmission regulates nicotine reward. Proc Natl Acad Sci U S A 105(49):19480–19485
37. Smith RJ, Aston-Jones G (2012) Orexin/hypocretin 1 receptor antagonist reduces heroin self-administration and cue-induced heroin seeking. Eur J Neurosci 35(5):798–804
38. LeSage MG, Perry JL, Kotz CM, Shelley D, Corrigall WA (2010) Nicotine self-administration in the rat: effects of hypocretin antagonists and changes in hypocretin mRNA. Psychopharmacology (Berl) 209(2):203–212
39. Steiner MA, Lecourt H, Jenck F (2013) The dual orexin receptor antagonist almorexant, alone and in combination with morphine, cocaine and amphetamine, on conditioned place preference and locomotor sensitization in the rat. Int J Neuropsychopharmacol 16(2):417–432
40. Sartor GC, Aston-Jones GS (2012) A septal-hypothalamic pathway drives orexin neurons, which is necessary for conditioned cocaine preference. J Neurosci 32(13):4623–4631
41. Rao Y, Mineur YS, Gan G, Wang AH, Liu ZW, Wu X, Suyama S, de Lecea L, Horvath TL, Picciotto MR, Gao XB (2013) Repeated in vivo exposure of cocaine induces long-lasting synaptic plasticity in hypocretin/orexin-producing neurons in the lateral hypothalamus in mice. J Physiol 591(Pt 7):1951–1966
42. Marchant NJ, Li X, Shaham Y (2013) Recent developments in animal models of drug relapse. Curr Opin Neurobiol 23(4):675–683
43. Zhou L, Ghee SM, Chan C, Lin L, Cameron MD, Kenny PJ, See RE (2012) Orexin-1 receptor mediation of cocaine seeking in male and female rats. J Pharmacol Exp Ther 340(3):801–809
44. Martin-Fardon R, Weiss F (2014a) Blockade of hypocretin receptor-1 preferentially prevents cocaine seeking: comparison with natural reward seeking. Neuroreport 25(7):485–488

45. Smith RJ, Tahsili-Fahadan P, Aston-Jones G (2010) Orexin/hypocretin is necessary for context-driven cocaine-seeking. Neuropharmacology 58(1):179–184
46. Martin-Fardon R, Weiss F (2014b) N-(2-methyl-6-benzoxazolyl)-N′-1,5-naphthyridin-4-yl urea (SB334867), a hypocretin receptor-1 antagonist, preferentially prevents ethanol seeking: comparison with natural reward seeking. Addict Biol 19(2):233–236
47. Moorman DE, James MH, Kilroy EA, Aston-Jones G (2016) Orexin/hypocretin-1 receptor antagonism reduces ethanol self-administration and reinstatement selectively in highly-motivated rats. Brain Res 1654:34–42. doi:10.1016/j.brainres.2016.10.018
48. Plaza-Zabala A, Martin-Garcia E, de Lecea L, Maldonado R, Berrendero F (2010) Hypocretins regulate the anxiogenic-like effects of nicotine and induce reinstatement of nicotine-seeking behavior. J Neurosci 30(6):2300–2310
49. Porter-Stransky KA, Bentzley BS, Aston-Jones G (2015) Individual differences in orexin-I receptor modulation of motivation for the opioid remifentanil. Addict Biol. doi:10.1111/adb.12323
50. Uslaner JM, Winrow CJ, Gotter AL, Roecker AJ, Coleman PJ, Hutson PH, Le AD, Renger JJ (2014) Selective orexin 2 receptor antagonism blocks cue-induced reinstatement, but not nicotine self-administration or nicotine-induced reinstatement. Behav Brain Res 269:61–65
51. Brown RM, Khoo SY-S, Lawrence AJ (2013) Central orexin (hypocretin) 2 receptor antagonism reduces ethanol self-administration, but not cue-conditioned ethanol-seeking, in ethanol-preferring rats. Int J Neuropsychopharmacol 16(9):2067–2079
52. Di Ciano P, Everitt BJ (2005) Neuropsychopharmacology of drug seeking: Insights from studies with second-order schedules of drug reinforcement. Eur J Pharmacol 526 (1–3):186–198
53. Schindler CW, Panlilio LV, Goldberg SR (2002) Second-order schedules of drug self-administration in animals. Psychopharmacology (Berl) 163(3–4):327–344
54. James MH, Campbell EJ, Dayas CV (2016) Role of the orexin/hypocretin system in stress-related psychiatric disorders. Curr Top Behav Neurosci. doi:10.1007/7854_2016_56
55. Chen X, Wang H, Lin Z, Li S, Li Y, Bergen HT, Vrontakis ME, Kirouac GJ (2013) Orexins (hypocretins) contribute to fear and avoidance in rats exposed to a single episode of footshocks. Brain Struct Funct 219:2103–2118
56. Furlong TM, Vianna DM, Liu L, Carrive P (2009) Hypocretin/orexin contributes to the expression of some but not all forms of stress and arousal. Eur J Neurosci 30(8):1603–1614
57. James MH, Campbell EJ, Walker FR, Smith DW, Richardson HN, Hodgson DM, Dayas CV (2014) Exercise reverses the effects of early life stress on orexin cell reactivity in male but not female rats. Front Behav Neurosci 8:244
58. Martins PJ, D'Almeida V, Pedrazzoli M, Lin L, Mignot E, Tufik S (2004) Increased hypocretin-1 (orexin-a) levels in cerebrospinal fluid of rats after short-term forced activity. Regul Pept 117(3):155–158
59. Bonnavion P, Jackson AC, Carter ME, de Lecea L (2015) Antagonistic interplay between hypocretin and leptin in the lateral hypothalamus regulates stress responses. Nat Commun 6:6266
60. Heydendael W, Sengupta A, Beck S, Bhatnagar S (2013) Optogenetic examination identifies a context-specific role for orexins/hypocretins in anxiety-related behavior. Physiol Behav 130:182–190
61. Chang H, Saito T, Ohiwa N, Tateoka M, Deocaris CC, Fujikawa T, Soya H (2007) Inhibitory effects of an orexin-2 receptor antagonist on orexin A- and stress-induced ACTH responses in conscious rats. Neurosci Res 57(3):462–466
62. Johnson PL, Truitt W, Fitz SD, Minick PE, Dietrich A, Sanghani S, Traskman-Bendz L, Goddard AW, Brundin L, Shekhar A (2010) A key role for orexin in panic anxiety. Nat Med 16(1):111–115
63. Shaham Y, Erb S, Stewart J (2000) Stress-induced relapse to heroin and cocaine seeking in rats: a review. Brain Res Brain Res Rev 33(1):13–33

64. Shaham Y, Shalev U, Lu L, De Wit H, Stewart J (2003) The reinstatement model of drug relapse: history, methodology and major findings. Psychopharmacology (Berl) 168(1–2):3–20
65. Shalev U, Marinelli M, Baumann MH, Piazza PV, Shaham Y (2003) The role of corticosterone in food deprivation-induced reinstatement of cocaine seeking in the rat. Psychopharmacology (Berl) 168(1–2):170–176
66. Le AD, Harding S, Juzytsch W, Funk D, Shaham Y (2005) Role of alpha-2 adrenoceptors in stress-induced reinstatement of alcohol seeking and alcohol self-administration in rats. Psychopharmacology (Berl) 179(2):366–373
67. Lee B, Tiefenbacher S, Platt DM, Spealman RD (2004) Pharmacological blockade of alpha2-adrenoceptors induces reinstatement of cocaine-seeking behavior in squirrel monkeys. Neuropsychopharmacology 29(4):686–693
68. Conrad KL, Davis AR, Silberman Y, Sheffler DJ, Shields AD, Saleh SA, Sen N, Matthies HJG, Javitch JA, Lindsley CW, Winder DG (2012) Yohimbine depresses excitatory transmission in BNST and impairs extinction of cocaine place preference through orexin-dependent, norepinephrine-independent processes. Neuropsychopharmacology 37 (10):2253–2266
69. Qi K, Wei C, Li Y, Sui N (2013) Orexin receptors within the nucleus accumbens shell mediate the stress but not drug priming-induced reinstatement of morphine conditioned place preference. Front Behav Neurosci 7:144
70. Ebrahimian F, Naghavi FS, Yazdi F, Sadeghzadeh F, Taslimi Z, Haghparast A (2016) Differential roles of orexin receptors within the dentate gyrus in stress- and drug priming-induced reinstatement of conditioned place preference in rats. Behav Neurosci 130 (1):91–102
71. Wang B, You ZB, Wise RA (2009) Reinstatement of cocaine seeking by hypocretin (orexin) in the ventral tegmental area: independence from the local corticotropin-releasing factor network. Biol Psychiatry 65(10):857–862
72. Ehrman RN, Robbins SJ, Childress AR, O'Brien CP (1992) Conditioned responses to cocaine-related stimuli in cocaine abuse patients. Psychopharmacology (Berl) 107 (4):523–529
73. Ferguson SG, Shiffman S (2009) The relevance and treatment of cue-induced cravings in tobacco dependence. J Subst Abuse Treat 36(3):235–243
74. O'Brien CP, Childress AR, McLellan AT, Ehrman R (1992) Classical conditioning in drug-dependent humans. Ann N Y Acad Sci 654:400–415
75. Sinha R, Fox HC, Hong KI, Hansen J, Tuit K, Kreek MJ (2011) Effects of adrenal sensitivity, stress- and cue-induced craving, and anxiety on subsequent alcohol relapse and treatment outcomes. Arch Gen Psychiatry 68(9):942–952
76. Sinha R, Li CS (2007) Imaging stress- and cue-induced drug and alcohol craving: association with relapse and clinical implications. Drug Alcohol Rev 26(1):25–31
77. Volkow ND, Wang GJ, Telang F, Fowler JS, Logan J, Childress AR, Jayne M, Ma Y, Wong C (2008) Dopamine increases in striatum do not elicit craving in cocaine abusers unless they are coupled with cocaine cues. Neuroimage 39(3):1266–1273
78. Cason AM, Aston-Jones G (2013) Role of orexin/hypocretin in conditioned sucrose-seeking in rats. Psychopharmacology (Berl) 226(1):155–165
79. Muschamp JW, Dominguez JM, Sato SM, Shen RY, Hull EM (2007) A role for hypocretin (orexin) in male sexual behavior. J Neurosci 27(11):2837–2845
80. Cruz HG, Hoever P, Chakraborty B, Schoedel K, Sellers EM, Dingemanse J (2014) Assessment of the abuse liability of a dual orexin receptor antagonist: a crossover study of almorexant and zolpidem in recreational drug users. CNS Drugs 28(4):361–372
81. Yeoh JW, Campbell EJ, James MH, Graham BA, Dayas CV (2014a) Orexin antagonists for neuropsychiatric disease: progress and potential pitfalls. Front Neurosci 8:36
82. Bentzley BS, Fender KM, Aston-Jones G (2013) The behavioral economics of drug self-administration: a review and new analytical approach for within-session procedures. Psychopharmacology (Berl) 226(1):113–125

83. Oleson EB, Roberts DCS (2008) Behavioral economic assessment of price and cocaine consumption following self-administration histories that produce escalation of either final ratios or intake. Neuropsychopharmacology 34(3):796–804

84. Hursh SR, Silberberg A (2008) Economic demand and essential value. Psychol Rev 115 (1):186–198

85. Bentzley BS, Jhou TC, Aston-Jones G (2014) Economic demand predicts addiction-like behavior and therapeutic efficacy of oxytocin in the rat. Proc Natl Acad Sci U S A 111 (32):11822–11827

86. Robinson TE, Yager LM, Cogan ES, Saunders BT (2014) On the motivational properties of reward cues: individual differences. Neuropharmacology 76(Pt B):450–459

87. Marcus JN, Aschkenasi CJ, Lee CE, Chemelli RM, Saper CB, Yanagisawa M, Elmquist JK (2001) Differential expression of orexin receptors 1 and 2 in the rat brain. J Comp Neurol 435 (1):6–25

88. Peyron C, Tighe DK, van den Pol AN, de Lecea L, Heller HC, Sutcliffe JG, Kilduff TS (1998) Neurons containing hypocretin (orexin) project to multiple neuronal systems. J Neurosci 18 (23):9996–10015

89. Sakurai T, Nagata R, Yamanaka A, Kawamura H, Tsujino N, Muraki Y, Kageyama H, Kunita S, Takahashi S, Goto K, Koyama Y, Shioda S, Yanagisawa M (2005) Input of orexin/hypocretin neurons revealed by a genetically encoded tracer in mice. Neuron 46 (2):297–308

90. Yoshida K, McCormack S, Espana RA, Crocker A, Scammell TE (2006) Afferents to the orexin neurons of the rat brain. J Comp Neurol 494(5):845–861

91. Hamlin AS, Clemens KJ, McNally GP (2008) Renewal of extinguished cocaine-seeking. Neuroscience 151(3):659–670

92. Marchant NJ, Hamlin AS, McNally GP (2009) Lateral hypothalamus is required for context-induced reinstatement of extinguished reward seeking. J Neurosci 29(5):1331–1342

93. Millan EZ, Furlong TM, McNally GP (2010) Accumbens shell-hypothalamus interactions mediate extinction of alcohol seeking. J Neurosci 30(13):4626–4635

94. Baldo BA, Gual-Bonilla L, Sijapati K, Daniel RA, Landry CF, Kelley AE (2004) Activation of a subpopulation of orexin/hypocretin-containing hypothalamic neurons by GABAA receptor-mediated inhibition of the nucleus accumbens shell, but not by exposure to a novel environment. Eur J Neurosci 19(2):376–386

95. Yeoh JW, James MH, Jobling P, Bains JS, Graham BA, Dayas CV (2012) Cocaine potentiates excitatory drive in the perifornical/lateral hypothalamus. J Physiol 590 (Pt 16):3677–3689

96. Swanson LW (1982) The projections of the ventral tegmental area and adjacent regions: a combined fluorescent retrograde tracer and immunofluorescence study in the rat. Brain Res Bull 9(1–6):321–353

97. Di Chiara G, Bassareo V, Fenu S, De Luca MA, Spina L, Cadoni C, Acquas E, Carboni E, Valentini V, Lecca D (2004) Dopamine and drug addiction: the nucleus accumbens shell connection. Neuropharmacology 47(Suppl 1):227–241

98. Ranaldi R, Wise RA (2001) Blockade of D1 dopamine receptors in the ventral tegmental area decreases cocaine reward: possible role for dendritically released dopamine. J Neurosci 21 (15):5841–5846

99. Wise RA (2008) Dopamine and reward: the anhedonia hypothesis 30 years on. Neurotox Res 14(2–3):169–183

100. Balcita-Pedicino JJ, Sesack SR (2007) Orexin axons in the rat ventral tegmental area synapse infrequently onto dopamine and gamma-aminobutyric acid neurons. J Comp Neurol 503 (5):668–684

101. Korotkova TM, Sergeeva OA, Eriksson KS, Haas HL, Brown RE (2003) Excitation of ventral tegmental area dopaminergic and nondopaminergic neurons by orexins/hypocretins. J Neurosci 23(1):7–11

102. James MH, Charnley JL, Levi EM, Jones E, Yeoh JW, Smith DW, Dayas CV (2011b) Orexin-1 receptor signalling within the ventral tegmental area, but not the paraventricular thalamus, is critical to regulating cue-induced reinstatement of cocaine-seeking. Int J Neuropsychopharmacol 14(5):684–690
103. Mahler SV, Smith RJ, Aston-Jones G (2013) Interactions between VTA orexin and glutamate in cue-induced reinstatement of cocaine seeking in rats. Psychopharmacology (Berl) 226(4):687–698
104. Brown RM, Kim AK, Khoo SY, Kim JH, Jupp B, Lawrence AJ (2016) Orexin-1 receptor signalling in the prelimbic cortex and ventral tegmental area regulates cue-induced reinstatement of ethanol-seeking in iP rats. Addict Biol 21(3):603–612
105. Berridge KC, Robinson TE (1998) What is the role of dopamine in reward: hedonic impact, reward learning, or incentive salience? Brain Res Rev 28(3):309–369
106. Cheer JF, Aragona BJ, Heien ML, Seipel AT, Carelli RM, Wightman RM (2007) Coordinated accumbal dopamine release and neural activity drive goal-directed behavior. Neuron 54 (2):237–244
107. Fiorillo CD, Tobler PN, Schultz W (2003) Discrete coding of reward probability and uncertainty by dopamine neurons. Science 299(5614):1898–1902
108. Ikemoto S, Panksepp J (1999) The role of nucleus accumbens dopamine in motivated behavior: a unifying interpretation with special reference to reward-seeking. Brain Res Brain Res Rev 31(1):6 41
109. Phillips PE, Stuber GD, Heien ML, Wightman RM, Carelli RM (2003) Subsecond dopamine release promotes cocaine seeking. Nature 422(6932):614–618
110. Salamone JD, Correa M, Farrar A, Mingote SM (2007) Effort-related functions of nucleus accumbens dopamine and associated forebrain circuits. Psychopharmacology (Berl) 191 (3):461–482
111. Satoh T, Nakai S, Sato T, Kimura M (2003) Correlated coding of motivation and outcome of decision by dopamine neurons. J Neurosci 23(30):9913–9923
112. Stuber GD, Roitman MF, Phillips PE, Carelli RM, Wightman RM (2005) Rapid dopamine signaling in the nucleus accumbens during contingent and noncontingent cocaine administration. Neuropsychopharmacology 30(5):853–863
113. Vittoz NM, Schmeichel B, Berridge CW (2008) Hypocretin/orexin preferentially activates caudomedial ventral tegmental area dopamine neurons. Eur J Neurosci 28(8):1629–1640
114. Chergui K, Charlety PJ, Akaoka H, Saunier CF, Brunet JL, Buda M, Svensson TH, Chouvet G (1993) Tonic activation of NMDA receptors causes spontaneous burst discharge of rat midbrain dopamine neurons in vivo. Eur J Neurosci 5(2):137–144
115. Harris GC, Aston-Jones G (2003) Critical role for ventral tegmental glutamate in preference for a cocaine-conditioned environment. Neuropsychopharmacology 28(1):73–76
116. Wang T, O'Connor WT, Ungerstedt U, French ED (1994) N-methyl-d-aspartic acid biphasically regulates the biochemical and electrophysiological response of A10 dopamine neurons in the ventral tegmental area: in vivo microdialysis and in vitro electrophysiological studies. Brain Res 666(2):255–262
117. Zellner MR, Ranaldi R (2010) How conditioned stimuli acquire the ability to activate VTA dopamine cells: a proposed neurobiological component of reward-related learning. Neurosci Biobehav Rev 34(5):769–780
118. Borgland SL, Taha SA, Sarti F, Fields HL, Bonci A (2006) Orexin A in the VTA is critical for the induction of synaptic plasticity and behavioral sensitization to cocaine. Neuron 49(4):589–601
119. Borgland SL, Storm E, Bonci A (2008) Orexin B/hypocretin 2 increases glutamatergic transmission to ventral tegmental area neurons. Eur J Neurosci 28(8):1545–1556
120. Geisler S, Marinelli M, Degarmo B, Becker ML, Freiman AJ, Beales M, Meredith GE, Zahm DS (2008) Prominent activation of brainstem and pallidal afferents of the ventral tegmental area by cocaine. Neuropsychopharmacology 33(11):2688–2700
121. Moorman DE, James MH, McGlinchey EM, Aston-Jones G (2015) Differential roles of medial prefrontal subregions in the regulation of drug seeking. Brain Res 1628 (Pt A):130–146

122. Gariano RF, Groves PM (1988) Burst firing induced in midbrain dopamine neurons by stimulation of the medial prefrontal and anterior cingulate cortices. Brain Res 462 (1):194–198
123. Tong ZY, Overton PG, Clark D (1996) Stimulation of the prefrontal cortex in the rat induces patterns of activity in midbrain dopaminergic neurons which resemble natural burst events. Synapse 22(3):195–208
124. Moorman DE, Aston-Jones G (2010) Orexin/hypocretin modulates response of ventral tegmental dopamine neurons to prefrontal activation: diurnal influences. J Neurosci 30 (46):15585–15599
125. Mahler SV, Aston-Jones GS (2012) Fos activation of selective afferents to ventral tegmental area during cue-induced reinstatement of cocaine seeking in rats. J Neurosci 32 (38):13309–13326
126. Ishibashi M, Takano S, Yanagida H, Takatsuna M, Nakajima K, Oomura Y, Wayner MJ, Sasaki K (2005) Effects of orexins/hypocretins on neuronal activity in the paraventricular nucleus of the thalamus in rats in vitro. Peptides 26(3):471–481
127. James MH, Charnley JL, Flynn JR, Smith DW, Dayas CV (2011a) Propensity to 'relapse' following exposure to cocaine cues is associated with the recruitment of specific thalamic and epithalamic nuclei. Neuroscience 199:235–242
128. Barson JR, Ho HT, Leibowitz SF (2015) Anterior thalamic paraventricular nucleus is involved in intermittent access ethanol drinking: role of orexin receptor 2. Addict Biol 20(3):469–481
129. Dayas CV, McGranahan TM, Martin-Fardon R, Weiss F (2008) Stimuli linked to ethanol availability activate hypothalamic CART and orexin neurons in a reinstatement model of relapse. Biol Psychiatry 63(2):152–157
130. James MH, Charnley JL, Jones E, Levi EM, Yeoh JW, Flynn JR, Smith DW, Dayas CV (2010) Cocaine- and amphetamine-regulated transcript (CART) signaling within the paraventricular thalamus modulates cocaine-seeking behaviour. PLoS One 5(9):e12980
131. James MH, Dayas CV (2013) What about me...? The PVT: a role for the paraventricular thalamus (PVT) in drug-seeking behavior. Front Behav Neurosci 7:18
132. Martin-Fardon R, Boutrel B (2012) Orexin/hypocretin (Orx/Hcrt) transmission and drug-seeking behavior: is the paraventricular nucleus of the thalamus (PVT) part of the drug seeking circuitry? Front Behav Neurosci 6:75
133. Yeoh JW, James MH, Graham BA, Dayas CV (2014b) Electrophysiological characteristics of paraventricular thalamic (PVT) neurons in response to cocaine and cocaine- and amphetamine-regulated transcript (CART). Front Behav Neurosci 8:280
134. Matzeu A, Kerr TM, Weiss F, Martin-Fardon R (2016) Orexin-A/hypocretin-1 mediates cocaine-seeking behavior in the posterior paraventricular nucleus of the thalamus via orexin/hypocretin receptor-2. J Pharmacol Exp Ther 359:273–279. doi:10.1124/jpet.116.235945
135. Trivedi P, Yu H, MacNeil DJ, Van der Ploeg LH, Guan XM (1998) Distribution of orexin receptor mRNA in the rat brain. FEBS Lett 438(1–2):71–75
136. Mori K, Kim J, Sasaki K (2011) Electrophysiological effects of orexin-B and dopamine on rat nucleus accumbens shell neurons in vitro. Peptides 32(2):246–252
137. Mukai K, Kim J, Nakajima K, Oomura Y, Wayner MJ, Sasaki K (2009) Electrophysiological effects of orexin/hypocretin on nucleus accumbens shell neurons in rats: an in vitro study. Peptides 30(8):1487–1496
138. Castro DC, Terry RA, Berridge KC (2016) Orexin in rostral hotspot of nucleus accumbens enhances sucrose 'liking' and intake but scopolamine in caudal shell shifts 'liking' toward 'disgust' and 'fear'. Neuropsychopharmacology 41:2101–2111
139. Ho CY, Berridge KC (2013) An orexin hotspot in ventral pallidum amplifies hedonic 'liking' for sweetness. Neuropsychopharmacology 38(9):1655–1664
140. Ubaldi M, Giordano A, Severi I, Li H, Kallupi M, de Guglielmo G, Ruggeri B, Stopponi S, Ciccocioppo R, Cannella N (2016) Activation of hypocretin-1/orexin-A neurons projecting to the bed nucleus of the stria terminalis and paraventricular nucleus is critical for reinstatement of alcohol seeking by neuropeptide S. Biol Psychiatry 79(6):452–462

141. Kastman HE, Blasiak A, Walker L, Siwiec M, Krstew EV, Gundlach AL, Lawrence AJ (2016) Nucleus incertus Orexin2 receptors mediate alcohol seeking in rats. Neuropharmacology 110(Part A):82–91

142. Srinivasan S, Simms JA, Nielsen CK, Lieske SP, Bito-Onon JJ, Yi H, Hopf FW, Bonci A, Bartlett SE (2012) The dual orexin/hypocretin receptor antagonist, almorexant, in the ventral tegmental area attenuates ethanol self-administration. PLoS One 7(9):e44726

143. Li J, Hu Z, de Lecea L (2014) The hypocretins/orexins: integrators of multiple physiological functions. Br J Pharmacol 171(2):332–350

144. Sakurai T (2014) The role of orexin in motivated behaviours. Nat Rev Neurosci 15 (11):719–731

145. Fadel J, Bubser M, Deutch AY (2002) Differential activation of orexin neurons by antipsychotic drugs associated with weight gain. J Neurosci 22(15):6742–6746

146. Estabrooke IV, McCarthy MT, Ko E, Chou TC, Chemelli RM, Yanagisawa M, Saper CB, Scammell TE (2001) Fos expression in orexin neurons varies with behavioral state. J Neurosci 21(5):1656–1662

147. Murphy JA, Deurveilher S, Semba K (2003) Stimulant doses of caffeine induce c-FOS activation in orexin/hypocretin containing neurons in rat. Neuroscience 121(2):269 275

148. Sakamoto F, Yamada S, Ueta Y (2004) Centrally administered orexin-A activates corticotropin-releasing factor-containing neurons in the hypothalamic paraventricular nucleus and central amygdaloid nucleus of rats: possible involvement of central orexins on stress-activated central CRF neurons. Regul Pept 118(3):183–191

149. Winsky-Sommerer R, Yamanaka A, Diano S, Borok E, Roberts AJ, Sakurai T, Kilduff TS, Horvath TL, de Lecea L (2004) Interaction between the corticotropin-releasing factor system and hypocretins (orexins): a novel circuit mediating stress response. J Neurosci 24 (50):11439–11448

150. Harris GC, Wimmer M, Randall-Thompson JF, Aston-Jones G (2007) Lateral hypothalamic orexin neurons are critically involved in learning to associate an environment with morphine reward. Behav Brain Res 183(1):43–51

151. Moorman DE, James MH, Kilroy EA, Aston-Jones G (2016) Orexin/hypocretin neuron activation is correlated with alcohol seeking and preference in a topographically specific manner. Eur J Neurosci 43:710–720

152. Richardson KA, Aston-Jones G (2012) Lateral hypothalamic orexin/hypocretin neurons that project to ventral tegmental area are differentially activated with morphine preference. J Neurosci 32(11):3809–3817

153. Gompf HS, Aston-Jones G (2008) Role of orexin input in the diurnal rhythm of locus coeruleus impulse activity. Brain Res 1224:43–52

154. Hamlin AS, Newby J, McNally GP (2007) The neural correlates and role of D1 dopamine receptors in renewal of extinguished alcohol-seeking. Neuroscience 146(2):525–536

155. Kallupi M, Cannella N, Economidou D, Ubaldi M, Ruggeri B, Weiss F, Massi M, Marugan J, Heilig M, Bonnavion P, de Lecea L, Ciccocioppo R (2010) Neuropeptide S facilitates cue-induced relapse to cocaine seeking through activation of the hypothalamic hypocretin system. Proc Natl Acad Sci U S A 107(45):19567–19572

156. Lasheras MC, Laorden ML, Milanes MV, Nunez C (2015) Corticotropin-releasing factor 1 receptor mediates the activity of the reward system evoked by morphine-induced conditioned place preference. Neuropharmacology 95:168–180

157. Hassani OK, Krause MR, Mainville L, Cordova CA, Jones BE (2016) Orexin neurons respond differentially to auditory cues associated with appetitive versus aversive outcomes. J Neurosci 36(5):1747–1757

158. Lee MG, Hassani OK, Jones BE (2005) Discharge of identified orexin/hypocretin neurons across the sleep-waking cycle. J Neurosci 25(28):6716–6720

159. Nollet M, Gaillard P, Minier F, Tanti A, Belzung C, Leman S (2011) Activation of orexin neurons in dorsomedial/perifornical hypothalamus and antidepressant reversal in a rodent model of depression. Neuropharmacology 61(1–2):336–346

160. Sharf R, Sarhan M, Dileone RJ (2008) Orexin mediates the expression of precipitated morphine withdrawal and concurrent activation of the nucleus accumbens shell. Biol Psychiatry 64(3):175–183

161. Plaza-Zabala A, Flores A, Maldonado R, Berrendero F (2012) Hypocretin/orexin signaling in the hypothalamic paraventricular nucleus is essential for the expression of nicotine withdrawal. Biol Psychiatry 71(3):214–223

162. Johnson PL, Molosh A, Fitz SD, Truitt WA, Shekhar A (2012a) Orexin, stress, and anxiety/panic states. Prog Brain Res 198:133–161

163. Johnson PL, Samuels BC, Fitz SD, Federici LM, Hammes N, Early MC, Truitt W, Lowry CA, Shekhar A (2012b) Orexin 1 receptors are a novel target to modulate panic responses and the panic brain network. Physiol Behav 107(5):733–742

164. Sunanaga J, Deng BS, Zhang W, Kanmura Y, Kuwaki T (2009) CO2 activates orexin-containing neurons in mice. Respir Physiol Neurobiol 166(3):184–186

165. Campbell EJ, Watters SM, Zouikr I, Hodgson DM, Dayas CV (2015) Recruitment of hypothalamic orexin neurons after formalin injections in adult male rats exposed to a neonatal immune challenge. Front Neurosci 9:65

166. Tung LW, Lu GL, Lee YH, Yu L, Lee HJ, Leishman E, Bradshaw H, Hwang LL, Hung MS, Mackie K, Zimmer A, Chiou LC (2016) Orexins contribute to restraint stress-induced cocaine relapse by endocannabinoid-mediated disinhibition of dopaminergic neurons. Nat Commun 7:12199

167. Johnson PL et al. (2011) Induction of c-Fos in 'panic/defence'-related brain circuits following brief hypercarbic gas exposure. J Psychopharmacol 25:26–36. doi:10.1177/0269881109353464

168. Webb IC, Patton DF, Hamson DK, Mistlberger RE (2008) Neural correlates of arousal-induced circadian clock resetting: hypocretin/orexin and the intergeniculate leaflet. Eur J Neurosci 27:828–835. doi:10.1111/j.1460-9568.2008.06074

169. Chemelli RM et al. (1999) Narcolepsy in orexin knockout mice: molecular genetics of sleep regulation. Cell 20:437–451

170. Plaza-Zabala A et al. (2013) A role for hypocretin/orexin receptor-1 in cue-induced reinstatement of nicotine-seeking behavior. Neuropsychopharmacology 38:1724–1736

171. Cornish JL et al. (2012) Regional c-Fos and FosB/ΔFosB expression associated with chronic methamphetamine self-administration and methamphetamine-seeking behavior in rat. Neuroscience 206:100–114. doi:10.1016/j.neuroscience.2012.01.004

172. Espana RA, Valentino RJ, Berridge CW (2003) Fos immunoreactivity in hypocretin-synthesizing and hypocretin-1 receptor-expressing neurons: effects of diurnal and nocturnal spontaneous waking, stress and hypocretin-1 administration. Neuroscience 121(1):201–217

173. McGregor R, Wu MF, Barber G, Ramanathan L, Siegel JM (2011) Highly specific role of hypocretin (orexin) neurons: differential activation as a function of diurnal phase, operant reinforcement versus operant avoidance and light level. J Neurosci 31(43):15455–15467

174. Cole S, Hobin MP, Petrovich GD (2015) Appetitive associative learning recruits a distinct network with cortical, striatal, and hypothalamic regions. Neuroscience 286:187–202

175. Petrovich GD, Hobin MP, Reppucci CJ (2012) Selective Fos induction in hypothalamic orexin/hypocretin, but not melanin-concentrating hormone neurons, by a learned food-cue that stimulates feeding in sated rats. Neuroscience 224:70–80

176. Choi DL, Davis JF, Fitzgerald ME, Benoit SC (2010) The role of orexin-A in food motivation, reward-based feeding behavior and food-induced neuronal activation in rats. Neuroscience 167(1):11–20

177. Mena JD, Selleck RA, Baldo BA (2013) Mu-opioid stimulation in rat prefrontal cortex engages hypothalamic orexin/hypocretin-containing neurons, and reveals dissociable roles of nucleus accumbens and hypothalamus in cortically driven feeding. J Neurosci 33(47):18540–18552

178. Zheng H, Patterson LM, Berthoud HR (2007) Orexin signaling in the ventral tegmental area is required for high-fat appetite induced by opioid stimulation of the nucleus accumbens. J Neurosci 27(41):11075–11082

179. Mileykovskiy BY, Kiyashchenko LI, Siegel JM (2005) Behavioral correlates of activity in identified hypocretin/orexin neurons. Neuron 46(5):787–798

180. Adamantidis AR, Zhang F, Aravanis AM, Deisseroth K, de Lecea L (2007) Neural substrates of awakening probed with optogenetic control of hypocretin neurons. Nature 450 (7168):420–424
181. Schone C, Cao ZF, Apergis-Schoute J, Adamantidis A, Sakurai T, Burdakov D (2012) Optogenetic probing of fast glutamatergic transmission from hypocretin/orexin to histamine neurons in situ. J Neurosci 32(36):12437–12443
182. Tsunematsu T, Kilduff TS, Boyden ES, Takahashi S, Tominaga M, Yamanaka A (2011) Acute optogenetic silencing of orexin/hypocretin neurons induces slow-wave sleep in mice. J Neurosci 31(29):10529–10539
183. Tsunematsu T, Tabuchi S, Tanaka KF, Boyden ES, Tominaga M, Yamanaka A (2013) Long-lasting silencing of orexin/hypocretin neurons using archaerhodopsin induces slow-wave sleep in mice. Behav Brain Res 255:64–74
184. Floyd NS, Price JL, Ferry AT, Keay KA, Bandler R (2001) Orbitomedial prefrontal cortical projections to hypothalamus in the rat. J Comp Neurol 432(3):307–328
185. Peters J, Kalivas PW, Quirk GJ (2009) Extinction circuits for fear and addiction overlap in prefrontal cortex. Learn Mem 16(5):279–288
186. Giardino WJ, de Lecea L (2014) Hypocretin (orexin) neuromodulation of stress and reward pathways. Curr Opin Neurobiol 29:103–108
187. Fadel J, Deutch AY (2002) Anatomical substrates of orexin-dopamine interactions: lateral hypothalamic projections to the ventral tegmental area. Neuroscience 111(2):379–387
188. Gonzalez JA, Jensen LT, Fugger L, Burdakov D (2012) Convergent inputs from electrically and topographically distinct orexin cells to locus coeruleus and ventral tegmental area. Eur J Neurosci 35(9):1426–1432
189. Narita M, Nagumo Y, Hashimoto S, Narita M, Khotib J, Miyatake M, Sakurai T, Yanagisawa M, Nakamachi T, Shioda S, Suzuki T (2006) Direct involvement of orexinergic systems in the activation of the mesolimbic dopamine pathway and related behaviors induced by morphine. J Neurosci 26(2):398–405
190. Schone C, Apergis-Schoute J, Sakurai T, Adamantidis A, Burdakov D (2014) Coreleased orexin and glutamate evoke nonredundant spike outputs and computations in histamine neurons. Cell Rep 7(3):697–704
191. Schone C, Burdakov D (2012) Glutamate and GABA as rapid effectors of hypothalamic "peptidergic" neurons. Front Behav Neurosci 6:81
192. Torrealba F, Yanagisawa M, Saper CB (2003) Colocalization of orexin a and glutamate immunoreactivity in axon terminals in the tuberomammillary nucleus in rats. Neuroscience 119(4):1033–1044
193. Chou TC, Lee CE, Lu J, Elmquist JK, Hara J, Willie JT, Beuckmann CT, Chemelli RM, Sakurai T, Yanagisawa M, Saper CB, Scammell TE (2001) Orexin (hypocretin) neurons contain dynorphin. J Neurosci 21(19):RC168
194. Muschamp JW, Hollander JA, Thompson JL, Voren G, Hassinger LC, Onvani S, Kamenecka TM, Borgland SL, Kenny PJ, Carlezon WA Jr (2014). Hypocretin (orexin) facilitates reward by attenuating the antireward effects of its cotransmitter dynorphin in ventral tegmental area. Proc Natl Acad Sci U S A 111(16):E1648–E1655
195. Williams RH, Alexopoulos H, Jensen LT, Fugger L, Burdakov D (2008) Adaptive sugar sensors in hypothalamic feeding circuits. Proc Natl Acad Sci U S A 105(33):11975–11980
196. Ahmed SH (2012) The science of making drug-addicted animals. Neuroscience 211:107–125
197. Ahmed SH, Walker JR, Koob GF (2000) Persistent increase in the motivation to take heroin in rats with a history of drug escalation. Neuropsychopharmacology 22(4):413–421
198. Zimmer BA, Oleson EB, Roberts DCS (2012) The motivation to self-administer is increased after a history of spiking brain levels of cocaine. Neuropsychopharmacology 37(8):1901–1910
199. Michelson D, Snyder E, Paradis E, Chengan-Liu M, Snavely DB, Hutzelmann J, Walsh JK, Krystal AD, Benca RM, Cohn M, Lines C, Roth T, Herring WJ (2014) Safety and efficacy of suvorexant during 1-year treatment of insomnia with subsequent abrupt treatment discontinuation: a phase 3 randomised, double-blind, placebo-controlled trial. Lancet Neurol 13 (5):461–471

Hypocretin/Orexin and Plastic Adaptations Associated with Drug Abuse

Corey Baimel and Stephanie L. Borgland

Abstract Dopamine neurons in the ventral tegmental area (VTA) are a critical part of the neural circuits that underlie reward learning and motivation. Dopamine neurons send dense projections throughout the brain and recent observations suggest that both the intrinsic properties and the functional output of dopamine neurons are dependent on projection target and are subject to neuromodulatory influences. Lateral hypothalamic hypocretin (also termed orexin) neurons project to the VTA and contain both hypocretin and dynorphin peptides in the same dense core vesicles suggesting they may be co-released. Hypocretin peptides act at excitatory $G_{\alpha q}$ protein-coupled receptors and dynorphin acts at inhibitory $G_{\alpha i/o}$ protein-coupled receptors, which are both expressed on subpopulations of dopamine neurons. This review describes a role for neuromodulation of dopamine neurons and the influence on motivated behaviour in response to natural and drug rewards.

Keywords AMPA · Dopamine · Hypocretin · Morphine · NMDA · Ventral tegmental area

Contents

C. Baimel
Department of Anesthesiology, Pharmacology and Therapeutics, University of British Columbia, 2176 Health Sciences Mall, Vancouver, BC, Canada, V6T 1Z3
Hotchkiss Brain Institute, University of Calgary, 3330 Hospital Dr. NW, Calgary, AB, Canada, T2N 4N1

S.L. Borgland (⊠)
Hotchkiss Brain Institute, University of Calgary, 3330 Hospital Dr. NW, Calgary, AB, Canada, T2N 4N1
e-mail: s.borgland@ucalgary.ca

© Springer International Publishing AG 2016
Curr Topics Behav Neurosci (2017) 33: 283–304
DOI 10.1007/7854_2016_44
Published Online: 8 March 2017

1   Introduction

Addiction is a disease that progresses in a step-wise manner, evolving from irregular recreational use, to sustained and escalated use that can transition into compulsive drug consumption with disregard for negative consequences [1, 2]. All addictive drugs target the mesocorticolimbic dopamine system originating in the ventral tegmental area (VTA) and have in common the ability to increase dopamine concentrations in the nucleus accumbens (NAc) [3], albeit by very different mechanisms. These drugs can be grouped into three categories: those that activate G-protein coupled receptors (GPCRs), those that alter the function of ion channels, which both signal primarily in the VTA, and those that target the monoamine transporters at terminal dopamine release sites [4]. It should however be noted that the effects of this third group, which includes the stimulants cocaine and amphetamine, are still dependent on modulatory mechanisms in the VTA. Specifically, cocaine enhances endocannabinoid signalling at presynaptic γ-aminobutyric acid (GABA) inputs to increase the firing rate of VTA dopamine neurons through disinhibition [5]. Drugs of abuse also share the ability to alter synaptic transmission in the mesocorticolimbic dopamine system. Addiction is a disorder of neural circuit dysfunction characterized by serial changes in synaptic transmission initiated in the VTA [6]. Synaptic plasticity mechanisms such as long-term potentiation (LTP) and long-term depression (LTD) are critically involved in learning, and a leading hypothesis suggests that addiction represents a powerful and aberrant form of learning and memory [7–10]. In fact, much evidence has accumulated demonstrating that beyond acute increases in dopamine concentration, addictive drugs also have in common the property of inducing synaptic plasticity at synapses in the VTA [11, 12]. This chapter focuses on how the lateral hypothalamic neuropeptide, hypocretin, modulates neuroplasticity in the VTA, an effect that may underlie this peptide's ability to promote motivated drug seeking.

Drug-induced plasticity in the VTA was initially demonstrated with the finding that a single, in vivo injection of cocaine potentiates the α-amino-3-hydroxy-5-methyl-4-isoxazolepropionic acid receptor (AMPAR) to N-methyl-D-aspartate receptor (NMDAR) ratio (AMPAR/NMDAR ratio) of dopamine neurons [13]. The AMPAR/NMDAR ratio is a normalized measure of basal synaptic strength for comparing between different cells and slice preparations which is independent of the positioning of electrodes or the number of synapses activated [9]. This plasticity is prolonged when animals self-administer cocaine, where increases in the AMPAR/NMDAR ratio are observed up to 3 months after the end of self-administration [14]. Similar drug-induced increases in excitatory synaptic strength have also been demonstrated with single injections of amphetamine [12, 15], nicotine, morphine, ethanol [12], benzodiazepines [16] and delta(9)-tetrahydrocannabinol, the main psychoactive ingredient of marijuana [17], suggesting a common mechanism for potentiating dopaminergic output. Drug-induced potentiation of AMPAR signalling in the VTA is expressed by a redistribution of AMPAR subunits. After drug exposure, GluA2 subunits are internalized and AMPAR excitatory postsynaptic currents (EPSCs) on dopamine neurons show an inwardly rectifying current–voltage relationship and are sensitive to polyamine block [11, 18]. Cocaine-induced changes at synaptic NMDARs have also been observed using two-photon uncaging of glutamate at single synapses. In single excitatory synapses from the VTA of cocaine-treated mice, the amplitude of AMPAR EPSCs was increased whereas NMDAR EPSC amplitude was decreased compared to saline-treated controls [19]. In fact, a small NMDAR EPSC was predictive of a large AMPAR EPSC that shows inward rectification, indicating that these changes do occur at the same synapse [19]. Moreover, the properties of NMDARs switched following cocaine exposure. Specifically, Ca^{2+} permeability of NMDARs, Mg^{2+} block and sensitivity to zinc decreased following cocaine exposure while current decay time and sensitivity to ifenprodil, an inhibitor of GluN2B-containing NMDARs, increased [20]. Taken together, drug-induced switches in both AMPAR and NMDAR subunit composition would drastically alter glutamatergic synaptic transmission and the regulation of dopamine neuron activity.

Synaptic plasticity in the VTA can be gated, strengthened or inhibited by various neuromodulators, including neuropeptides. Hypocretins, also known as orexins, are neuropeptides synthesized in the perifornical region and the lateral hypothalamus (LH) and project widely throughout the brain [21], including to the VTA. Importantly, hypocretin receptors 1 (hcrt-R1) and 2 (hcrt-R2) are expressed in the VTA and appear to play a role in promoting motivated behaviour [22]. This review focuses on how hypocretin modulates drug-induced synaptic transmission in the VTA and related behavioural outcomes.

2 Lateral Hypothalamic Hypocretin Neurons and Motivated Behaviour

The LH has long been studied for its role in motivated behaviour. Electrical stimulation of the LH can evoke behaviours including locomotion and goal-directed activities like feeding, drinking and copulation [23]. Because both rodents and humans will perform operant tasks for electrical stimulation of the LH [24, 25], the LH is now recognized as a critical part of the complex circuitry that regulates reward and reinforcement. Accordingly, addictive drugs modulate operant responding for self-stimulation of the LH [26, 27], and animals will perform operant tasks for direct administration of drugs into the LH [28].

The hypocretin neuropeptides were independently discovered by two groups nearly two decades ago [29, 30]. The hypocretin system consists of two peptides, hypocretin 1 and 2, both cleaved from the same pre-pro-hypocretin precursor molecule [29, 30]. Hypocretin neurons reside in the portions of the LH that support self-stimulation [21, 23, 31], and the functional effects of hypocretin receptor activation mimic those of LH stimulation. Central hypocretin administration evokes feeding [30, 32, 33], drinking [34], copulation [35] and locomotion [36]. Together this suggests that hypocretin neurons may be one of the critical substrates in the LH involved in the regulation of motivated behaviour. Much literature supports this view and proposes that the hypocretin system is critical to the integration of both external cues and internal states to coordinate arousal levels and drive motivational action [22, 37]. These behaviours are partly engaged through hypocretin projections to reward-responsive dopamine neurons in the VTA [38, 39].

Hypocretins bind and activate two GPCRs: hcrt-R1 and hcrt-R2 [30]. Although most of the functional roles of hypocretin receptor activation have been linked to the G_q pathway [40–43], both receptors have been shown to interact with G_q, G_s and $G_{i/o}$ proteins [44]. A lack of direct methods to measure G-protein activation has prevented a definitive evaluation of hypocretin receptor G-protein coupling [44]. Despite being small in number, the hypocretin cell population sends dense projections throughout the brain, including a dense projection to the VTA [21, 45]. Hcrt-R1 and Hcrt-R2 are similarly widely distributed within the brain [46], but their expression patterns are not uniform. For example, there is higher expression of Hcrt-R1 in cortical regions, the bed nucleus of the stria terminalis and the locus coeruleus, whereas Hcrt-R2 density is enriched in the NAc and in subregions of the thalamus and hypothalamus; there are relatively similar levels of expression of both receptors in the VTA [30, 46].

Hypocretin peptides were discovered for their role in regulating feeding and arousal [30, 47–51]. Accordingly, the loss of hypocretin neurons results in the neurological syndrome of narcolepsy, characterized by excessive daytime sleepiness [48, 49, 52]. Interestingly, narcoleptic patients have a lower prevalence of drug abuse [53–55], although it is unclear if and how reward processing is altered in narcoleptic patients [56]. Hypocretin peptides are released in a circadian rhythm and are elevated in the cerebrospinal fluid during active phases and are low during the resting phase [57, 58]. These fluctuations in peptide levels are consistent with the activity levels of hypocretin neurons, which show a similar pattern [59, 60]. Hypocretin signalling has been linked to many different behavioural phenotypes, but a recent hypothesis suggests that across all domains hypocretin plays a critical role in driving motivational activation [22]. Hypocretin neurons are activated by internal, homeostatic or external, motivationally relevant signals and coordinate both psychological and physiological processes to facilitate adaptive behaviours [22].

Hypocretin neurons are so named by way of the fact that they are the only known source of hypocretin 1 and hypocretin 2. That being said, this label is potentially misleading due to the fact that these cells also contain dynorphin [61, 62] and release glutamate [63–65]. Although it has been largely overlooked, potential simultaneous or co-release of these transmitters may have important functional consequences. Moreover, although hypocretin neurons are not thought to be GABAergic [63], recent evidence suggests that hypocretin neurons may be capable of releasing GABA at certain sites [66]. Optogenetic stimulation of hypocretin neurons evokes currents in melanin-concentrating hormone (MCH) cells that persist in the presence of CNQX and APV, blockers of AMPAR and NMDARs, and which decrease the firing of postsynaptic MCH neurons in a $GABA_A$-dependent manner [66]. Together, this suggests that hypocretin neurons are capable of complex regulation of downstream targets.

3 Drug-Induced Activation of Hypocretin Neurons

In terms of anatomical connectivity, hypocretin neurons are well situated to mediate reward and motivation. Many studies have examined the role of hypocretin neurons in drug-seeking behaviour. In conditioned place preference (CPP) tasks, hypocretin neurons in the LH, but not the dorsomedial hypothalamus (DMH) or the perifornical area (PFA), and that project to the VTA but not the locus coeruleus [67], are activated by cues associated with both drug (cocaine and morphine) and food rewards [68]. Moreover, the preference for the drug-paired environment positively correlates with expression levels of the immediate early gene c-Fos in hypocretin neurons [68], a surrogate marker of neuronal activation. Interestingly, following extinction training, stimulation of the LH with the neuropeptide Y receptor type 4 agonist (rat pancreatic polypeptide) or microinfusion of hypocretin A into the VTA reinstates CPP, both of which are inhibited by administration of the Hcrt-R1 antagonist N-(2-Methyl-6-benzoxazolyl)-N′-1,5-naphthyridin-4-yl urea (SB 334867) [68].

In addition to morphine and cocaine, other addictive drugs can also activate hypocretin neurons. Acute application of nicotine increases the firing of hypocretin containing neurons [69]. Similarly, nicotine treatment induces c-Fos expression in hypocretin neurons that project to the basal forebrain and the paraventricular nucleus of the dorsal thalamus [70, 71]. Acute methamphetamine also increases c-Fos activation in perifornical hypocretin neurons [58], whereas acute amphetamine induces c-Fos in medial but not lateral LH/PFA hypocretin neurons [72]. Finally, alcohol seeking [73] or cues previously paired with alcohol [74] increase c-Fos expression in hypocretin neurons. Taken together, drugs of abuse activate hypocretin neurons although the subregional dependence of c-Fos expression varies.

Stressful stimuli like foot shock can also reinstate drug seeking, but do not activate LH hypocretin neurons, but rather activate those in the DMH and the PFA [68]. These results, coupled with data demonstrating that morphine CPP increases c-Fos activation in LH hypocretin neurons [68], suggest that there may be a functional dichotomy of hypocretin neurons in terms of their response to pharmacological and environmental stimuli and how they in turn organize behaviour [75]. It was proposed that DMH and PFA hypocretin neurons were involved in promoting arousal [75], whereas LH hypocretin neurons stimulated reward seeking [75, 76]. This dichotomy was thought to result from discrete targeting of hypocretin projections within each subregion; such that DMH and PFA hypocretin neurons projected to regions like the locus coeruleus to induce arousal; and LH hypocretin neurons likely influenced reward through projections to the reward circuit including, but not limited to, the VTA [75]. However, hypocretin neurons that project to both the locus coeruleus and the VTA are present and spread throughout the hypocretin field, and VTA projecting hypocretin neurons outnumber locus coeruleus projecting neurons in medial areas [77]. In addition, a sensitizing regimen of amphetamine induces c-Fos hypocretin double-labelling throughout the hypocretin field [78], as does naloxone-induced morphine withdrawal [79, 80]. Therefore, an alternative hypothesis is that differential recruitment of hypocretin neurons results from the activation of distinct upstream afferent projections to these neurons rather than anatomically defined segregation within the hypothalamus. Accordingly, hypocretin neurons in the medial regions receive preferential innervation from hypothalamic neurons while those in the LH are targeted by inputs from the brain stem and reward-related regions [81]. Moreover, neurons in the rostral lateral septum, which are activated by addictive drugs and cues in the environment linked to the drug taking experience [82, 83], send projections to the LH, but not to the DMH or PFA [84].

4 Drug-Induced Plasticity at Hypocretin Neurons

Hypocretin neurons are somewhat unique in terms of their regulation by synaptic input [85]. In most long projecting neurons, inhibitory synapses are clustered around the cell body and excitatory inputs are found on more distal dendrites [86, 87]. However, hypocretin neurons have more asymmetric, presumably excitatory, synapses than symmetric inhibitory synapses on their cell bodies [87]. Accordingly, the frequency of excitatory mEPSCs is tenfold higher than that of mIPSCs in these neurons [87]. Furthermore, blocking glutamatergic transmission onto hypocretin neurons significantly inhibits firing of these neurons while inhibition of $GABA_A$ receptors has little effect [88, 89]. Interestingly under basal conditions, hypocretin neurons express GluA2-lacking AMPARs and the amplitude of AMPAR EPSCs is far greater than that of NMDAR EPSCs, giving them a high AMPAR/NMDAR ratio [85, 90–92]. Together this suggests that hypocretin neurons are likely highly responsive to presynaptic glutamate input and that Ca^{2+} regulation of synaptic plasticity in these cells may be dependent on AMPARs, rather than NMDARs [85]. Moreover, efficient glutamatergic transmission onto hypocretin neurons may facilitate excitation of these neurons in response to motivationally relevant sensory input [85].

The critical role of drug-induced plasticity within the mesocorticolimbic dopamine system in the development and the expression of addiction-like behaviours has been well characterized [6, 9, 93], but drug-induced modifications in synaptic transmission are not limited to these brain regions. In fact, given the unique synaptic regulation of hypocretin neurons, and their connections to the mesocorticolimbic dopamine system, drug-induced alterations in excitatory synaptic transmission onto hypocretin neurons may have important functional consequences towards motivated behaviour.

In vivo repeated cocaine administration induces synaptic plasticity at excitatory synapses onto hypocretin neurons. In mice that express a CPP for cocaine, both the AMPAR/NMDAR ratio and the amplitude of AMPAR mEPSCs are increased relative to saline control mice [92]. This plasticity was not induced with a single injection of cocaine, nor was it dependent on Hcrt-R1 signalling and was observed

5, but not 10 days after exposure [92]. Interestingly, this up-regulation of postsynaptic AMPARs did not occlude but rather facilitated the induction of high frequency stimulation induced LTP [92]. This suggests that cocaine-induced plasticity at these synapses, similar to that at synapses in the VTA, may serve as a type of metaplasticity lowering the threshold for plasticity induced by other synaptic stimulations [85]. In contrast, others have reported presynaptic cocaine-induced plasticity including an increase in the frequency of mEPSCs, paired pulse-depression, and an increase in the number of vesicular glutamate transporter 2 (VGLUT-2) positive puncta onto hypocretin neurons [94]. Nevertheless, it should be noted that the behavioural implication of cocaine-induced plasticity at hypocretin neurons has not yet been examined, and it is unclear whether or not synaptic plasticity at these synapses is involved in the development of drug-seeking behaviours. In fact, the Hcrt-R1 antagonist SB 334867 blocks cocaine CPP despite the persistence of cocaine-induced plasticity at synapses onto hypocretin neurons [92]. Therefore, downstream release of hypocretin peptides is likely the critical factor driving motivated behaviour. Plasticity at hypocretin neurons may contribute to motivated behaviour by increasing the probability of hypocretin release.

Given the role of hypocretin receptor signalling in morphine CPP [68], it is surprising that application of MOR agonists decreases the firing rate and the frequency of mEPSCs onto hypocretin neurons [37, 95]. That being said, the effect of in vivo morphine exposure on the activity of hypocretin neurons has not been directly determined. Moreover, because hypocretin neurons are a functionally segregated cell population [75] and only half of hypocretin cells express MORs [79], it is possible that the population of cells inhibited by morphine do not project to reward-related brain areas but may mediate the sedative effects of the drug through projections to arousal-related regions.

5 Hypocretin Release in the Ventral Tegmental Area

Because hypocretin neurons synthesize and release multiple transmitters and peptides [61–66], it is important to consider the conditions under which each would be released. Although there is substantial overlap in the processes that regulate neurotransmitter and neuropeptide release [96], the differences confer important functional considerations. Interestingly, both hypocretin in dense-core vesicles and synaptic vesicles containing glutamate are found in the same terminals [29, 38]. Nevertheless, amino acid neurotransmitters are released by Ca^{2+} dependent exocytosis at specialized sites in the presynaptic "active zone" [97], where docked and primed vesicles are stored in close proximity to voltage-gated Ca^{2+} channels to allow fast synchronous excitation-release coupling [96]. Neuropeptides, on the other hand, are less restrictive and are released from many sites void of synaptic specializations away from the active zone [98]. The dispersed nature of peptide release sites creates an interesting paradox in release requirements. Although dense-core vesicles have a lower Ca^{2+} threshold for vesicle fusion (~1 μM compared to 10–100 μM for synaptic vesicles) [99], they require a greater Ca^{2+} signal to be released, such that cytosolic Ca^{2+} concentrations need to be high enough to permit diffusion of Ca^{2+} to the extrasynaptic sites [98, 99]. Therefore, neuropeptide release likely requires high frequency trains of action potentials. In line with this is the observation that optogenetic stimulation of hypocretin terminals in the tuberomammillary nucleus at 1, 5, 10 and 20 Hz induces glutamate receptor currents in postsynaptic histamine neurons, but hypocretin release, measured by an increase in hypocretin receptor dependent histamine neuronal firing rate, is only observed with 20 Hz stimulations [64]. Neuropeptide release from extrasynaptic sites does not imply that neuropeptides cannot act on cells immediately postsynaptic to the axon, but rather that neuropeptide release and activity is likely much less confined in space [98]. In fact, the diffusion of amino acid neurotransmitters is somewhat limited by the structure of the synapse itself and so they only diffuse small distances in the tens of nanometres, before they undergo rapid degradation or reuptake [98]. Neuropeptides, on the other hand, often lack specific reuptake mechanisms and rather are subject to degradation by peptidases that are present in the extracellular space [98]. Because of this, peptides often have longer extracellular half-lives, which increases the time available for diffusion and are said to signal by volume transmission [100–102].

The extrasynaptic nature of peptide release is of particular relevance for the hypocretin input to the VTA. Although hypocretin fibres are present throughout the VTA [39], only 15% of hypocretin axons make appositional contacts in the VTA and even less (5%) show identifiable synaptic specializations [38]. However, most axons stain heavily for dense-core vesicles suggesting that hypocretin release in this region is largely extrasynaptic [38]. Given that neuropeptide release can occur at sites void of synaptic specializations, the low synaptic incidence of hypocretin neurons in the VTA likely does not impede its neuromodulatory influence over VTA dopaminergic and GABAergic neurons [103].

It is also important to consider that peptides are generated through protein synthesis in the endoplasmic reticulum and are processed in the Golgi apparatus before being packaged in dense-core vesicles and transported to presynaptic terminals through axonal transport. Because of the large energy expenditure required for this process, neuropeptides may be released in smaller numbers of molecules, although this can be compensated for by amplification at GPCRs and increased potency at their receptors [98]. That being said, the time scale required to replenish pools of dense-core vesicles is much greater than that of synaptic vesicles, which are regulated within the terminals [98]. Because of this, hypocretin release may occur in intermittent bouts, and thus be suited to induce longer-lasting synaptic changes to mediate their effects.

6 Hypocretin Modulates Synaptic Transmission in the Ventral Tegmental Area

In whole-cell patch clamp recordings from VTA dopamine neurons, hypocretin 1 induces a dose-dependent transient increase in NMDAR EPSCs [41]. This potentiation is due to the activation of protein kinase C (PKC)/phospholipase C (PLC) intracellular signalling cascades and results in increased trafficking of NMDARs containing the

GluN2A subunit to the membrane [41]. Additionally, NMDAR mEPSC amplitude is increased 15 min, but not 3–4 h after hypocretin 1 application, further indicating the short-term nature of this potentiation [41]. One of the major roles of NMDARs in dopamine neurons is the promotion of burst firing [104–106] a phenomenon associated with more efficient dopamine release in downstream target structures. Therefore, hypocretin 1-induced potentiation of NMDAR may result in increased dopamine neuron activity and increased dopamine release in projecting regions. Hypocretin 1 application to the VTA increases the firing rate of dopamine neurons and in some cases induces burst firing [107, 108]. Moreover, intra-VTA hypocretin 1 facilitates firing induced by stimulation of medial PFC inputs to the VTA [109] and can increase dopamine release in both the NAc and the PFC [110, 111]. Behaviourally, intra-VTA hypocretin 1 induces a CPP, which is blocked with PKC and PLC inhibitors [112].

NMDARs are involved in the induction of certain forms of LTP. It is therefore possible that hypocretin 1-induced potentiation of NMDAR EPSCs facilitates the induction of synaptic plasticity in dopamine neurons [41]. Interestingly, there are no acute effects of hypocretin 1 application on AMPAR EPSCs [41]. However, hypocretin 1 application increases the AMPAR/NMDAR ratio when measured 3–4 h later, an effect that is associated with an increase in both the frequency and the amplitude of AMPAR mEPSCs [41]. The increase in mEPSC amplitude, but not the frequency, was inhibited with the NMDAR antagonist APV, suggesting that AMPAR EPSCs were potentiated through an NMDAR-dependent postsynaptic mechanism as well as an NMDAR-independent increase in glutamate release [41].

Hypocretin 2 is presumably co-released with hypocretin 1 and increases the firing rate of dopamine neurons [107], as well as potentiating NMDAR and AMPAR EPSCs [40]. Hypocretin 2 in the VTA signals primarily through Hcrt-R2 and potentiates NMDAR EPSCs via a PKC-dependent mechanism [40]. Additionally, hypocretin 2 induces a presynaptic increase in glutamate release, which surprisingly is not blocked by antagonists for either Hcrt-R1 or Hcrt-R2 suggesting that the enhanced presynaptic glutamate release may be due to hypocretin 2 or one of its metabolites acting at an unidentified PKC-coupled receptor [40]. Interestingly, although hypocretin 2 infusions in the VTA induce a place preference and increase dopamine release in the NAc [112], inhibition of Hcrt-R2 does not alter reward-seeking behaviours [113–115]. Therefore, hypocretin 2 likely mediates reward seeking by acting on Hcrt-R1, while Hcrt-R2 in the VTA may primarily mediate arousal-related functions [116].

Although there are multiple reports demonstrating the ability of hypocretin to increase the firing rate of dopamine neurons, the effect is quite variable in that many dopamine neurons do not respond to hypocretin application (~50%) [107]. Given the newfound heterogeneity of the dopamine system, both in terms of cellular properties and circuit level connections ([117–120]; Lammel et al. 2012), the inconsistent response to hypocretin application within the VTA might suggest that specific subsets of dopamine neurons are sensitive to the activating effects of hypocretin. Accordingly, hypocretin 1 preferentially induces c-Fos in VTA dopamine neurons that project to the medial PFC and the NAc shell [111]. Because most electrophysiological studies recording dopamine neurons primarily focus on lateral VTA dopamine neurons surrounding the medial terminalis of the optic nucleus that primarily project to the NAc [117], it will be interesting to determine whether subpopulations of VTA dopamine neurons defined by their projection target are differentially responsive to hypocretin.

7 Hypocretin Modulates Drug-Induced Plasticity in the Ventral Tegmental Area

Activation of hypocretin neurons by drugs of abuse may modulate plasticity in the VTA. Hypocretin neurons show high levels of c-Fos expression in drug CPP paradigms [68] and both systemic or intra-VTA SB 334867 attenuates cocaine- or amphetamine-induced increases in dopamine concentration in the NAc [121, 122]. Moreover, intra-VTA infusion of hypocretin 1 potentiates cocaine-induced increases in dopamine concentration and the efficacy of cocaine-mediated dopamine reuptake inhibition [123]. Because hypocretin neurons are likely activated by addictive drugs, and hypocretin potentiates NMDAR EPSCs in the VTA [41], signalling at Hcrt-R1 may be involved in the NMDAR-dependent increase in excitatory transmission induced by addictive drugs [11–13, 124]. Interestingly, when animals are pretreated with the Hcrt-R1 antagonist SB 334867, cocaine-induced potentiation of excitatory synaptic transmission is blocked [41]. Moreover, intra-VTA SB 334867 blocks the development, but not the expression, of cocaine sensitization [41] as well as sensitization induced by morphine [125] and amphetamine [122].

Hypocretin signalling is also required for morphine-induced synaptic plasticity in the VTA. Both systemic and intra-VTA administration of the hcrt-R1 antagonist SB 334867 inhibited morphine-induced potentiation of excitatory synaptic transmission [103]. Interestingly, hypocretin signalling was required for both increased presynaptic glutamate release as well as postsynaptic changes in AMPAR subunit composition induced by morphine [103]. Additionally, SB 334867 blocked a long-lasting decrease in the probability of presynaptic GABA release at inhibitory synapses onto dopamine neurons, indicating that Hcrt-R1 signalling is required for a morphine-induced shift in the balance of excitatory and inhibitory inputs onto dopamine neurons. Thus, hypocretin signalling in the VTA appears to be required for cocaine- or morphine-induced plasticity. Further experiments should assess if this is common for all drugs of abuse that potentiate excitatory synapses in the VTA.

Interestingly, self-administration of drugs or other highly salient reinforcers enhances the functional circuit between hypocretin and dopamine neurons. Following cocaine or high-fat food pellet self-administration, the magnitude of hypocretin-induced potentiation of NMDAR EPSCs is greater than that in naïve animals, or animals that self-administered regular food [126]. Moreover, in these animals hypocretin potentiates NMDAR EPSCs at doses that are subthreshold in control animals [126]. Therefore, exposure to salient reinforcers may facilitate the ability of hypocretin neurons to alter the output of VTA dopamine neurons.

8 Hypocretin in the Ventral Tegmental Area Modulates Motivated Behaviour

Hypocretins modulate many aspects of reward seeking, but several lines of evidence point to a particular role for hypocretin in motivated drug seeking. In self-administration studies, hcrt-R1 antagonists reduce ethanol [127] or cocaine [121, 126] self-administration under progressive ratio schedules of reinforcement, where animals must exert progressively greater effort to obtain the same reinforcement. Consistent with this, hcrt-1 antagonists reduce cocaine infusions on high fixed ratio schedules of reinforcement such as a fixed ratio 20, where 20 lever presses are required per reinforcement [128], but not on low fixed ratio schedules when drug infusions are easily obtained [114, 121]. Similarly, in an experiment where the dose of cocaine progressively declines within a session, such that the animal must progressively increase responding to maintain preferred levels of cocaine, SB 334867 has little effect in the early portion of the session when the response requirement is low, but attenuates responding later in the session as the response requirement increases [121]. Moreover, local infusion of hypocretin 1 in the VTA increases the breakpoint for cocaine self-administration on progressive ratio schedules of reinforcement [123], suggesting that VTA hypocretin 1 increases effortful responding for cocaine. Consistent with this, transgenic mice lacking hypocretin peptides have decreased cocaine self-administration in a fixed ratio 5–20 schedule [128]. Unlike cocaine reinforcement, Hcrt-R1 antagonists do reduce heroin self-administration on low effort fixed ratio 1 schedules of reinforcement [129], suggesting that the role of hypocretin in reinforcement may differ across drugs or perhaps that the motivational aspect of heroin self-administration may differ from that of cocaine. Interestingly, in a behavioural economics task, SB 334867 only reduces cocaine demand when cocaine-associated cues are present [130], further supporting the idea that hypocretin neurons integrate sensory stimuli to drive motivated behaviour [22].

Relapse to drug seeking is a common characteristic of addiction that can be modelled in rodents. Because cues in the environment hold salient motivational properties [131], they are capable of reinstating drug seeking. Hcrt-R1 antagonists block cue-induced reinstatement of drug seeking for cocaine [114, 132], alcohol [133, 134] and heroin [129]. Interestingly, cue-induced reinstatement of cocaine seeking requires both Hcrt-R1 and AMPAR signalling in the VTA [135]. Treatment with a positive allosteric modulator of AMPAR reverses SB 334867 inhibition of reinstatement of drug seeking [135]. Moreover, intracerebroventricular hypocretin administration is sufficient to reinstate cocaine seeking [136]. Although intra-VTA infusions of an Hcrt-R1 antagonist reduce cocaine-induced increases in dopamine concentration [121], it does not block drug-primed reinstatement of cocaine seeking [114]. Together, this suggests that hypocretin in the VTA promotes motivated behaviour when the effort requirement is high and drives reinstatement of drug seeking by modulating signalling at AMPARs.

9 Hypocretin Neurons Contain the Kappa Opioid Receptor Agonist Dynorphin

The neuropeptide dynorphin is widely distributed in the brain and is the endogenous ligand for the kappa-opioid receptor (KOR), but also binds with low affinity to both mu opioid and delta opioid receptors [137]. Dynorphin is cleaved from the precursor prodynorphin [138, 139] and is widely known to mediate negative emotional states. KOR agonists induce place aversions, depression-like behaviour and dysphoria in both humans and animals [140–142].

In the LH, dynorphin is almost exclusively expressed in hypocretin neurons, and nearly all hypocretin neurons contain dynorphin at both the mRNA and protein level [61]. Although dynorphin release from hypocretin neurons has been largely overlooked, it likely has important functional consequences. Both hypocretin and dynorphin levels in the LH increase at night, when animals are awake [143, 144], suggesting that both peptides share a circadian rhythm. Moreover, human narcoleptic patients show a marked reduction in both hypocretin and dynorphin in the hypothalamus [145], and ablating hypocretin neurons yields a phenotype very similar to that of human narcolepsy [146], while this phenotype is only mildly evident in hypocretin peptide knockout mice [48, 147]. Together this suggests that hypocretin and dynorphin may act together to regulate arousal and other behaviours and highlights the importance of taking into account the full peptide and neurotransmitter profile of a neuron when examining function.

Hypocretin receptors [30] and KORs [148] are expressed widely throughout the brain and have opposing actions on cellular excitability. Hypocretin binds to excitatory, primarily G_q coupled receptors, and increases the firing rate of postsynaptic neurons, while dynorphin binds to G_i/G_o-coupled KORs [138] and inhibits neural activity by activating GIRK channels. Nevertheless, recent evidence suggests that hypocretin and dynorphin are co-released [62, 149]. Hypocretin and dynorphin colocalize in 94% of neurons in the LH and are found together within the Golgi apparatus, in axonal processes, and importantly in axonal terminals at asymmetric synapses, where both peptides are located in the same large vesicles [62].

There are many ways in which co-release of hypocretin and dynorphin could modulate the activity of downstream target neurons. Because hypocretin and dynorphin have opposing effects on cellular excitability, the obvious notion is that they would counteract each other, but this is not necessarily the case. In the arcuate nucleus, hypocretin and dynorphin have synergistic effects in that hypocretin directly excites neuropeptide Y containing cells and dynorphin inhibits GABA release onto these neurons [149]. The same is true of hypocretin and dynorphin release onto histaminergic tuberomammillary neurons [150]. Both of these examples highlight the fact that the effect of co-release of hypocretin and dynorphin is very much dependent on the receptor expression of the target neurons [149]. It is possible that at terminal release sites, cells may only express the receptor for hypocretin or dynorphin, and therefore would only respond to one of the peptides [149]. However, even if target cells express both receptors, receptor activity may be differentially regulated. For example, MCH neurons in the hypothalamus are more responsive to dynorphin than they are to hypocretin;

therefore, co-application of these peptides induces a hyperpolarizing current [149]. However, with repeated application of these peptides, the current becomes depolarizing and dominated by hypocretin signalling due to more rapid desensitization of KORs in these cells [149]. Therefore, in some cases the level and the pattern of release may determine the response. Alternatively, the peptides may act with different time courses due to latencies in signalling or in duration of action. In this scenario, one peptide may limit the effect of the other over time [149]. Moreover, the activation of one receptor may alter the signalling of the other. For example, in cells heterologously expressing both Hcrt-R1s and KORs, Hcrt-R1 activation promotes preferential β-arrestin/p38 signalling over G_i of KORs [151]. Finally, and perhaps the most intriguing notion is that although these peptides are co-packaged in vesicles, they are encoded by different genes whose transcription is likely regulated by different factors [62, 149, 152]. Therefore, it is possible that under certain circumstances or physiological states, hypocretin and dynorphin may be differentially synthesized. This may drastically alter the modulatory effect of hypocretin neuron activity on downstream structures like the VTA.

10 Hypocretin and Dynorphin Have Opposing Effects on Motivated Behaviour

Recently, the interplay between hypocretin and dynorphin on motivated behaviour was examined for the first time. It was proposed that a key function of hypocretin is to negate the anti-reward effects of dynorphin [62]. Intra-VTA administration of an Hcrt-R1 antagonist increases reward thresholds for electrical self-stimulation of the LH, decreases impulsive responding in the five choice serial reaction time task and decreases cocaine self-administration on a fixed ratio 5 schedule of reinforcement [62]. Interestingly, all of these effects are blocked by pretreatment with norbinaltorphimine (norBNI), which suggests that in the absence of hypocretin signalling, the inhibitory effects of dynorphin on motivated behaviour are unopposed [62]. However, the effect is not reciprocal because unopposed hypocretin signalling does not increase reward-seeking behaviour [62]; the reason for this remains to be determined. Nevertheless, these effects appear to be due to equal but opposing actions of hypocretin and dynorphin on the firing rate of VTA dopamine neurons, although other structures may also be involved [62]. Although this work suggests that hypocretin and dynorphin signalling may oppose one another, this may be behaviour specific. For example, both SB 334867 and norBNI block stress-induced reinstatement of cocaine seeking [136, 153]. Moreover, morphine withdrawal increases c-Fos levels in hypocretin neurons [79, 80]. It is interesting to consider that under these conditions hypocretin and dynorphin may work synergistically to promote drug seeking, with hypocretin motivating drug seeking to alleviate some of the negative affect induced by dynorphin.

11　Summary

Dopamine signalling is linked to motivated behaviour and dopamine release is dependent on both the level and the pattern of activity of VTA dopamine neurons. These patterns of activity are determined by intrinsic conductances and afferent synaptic input, both of which are subject to neuromodulatory influences. Hypocretin neurons project to the VTA and promote motivated behaviour through Hcrt-R1 dependent increases in firing and changes in synaptic transmission.

Drugs of abuse target the mesocorticolimbic dopamine system and induce synaptic plasticity at both excitatory and inhibitory synapses onto these neurons. Despite much research, it remains unclear how this pharmacologically diverse group of drugs all converge to induce synaptic plasticity in the VTA. Hypocretin in the VTA is required for both cocaine- and morphine-induced potentiation of excitatory transmission [41, 103], suggesting it may be one mechanism by which drugs of differing mechanisms promote synaptic plasticity in the VTA. Presumably, increased drug-induced excitatory synaptic transmission onto dopamine neuron increases phasic dopamine release in target regions of the VTA. Ultimately, this may underlie enhanced drug-seeking behaviour and why hypocretin receptor antagonists are effective at inhibiting reinstatement of drug seeking.

Recent evidence highlights the complexity of the VTA dopamine system, such that distinct subpopulations of dopamine neurons differ in their intrinsic electrophysiological properties and their afferent and efferent connection [117, 118, 154]. Moreover, these different subpopulations may be subject to separate neuromodulatory influences [111, 119, 155, 156]. Because hypocretin neurons contain both excitatory hypocretin and inhibitory dynorphin not only in the same cell [61] but also within the same vesicles [62], potential co-release of these opposing peptides may have distinct effects on different subpopulations of dopamine neurons. Thus, the neuromodulatory influences of hypocretin and dynorphin on plasticity of dopamine neurons may vary according to their afferent and efferent connections. Future directions should be aimed at parsing out how these neuromodulators can tune dopamine neurons and their related circuits to drive motivated behaviours.

References

1. Deroche-Gamonet V, Piazza PV (2014) Psychobiology of cocaine addiction: contribution of a multi-symptomatic animal model of loss of control. Neuropharmacology 76(Pt B):437–449
2. Koob GF, Volkow ND (2010) Neurocircuitry of addiction. Neuropsychopharmacology 35:217–238
3. Di Chiara G, Imperato A (1988) Drugs abused by humans preferentially increase synaptic dopamine concentrations in the mesolimbic system of freely moving rats. Proc Natl Acad Sci U S A 85:5274–5278
4. Lüscher C, Ungless MA (2006) The mechanistic classification of addictive drugs. PLoS Med 3:e437

5. Wang H, Treadway T, Covey DP, Cheer JF, Lupica CR (2015) Cocaine-induced endocannabinoid mobilization in the ventral tegmental area. Cell Rep 12:1997–2008
6. Lüscher C, Pascoli V, Creed M (2015) Optogenetic dissection of neural circuitry: from synaptic causalities to blue prints for novel treatments of behavioral diseases. Curr Opin Neurobiol 35:95–100
7. Hyman SE, Malenka RC, Nestler EJ (2006) Neural mechanisms of addiction: the role of reward-related learning and memory. Annu Rev Neurosci 29:565–598
8. Kalivas PW, Volkow ND (2005) The neural basis of addiction: a pathology of motivation and choice. Am J Psychiatry 162:1403–1413
9. Kauer JA, Malenka RC (2007) Synaptic plasticity and addiction. Nat Rev Neurosci 8:844–858
10. Kelley AE (2004) Memory and addiction: shared neural circuitry and molecular mechanisms. Neuron 44:161–179
11. Brown MTC, Bellone C, Mameli M, Labouèbe G, Bocklisch C, Balland B, Dahan L, Luján R, Deisseroth K, Lüscher C (2010) Drug-driven AMPA receptor redistribution mimicked by selective dopamine neuron stimulation. PLoS One 5:e15870
12. Saal D, Dong Y, Bonci A, Malenka RC (2003) Drugs of abuse and stress trigger a common synaptic adaptation in dopamine neurons. Neuron 37:577–582
13. Ungless MA, Whistler JL, Malenka RC, Bonci A (2001) Single cocaine exposure in vivo induces long-term potentiation in dopamine neurons. Nature 411:583–587
14. Chen BT, Bowers MS, Martin M, Hopf FW, Guillory AM, Carelli RM, Chou JK, Bonci A (2008) Cocaine but not natural reward self-administration nor passive cocaine infusion produces persistent LTP in the VTA. Neuron 59:288–297
15. Faleiro LJ, Jones S, Kauer JA (2003) Rapid AMPAR/NMDAR response to amphetamine: a detectable increase in AMPAR/NMDAR ratios in the ventral tegmental area is detectable after amphetamine injection. Ann N Y Acad Sci 1003:391–394
16. Heikkinen AE, Möykkynen TP, Korpi ER (2009) Long-lasting modulation of glutamatergic transmission in VTA dopamine neurons after a single dose of benzodiazepine agonists. Neuropsychopharmacology 34:290–298
17. Good CH, Lupica CR (2010) Afferent-specific AMPA receptor subunit composition and regulation of synaptic plasticity in midbrain dopamine neurons by abused drugs. J Neurosci 30:7900–7909
18. Bellone C, Lüscher C (2006) Cocaine triggered AMPA receptor redistribution is reversed in vivo by mGluR-dependent long-term depression. Nat Neurosci 9:636–641
19. Mameli M, Bellone C, Brown MTC, Lüscher C (2011) Cocaine inverts rules for synaptic plasticity of glutamate transmission in the ventral tegmental area. Nat Neurosci 14:414–416
20. Yuan T, Mameli M, O' Connor EC, Dey PN, Verpelli C, Sala C, Perez-Otano I, Lüscher C, Bellone C (2013) Expression of cocaine-evoked synaptic plasticity by GluN3A-containing NMDA receptors. Neuron 80:1025–1038
21. Peyron C, Tighe DK, van den Pol AN, de Lecea L, Heller HC, Sutcliffe JG, Kilduff TS (1998) Neurons containing hypocretin (orexin) project to multiple neuronal systems. J Neurosci 18:9996–10015
22. Mahler SV, Moorman DE, Smith RJ, James MH, Aston-Jones G (2014) Motivational activation: a unifying hypothesis of orexin/hypocretin function. Nat Neurosci 17:1298–1303
23. Valenstein ES, Cox VC, Kakolewski JW (1970) Reexamination of the role of the hypothalamus in motivation. Psychol Rev 77:16–31
24. Heath RG (1963) Electrical self-stimulation of the brain in man. Am J Psychiatry 120:571–577
25. Olds J, Milner P (1954) Positive reinforcement produced by electrical stimulation of septal area and other regions of rat brain. J Comp Physiol Psychol 47:419–427
26. Adams WJ, Lorens SA, Mitchell CL (1972) Morphine enhances lateral hypothalamic self-stimulation in the rat. Proc Soc Exp Biol Med 140:770–771

27. Carey RJ, Goodal E (1975) Differential effects of amphetamine and food deprivation of self-stimulation of the lateral hypothalamus and medial frontal cortex. J Comp Physiol Psychol 88:224–230
28. Cazala P, Darracq C, Saint-Marc M (1987) Self-administration of morphine into the lateral hypothalamus in the mouse. Brain Res 416:283–288
29. de Lecea L, Kilduff TS, Peyron C, Gao X, Foye PE, Danielson PE, Fukuhara C, Battenberg EL, Gautvik VT, Bartlett FS, Frankel WN, van den Pol AN, Bloom FE, Gautvik KM, Sutcliffe JG (1998) The hypocretins: hypothalamus-specific peptides with neuroexcitatory activity. Proc Natl Acad Sci U S A 95:322–327
30. Sakurai T et al (1998) Orexins and orexin receptors: a family of hypothalamic neuropeptides and G protein-coupled receptors that regulate feeding behavior. Cell 92:573–585
31. Hollander JA, Lu Q, Cameron MD, Kamenecka TM, Kenny PJ (2008) Insular hypocretin transmission regulates nicotine reward. Proc Natl Acad Sci U S A 105:19480–19485
32. Dube MG, Kalra SP, Kalra PS (1999) Food intake elicited by central administration of orexins/hypocretins: identification of hypothalamic sites of action. Brain Res 842:473–477
33. Haynes AC, Jackson B, Overend P, Buckingham RE, Wilson S, Tadayyon M, Arch JR (1999) Effects of single and chronic intracerebroventricular administration of the orexins on feeding in the rat. Peptides 20:1099–1105
34. Kunii K, Yamanaka A, Nambu T, Matsuzaki I, Goto K, Sakurai T (1999) Orexins/hypocretins regulate drinking behaviour. Brain Res 842:256–261
35. Gulia KK, Mallick HN, Kumar VM (2003) Orexin A (hypocretin-1) application at the medial preoptic area potentiates male sexual behavior in rats. Neuroscience 116:921–923
36. Kotz CM, Teske JA, Levine JA, Wang C (2002) Feeding and activity induced by orexin A in the lateral hypothalamus in rats. Regul Pept 104:27–32
37. Baimel C, Bartlett SE, Chiou L-C, Lawrence AJ, Muschamp JW, Patkar O, Tung L-W, Borgland SL (2015) Orexin/hypocretin role in reward: implications for opioid and other addictions. Br J Pharmacol 172:334–348
38. Balcita-Pedicino JJ, Sesack SR (2007) Orexin axons in the rat ventral tegmental area synapse infrequently onto dopamine and gamma-aminobutyric acid neurons. J Comp Neurol 503:668–684
39. Fadel J, Deutch AY (2002) Anatomical substrates of orexin-dopamine interactions: lateral hypothalamic projections to the ventral tegmental area. Neuroscience 111:379–387
40. Borgland SL, Storm E, Bonci A (2008) Orexin B/hypocretin 2 increases glutamatergic transmission to ventral tegmental area neurons. Eur J Neurosci 28:1545–1556
41. Borgland SL, Taha SA, Sarti F, Fields HL, Bonci A (2006) Orexin A in the VTA is critical for the induction of synaptic plasticity and behavioral sensitization to cocaine. Neuron 49:589–601
42. Uramura K, Funahashi H, Muroya S, Shioda S, Takigawa M, Yada T (2001) Orexin-a activates phospholipase C- and protein kinase C-mediated Ca2+ signaling in dopamine neurons of the ventral tegmental area. Neuroreport 12:1885–1889
43. Yang B, Samson WK, Ferguson AV (2003) Excitatory effects of orexin-A on nucleus tractus solitarius neurons are mediated by phospholipase C and protein kinase C. J Neurosci 23:6215–6222
44. Kukkonen JP, Leonard CS (2014) Orexin/hypocretin receptor signalling cascades. Br J Pharmacol 171:314–331
45. Baldo BA, Daniel RA, Berridge CW, Kelley AE (2003) Overlapping distributions of orexin/hypocretin- and dopamine-beta-hydroxylase immunoreactive fibers in rat brain regions mediating arousal, motivation, and stress. J Comp Neurol 464:220–237
46. Trivedi P, Yu H, MacNeil DJ, Van der Ploeg LH, Guan XM (1998) Distribution of orexin receptor mRNA in the rat brain. FEBS Lett 438:71–75
47. Adamantidis AR, Zhang F, Aravanis AM, Deisseroth K, de Lecea L (2007) Neural substrates of awakening probed with optogenetic control of hypocretin neurons. Nature 450:420–424
48. Chemelli RM, Willie JT, Sinton CM, Elmquist JK, Scammell T, Lee C, Richardson JA, Williams SC, Xiong Y, Kisanuki Y, Fitch TE, Nakazato M, Hammer RE, Saper CB, Yanagisawa M (1999) Narcolepsy in orexin knockout mice: molecular genetics of sleep regulation. Cell 98:437–451

49. Lin L, Faraco J, Li R, Kadotani H, Rogers W, Lin X, Qiu X, de Jong PJ, Nishino S, Mignot E (1999) The sleep disorder canine narcolepsy is caused by a mutation in the hypocretin (orexin) receptor 2 gene. Cell 98:365–376
50. Peyron C et al (2000) A mutation in a case of early onset narcolepsy and a generalized absence of hypocretin peptides in human narcoleptic brains. Nat Med 6:991–997
51. Thannickal TC, Moore RY, Nienhuis R, Ramanathan L, Gulyani S, Aldrich M, Cornford M, Siegel JM (2000) Reduced number of hypocretin neurons in human narcolepsy. Neuron 27:469–474
52. Burgess CR, Scammell TE (2012) Narcolepsy: neural mechanisms of sleepiness and cataplexy. J Neurosci 32:12305–12311
53. Akimoto H, Honda Y, Takahashi Y (1960) Pharmacotherapy in narcolepsy. Dis Nerv Syst 21:704–706
54. Guilleminault C, Carskadon M, Dement WC (1974) On the treatment of rapid eye movement narcolepsy. Arch Neurol 30:90–93
55. Nishino S, Mignot E (1997) Pharmacological aspects of human and canine narcolepsy. Prog Neurobiol 52:27–78
56. Dimitrova A, Fronczek R, Van der Ploeg J, Scammell T, Gautam S, Pascual-Leone A, Lammers GJ (2011) Reward-seeking behavior in human narcolepsy. J Clin Sleep Med 7:293–300
57. Blouin AM, Fried I, Wilson CL, Staba RJ, Behnke EJ, Lam HA, Maidment NT, Karlsson KÆ, Lapierre JL, Siegel JM (2013) Human hypocretin and melanin-concentrating hormone levels are linked to emotion and social interaction. Nat Commun 4:1547
58. Estabrooke IV, McCarthy MT, Ko E, Chou TC, Chemelli RM, Yanagisawa M, Saper CB, Scammell TE (2001) Fos expression in orexin neurons varies with behavioral state. J Neurosci 21:1656–1662
59. Lee MG, Hassani OK, Jones BE (2005) Discharge of identified orexin/hypocretin neurons across the sleep-waking cycle. J Neurosci 25:6716–6720
60. Mileykovskiy BY, Kiyashchenko LI, Siegel JM (2005) Behavioral correlates of activity in identified hypocretin/orexin neurons. Neuron 46:787–798
61. Chou TC, Lee CE, Lu J, Elmquist JK, Hara J, Willie JT, Beuckmann CT, Chemelli RM, Sakurai T, Yanagisawa M, Saper CB, Scammell TE (2001) Orexin (hypocretin) neurons contain dynorphin. J Neurosci 21:RC168
62. Muschamp JW, Hollander JA, Thompson JL, Voren G, Hassinger LC, Onvani S, Kamenecka TM, Borgland SL, Kenny PJ, Carlezon WA (2014) Hypocretin (orexin) facilitates reward by attenuating the antireward effects of its cotransmitter dynorphin in ventral tegmental area. Proc Natl Acad Sci U S A 111:E1648–E1655
63. Rosin DL, Weston MC, Sevigny CP, Stornetta RL, Guyenet PG (2003) Hypothalamic orexin (hypocretin) neurons express vesicular glutamate transporters VGLUT1 or VGLUT2. J Comp Neurol 465:593–603
64. Schöne C, Apergis-Schoute J, Sakurai T, Adamantidis A, Burdakov D (2014) Coreleased orexin and glutamate evoke nonredundant spike outputs and computations in histamine neurons. Cell Rep 7:697–704
65. Schöne C, Cao ZFH, Apergis-Schoute J, Adamantidis A, Sakurai T, Burdakov D (2012) Optogenetic probing of fast glutamatergic transmission from hypocretin/orexin to histamine neurons in situ. J Neurosci 32:12437–12443
66. Apergis-Schoute J, Iordanidou P, Faure C, Jego S, Schöne C, Aitta-Aho T, Adamantidis A, Burdakov D (2015) Optogenetic evidence for inhibitory signaling from orexin to MCH neurons via local microcircuits. J Neurosci 35:5435–5441
67. Richardson KA, Aston-Jones G (2012) Lateral hypothalamic orexin/hypocretin neurons that project to ventral tegmental area are differentially activated with morphine preference. J Neurosci 32:3809–3817
68. Harris GC, Wimmer M, Aston-Jones G (2005) A role for lateral hypothalamic orexin neurons in reward seeking. Nature 437:556–559

69. Huang H, Xu Y, van den Pol AN (2011) Nicotine excites hypothalamic arcuate anorexigenic proopiomelanocortin neurons and orexigenic neuropeptide Y neurons: similarities and differences. J Neurophysiol 106:1191–1202

70. Pasumarthi RK, Fadel J (2008) Activation of orexin/hypocretin projections to basal forebrain and paraventricular thalamus by acute nicotine. Brain Res Bull 77:367–373

71. Plaza-Zabala A, Martín-García E, de Lecea L, Maldonado R, Berrendero F (2010) Hypocretins regulate the anxiogenic-like effects of nicotine and induce reinstatement of nicotine-seeking behavior. J Neurosci 30:2300–2310

72. Fadel J, Bubser M, Deutch AY (2002) Differential activation of orexin neurons by antipsychotic drugs associated with weight gain. J Neurosci 22:6742–6746

73. Moorman DE, James MH, Kilroy EA, Aston-Jones G (2016) Orexin/hypocretin neuron activation is correlated with alcohol seeking and preference in a topographically specific manner. Eur J Neurosci 43:710–720

74. Dayas CV, McGranahan TM, Martin-Fardon R, Weiss F (2008) Stimuli linked to ethanol availability activate hypothalamic CART and orexin neurons in a reinstatement model of relapse. Biol Psychiatry 63:152–157

75. Harris GC, Aston-Jones G (2006) Arousal and reward: a dichotomy in orexin function. Trends Neurosci 29:571–577

76. Willie JT, Chemelli RM, Sinton CM, Tokita S, Williams SC, Kisanuki YY, Marcus JN, Lee C, Elmquist JK, Kohlmeier KA, Leonard CS, Richardson JA, Hammer RE, Yanagisawa M (2003) Distinct narcolepsy syndromes in Orexin receptor-2 and Orexin null mice: molecular genetic dissection of Non-REM and REM sleep regulatory processes. Neuron 38:715–730

77. González JA, Jensen LT, Fugger L, Burdakov D (2012) Convergent inputs from electrically and topographically distinct orexin cells to locus coeruleus and ventral tegmental area. Eur J Neurosci 35:1426–1432

78. McPherson CS, Featherby T, Krstew E, Lawrence AJ (2007) Quantification of phosphorylated cAMP-response element-binding protein expression throughout the brain of amphetamine-sensitized rats: activation of hypothalamic orexin A-containing neurons. J Pharmacol Exp Ther 323:805–812

79. Georgescu D, Zachariou V, Barrot M, Mieda M, Willie JT, Eisch AJ, Yanagisawa M, Nestler EJ, DiLeone RJ (2003) Involvement of the lateral hypothalamic peptide orexin in morphine dependence and withdrawal. J Neurosci 23:3106–3111

80. Laorden ML, Ferenczi S, Pintér-Kübler B, González-Martín LL, Lasheras MC, Kovács KJ, Milanés MV, Núñez C (2012) Hypothalamic orexin – a neurons are involved in the response of the brain stress system to morphine withdrawal. PLoS One 7:e36871

81. Yoshida K, McCormack S, España RA, Crocker A, Scammell TE (2006) Afferents to the orexin neurons of the rat brain. J Comp Neurol 494:845–861

82. Shoji S, Simms D, McDaniel WC, Gallagher JP (1997) Chronic cocaine enhances gamma-aminobutyric acid and glutamate release by altering presynaptic and not postsynaptic gamma-aminobutyric acidB receptors within the rat dorsolateral septal nucleus. J Pharmacol Exp Ther 280:129–137

83. Shoji S, Simms D, Yamada K, Gallagher JP (1998) Cocaine administered in vitro to brain slices from rats treated with cocaine chronically in vivo results in a gamma-aminobutyric acid receptor-mediated hyperpolarization recorded from the dorsolateral septum. J Pharmacol Exp Ther 286:509–518

84. Sartor GC, Aston-Jones GS (2012) A septal-hypothalamic pathway drives orexin neurons, which is necessary for conditioned cocaine preference. J Neurosci 32:4623–4631

85. Gao X-B, Hermes G (2015) Neural plasticity in hypocretin neurons: the basis of hypocretinergic regulation of physiological and behavioral functions in animals. Front Syst Neurosci 9:142

86. Douglas RJ, Martin KAC (2004) Neuronal circuits of the neocortex. Annu Rev Neurosci 27:419–451

87. Horvath TL, Gao X-B (2005) Input organization and plasticity of hypocretin neurons: possible clues to obesity's association with insomnia. Cell Metab 1:279–286

88. Li Y, Gao XB, Sakurai T, van den Pol AN (2002) Hypocretin/orexin excites hypocretin neurons via a local glutamate neuron – a potential mechanism for orchestrating the hypothalamic arousal system. Neuron 36:1169–1181
89. Xie X, Crowder TL, Yamanaka A, Morairty SR, Lewinter RD, Sakurai T, Kilduff TS (2006) GABA(B) receptor-mediated modulation of hypocretin/orexin neurones in mouse hypothalamus. J Physiol 574:399–414
90. Rao Y, Liu Z-W, Borok E, Rabenstein RL, Shanabrough M, Lu M, Picciotto MR, Horvath TL, Gao X-B (2007) Prolonged wakefulness induces experience-dependent synaptic plasticity in mouse hypocretin/orexin neurons. J Clin Invest 117:4022–4033
91. Rao Y, Lu M, Ge F, Marsh DJ, Qian S, Wang AH, Picciotto MR, Gao X-B (2008) Regulation of synaptic efficacy in hypocretin/orexin-containing neurons by melanin concentrating hormone in the lateral hypothalamus. J Neurosci 28:9101–9110
92. Rao Y, Mineur YS, Gan G, Wang AH, Liu Z-W, Wu X, Suyama S, de Lecea L, Horvath TL, Picciotto MR, Gao X-B (2013) Repeated in vivo exposure of cocaine induces long-lasting synaptic plasticity in hypocretin/orexin-producing neurons in the lateral hypothalamus in mice. J Physiol 591:1951–1966
93. Creed MC, Lüscher C (2013) Drug-evoked synaptic plasticity: beyond metaplasticity. Curr Opin Neurobiol 23:553–558
94. Yeoh JW, James MH, Jobling P, Bains JS, Graham BA, Dayas CV (2012) Cocaine potentiates excitatory drive in the perifornical/lateral hypothalamus. J Physiol 590:3677–3689
95. Li Y, van den Pol AN (2008) Mu-opioid receptor-mediated depression of the hypothalamic hypocretin/orexin arousal system. J Neurosci 28:2814–2819
96. Südhof TC (2012) The presynaptic active zone. Neuron 75:11–25
97. Couteaux R, Pécot-Dechavassine M (1970) Synaptic vesicles and pouches at the level of "active zones" of the neuromuscular junction. C R Acad Sci Hebd Seances Acad Sci D 271:2346–2349
98. van den Pol AN (2012) Neuropeptide transmission in brain circuits. Neuron 76:98–115
99. Südhof TC (2012) Calcium control of neurotransmitter release. Cold Spring Harb Perspect Biol 4:a011353
100. Agnati LF, Fuxe K, Zoli M, Ozini I, Toffano G, Ferraguti F (1986) A correlation analysis of the regional distribution of central enkephalin and beta-endorphin immunoreactive terminals and of opiate receptors in adult and old male rats. Evidence for the existence of two main types of communication in the central nervous system: the volume transmission and the wiring transmission. Acta Physiol Scand 128:201–207
101. Agnati LF, Guidolin D, Guescini M, Genedani S, Fuxe K (2010) Understanding wiring and volume transmission. Brain Res Rev 64:137–159
102. Agnati LF, Zoli M, Strömberg I, Fuxe K (1995) Intercellular communication in the brain: wiring versus volume transmission. Neuroscience 69:711–726
103. Baimel C, Borgland SL (2015) Orexin signaling in the VTA gates morphine-induced synaptic plasticity. J Neurosci 35:7295–7303
104. Overton P, Clark D (1992) Iontophoretically administered drugs acting at the N-methyl-D-aspartate receptor modulate burst firing in A9 dopamine neurons in the rat. Synapse 10:131–140
105. Suaud-Chagny MF, Chergui K, Chouvet G, Gonon F (1992) Relationship between dopamine release in the rat nucleus accumbens and the discharge activity of dopaminergic neurons during local in vivo application of amino acids in the ventral tegmental area. Neuroscience 49:63–72
106. Tong ZY, Overton PG, Clark D (1996) Antagonism of NMDA receptors but not AMPA/kainate receptors blocks bursting in dopaminergic neurons induced by electrical stimulation of the prefrontal cortex. J Neural Transm (Vienna) 103:889–904
107. Korotkova TM, Sergeeva OA, Eriksson KS, Haas HL, Brown RE (2003) Excitation of ventral tegmental area dopaminergic and nondopaminergic neurons by orexins/hypocretins. J Neurosci 23:7–11
108. Muschamp JW, Dominguez JM, Sato SM, Shen R-Y, Hull EM (2007) A role for hypocretin (orexin) in male sexual behavior. J Neurosci 27:2837–2845

109. Moorman DE, Aston-Jones G (2010) Orexin/hypocretin modulates response of ventral tegmental dopamine neurons to prefrontal activation: diurnal influences. J Neurosci 30:15585–15599
110. Vittoz NM, Berridge CW (2006) Hypocretin/orexin selectively increases dopamine efflux within the prefrontal cortex: involvement of the ventral tegmental area. Neuropsychopharmacology 31:384–395
111. Vittoz NM, Schmeichel B, Berridge CW (2008) Hypocretin/orexin preferentially activates caudomedial ventral tegmental area dopamine neurons. Eur J Neurosci 28:1629–1640
112. Narita M, Nagumo Y, Miyatake M, Ikegami D, Kurahashi K, Suzuki T (2007) Implication of protein kinase C in the orexin-induced elevation of extracellular dopamine levels and its rewarding effect. Eur J Neurosci 25:1537–1545
113. Prince CD, Rau AR, Yorgason JT, España RA (2015) Hypocretin/orexin regulation of dopamine signaling and cocaine self-administration is mediated predominantly by hypocretin receptor 1. ACS Chem Nerosci 6:138–146
114. Smith RJ, See RE, Aston-Jones G (2009) Orexin/hypocretin signaling at the orexin 1 receptor regulates cue-elicited cocaine-seeking. Eur J Neurosci 30:493–503
115. Wang B, You Z-B, Wise RA (2009) Reinstatement of cocaine seeking by hypocretin (orexin) in the ventral tegmental area: independence from the local corticotropin-releasing factor network. Biol Psychiatry 65:857–862
116. Aston-Jones G, Smith RJ, Sartor GC, Moorman DE, Massi L, Tahsili-Fahadan P, Richardson KA (2010) Lateral hypothalamic orexin/hypocretin neurons: a role in reward-seeking and addiction. Brain Res 1314:74–90
117. Lammel S, Hetzel A, Häckel O, Jones I, Liss B, Roeper J (2008) Unique properties of mesoprefrontal neurons within a dual mesocorticolimbic dopamine system. Neuron 57:760–773
118. Lammel S, Ion DI, Roeper J, Malenka RC (2011) Projection-specific modulation of dopamine neuron synapses by aversive and rewarding stimuli. Neuron 70:855–862
119. Margolis EB, Mitchell JM, Ishikawa J, Hjelmstad GO, Fields HL (2008) Midbrain dopamine neurons: projection target determines action potential duration and dopamine D(2) receptor inhibition. J Neurosci 28:8908–8913
120. Lammel S, Lim BK, Ran C, Huang KW, Betley MJ, Tye KM, Deisseroth K, Malenka RC (2012) Input-specific control of reward and aversion in the ventral tegmental area. Nature 491:212–217
121. España RA, Oleson EB, Locke JL, Brookshire BR, Roberts DCS, Jones SR (2010) The hypocretin-orexin system regulates cocaine self-administration via actions on the mesolimbic dopamine system. Eur J Neurosci 31:336–348
122. Quarta D, Valerio E, Hutcheson DM, Hedou G, Heidbreder C (2010) The orexin-1 receptor antagonist SB-334867 reduces amphetamine-evoked dopamine outflow in the shell of the nucleus accumbens and decreases the expression of amphetamine sensitization. Neurochem Int 56:11–15
123. España RA, Melchior JR, Roberts DCS, Jones SR (2011) Hypocretin 1/orexin A in the ventral tegmental area enhances dopamine responses to cocaine and promotes cocaine self-administration. Psychopharmacology (Berl) 214:415–426
124. Borgland SL, Malenka RC, Bonci A (2004) Acute and chronic cocaine-induced potentiation of synaptic strength in the ventral tegmental area: electrophysiological and behavioral correlates in individual rats. J Neurosci 24:7482–7490
125. Narita M, Nagumo Y, Hashimoto S, Narita M, Khotib J, Miyatake M, Sakurai T, Yanagisawa M, Nakamachi T, Shioda S, Suzuki T (2006) Direct involvement of orexinergic systems in the activation of the mesolimbic dopamine pathway and related behaviors induced by morphine. J Neurosci 26:398–405
126. Borgland SL, Chang S-J, Bowers MS, Thompson JL, Vittoz N, Floresco SB, Chou J, Chen BT, Bonci A (2009) Orexin A/hypocretin-1 selectively promotes motivation for positive reinforcers. J Neurosci 29:11215–11225
127. Jupp B, Krivdic B, Krstew E, Lawrence AJ (2011) The orexin$_1$ receptor antagonist SB-334867 dissociates the motivational properties of alcohol and sucrose in rats. Brain Res 1391:54–59

128. Hollander JA, Pham D, Fowler CD, Kenny PJ (2012) Hypocretin-1 receptors regulate the reinforcing and reward-enhancing effects of cocaine: pharmacological and behavioral genetics evidence. Front Behav Neurosci 6:47
129. Smith RJ, Aston-Jones G (2012) Orexin/hypocretin 1 receptor antagonist reduces heroin self-administration and cue-induced heroin seeking. Eur J Neurosci 35:798–804
130. Bentzley BS, Aston-Jones G (2015) Orexin-1 receptor signaling increases motivation for cocaine-associated cues. Eur J Neurosci 41:1149–1156
131. Robinson TE, Berridge KC (1993) The neural basis of drug craving: an incentive-sensitization theory of addiction. Brain Res Brain Res Rev 18:247–291
132. James MH, Charnley JL, Levi EM, Jones E, Yeoh JW, Smith DW, Dayas CV (2011) Orexin-1 receptor signalling within the ventral tegmental area, but not the paraventricular thalamus, is critical to regulating cue-induced reinstatement of cocaine-seeking. Int J Neuropsychopharmacol 14:684–690
133. Brown RM, Kim AK, Khoo SY-S, Kim JH, Jupp B, Lawrence AJ (2016) Orexin-1 receptor signalling in the prelimbic cortex and ventral tegmental area regulates cue-induced reinstatement of ethanol-seeking in iP rats. Addict Biol 21:603–612
134. Lawrence AJ, Cowen MS, Yang H-J, Chen F, Oldfield B (2006) The orexin system regulates alcohol-seeking in rats. Br J Pharmacol 148:752–759
135. Mahler SV, Smith RJ, Aston-Jones G (2013) Interactions between VTA orexin and glutamate in cue-induced reinstatement of cocaine seeking in rats. Psychopharmacology (Berl) 226:687–698
136. Boutrel B, Kenny PJ, Specio SE, Martin-Fardon R, Markou A, Koob GF, de Lecea L (2005) Role for hypocretin in mediating stress-induced reinstatement of cocaine-seeking behavior. Proc Natl Acad Sci U S A 102:19168–19173
137. Zhang S, Tong Y, Tian M, Dehaven RN, Cortesburgos L, Mansson E, Simonin F, Kieffer B, Yu L (1998) Dynorphin A as a potential endogenous ligand for four members of the opioid receptor gene family. J Pharmacol Exp Ther 286:136–141
138. Chavkin C, James IF, Goldstein A (1982) Dynorphin is a specific endogenous ligand of the kappa opioid receptor. Science 215:413–415
139. Corbett AD, Paterson SJ, McKnight AT, Magnan J, Kosterlitz HW (1982) Dynorphin and dynorphin are ligands for the kappa-subtype of opiate receptor. Nature 299:79–81
140. Mucha RF, Herz A (1985) Motivational properties of kappa and mu opioid receptor agonists studied with place and taste preference conditioning. Psychopharmacology (Berl) 86:274–280
141. Pfeiffer A, Brantl V, Herz A, Emrich HM (1986) Psychotomimesis mediated by kappa opiate receptors. Science 233:774–776
142. Shippenberg TS, Zapata A, Chefer VI (2007) Dynorphin and the pathophysiology of drug addiction. Pharmacol Ther 116:306–321
143. Fujiki N, Yoshida Y, Ripley B, Honda K, Mignot E, Nishino S (2001) Changes in CSF hypocretin-1 (orexin A) levels in rats across 24 hours and in response to food deprivation. Neuroreport 12:993–997
144. Przewłocki R, Lasón W, Konecka AM, Gramsch C, Herz A, Reid LD (1983) The opioid peptide dynorphin, circadian rhythms, and starvation. Science 219:71–73
145. Crocker A, España RA, Papadopoulou M, Saper CB, Faraco J, Sakurai T, Honda M, Mignot E, Scammell TE (2005) Concomitant loss of dynorphin, NARP, and orexin in narcolepsy. Neurology 65:1184–1188
146. Hara J, Beuckmann CT, Nambu T, Willie JT, Chemelli RM, Sinton CM, Sugiyama F, Yagami K, Goto K, Yanagisawa M, Sakurai T (2001) Genetic ablation of orexin neurons in mice results in narcolepsy, hypophagia, and obesity. Neuron 30:345–354
147. Willie JT, Chemelli RM, Sinton CM, Yanagisawa M (2001) To eat or to sleep? Orexin in the regulation of feeding and wakefulness. Annu Rev Neurosci 24:429–458
148. DePaoli AM, Hurley KM, Yasada K, Reisine T, Bell G (1994) Distribution of kappa opioid receptor mRNA in adult mouse brain: an in situ hybridization histochemistry study. Mol Cell Neurosci 5:327–335

149. Li Y, van den Pol AN (2006) Differential target-dependent actions of coexpressed inhibitory dynorphin and excitatory hypocretin/orexin neuropeptides. J Neurosci 26:13037–13047
150. Eriksson KS, Sergeeva OA, Selbach O, Haas HL (2004) Orexin (hypocretin)/dynorphin neurons control GABAergic inputs to tuberomammillary neurons. Eur J Neurosci 19:1278–1284
151. Robinson JD, McDonald PH (2015) The orexin 1 receptor modulates kappa opioid receptor function via a JNK-dependent mechanism. Cell Signal 27:1449–1456
152. Li X, Marchant NJ, Shaham Y (2014) Opposing roles of cotransmission of dynorphin and hypocretin on reward and motivation. Proc Natl Acad Sci U S A 111:5765–5766
153. Graziane NM, Polter AM, Briand LA, Pierce RC, Kauer JA (2013) Kappa opioid receptors regulate stress-induced cocaine seeking and synaptic plasticity. Neuron 77:942–954
154. Beier KT, Steinberg EE, DeLoach KE, Xie S, Miyamichi K, Schwarz L, Gao XJ, Kremer EJ, Malenka RC, Luo L (2015) Circuit architecture of VTA dopamine neurons revealed by systematic input-output mapping. Cell 162:622–634
155. Ford CP, Mark GP, Williams JT (2006) Properties and opioid inhibition of mesolimbic dopamine neurons vary according to target location. J Neurosci 26:2788–2797
156. Margolis EB, Lock H, Chefer VI, Shippenberg TS, Hjelmstad GO, Fields HL (2006) Kappa opioids selectively control dopaminergic neurons projecting to the prefrontal cortex. Proc Natl Acad Sci U S A 103:2938–2942

Orexin and Alzheimer's Disease

Claudio Liguori

Abstract Alzheimer's disease (AD) is the most frequent age-related dementia. It prevalently causes cognitive decline, although it is frequently associated with secondary behavioral disturbances. AD neurodegeneration characteristically produces a remarkable destruction of the sleep–wake cycle, with diurnal napping, nighttime arousals, sleep fragmentation, and REM sleep impairment. It was recently hypothesized that the orexinergic system was involved in AD pathology. Accordingly, recent papers showed the association between orexinergic neurotransmission dysfunction, sleep impairment, and cognitive decline in AD. Orexin is a hypothalamic neurotransmitter which physiologically produces wakefulness and reduces REM sleep and may alter the sleep–wake cycle in AD patients. Furthermore, the orexinergic system seems to interact with CSF AD biomarkers, such as beta-amyloid and tau proteins. Beta-amyloid accumulation is the main hallmark of AD pathology, while tau proteins mark brain neuronal injury due to AD pathology. Investigations so far suggest that orexinergic signaling overexpression alters the sleep–wake cycle and secondarily induces beta-amyloid accumulation and tau-mediated neurodegeneration. Therefore, considering that orexinergic system dysregulation impairs sleep–wake rhythms and may influence AD pathology, it is hypothesized that orexin receptor antagonists are likely potential preventive/therapeutic options in AD patients.

Keywords Alzheimer's disease · Beta-amyloid · Orexin · Polysomnography · REM · Sleep disturbances · Sleep–wake cycle · Tau

C. Liguori (✉)
Sleep Medicine Centre, Neurophysiopathology Unit, Department of Systems Medicine, University of Rome "Tor Vergata", Rome, Italy
e-mail: dott.claudioliguori@yahoo.it

© Springer International Publishing AG 2016
Curr Topics Behav Neurosci (2017) 33: 305–322
DOI 10.1007/7854_2016_50
Published Online: 20 December 2016

Contents

1 Alzheimer's Disease

Alzheimer's disease (AD) is a progressive neurodegenerative disorder, which has been identified as the main cause of cognitive decline in the elderly [1, 2]. The neuropathological hallmarks of AD are the accumulation of both amyloid-containing neuritic plaques and neurofibrillary tangles (NFTs) of tau proteins [3]. These neuropathological brain changes have been suggested to occur 15–20 years before the onset of AD symptoms [4]. Accordingly, the preclinical AD concept has been recently established as the presence of AD markers in cognitively normal individuals. In this regard, the expectations for disease-modifying therapeutic strategies are highly relevant for preclinical AD patients, since this condition may last years before the appearance of clinical AD. Therefore, the identification of risk factors for developing AD pathology is a current significant biological issue. Accordingly, researchers are presently highly focused at recognizing causes of AD, in order to counterbalance the pathological mechanisms triggering AD pathology. Therefore, substantial efforts have been made in the last decade to understand the pathophysiological processes underpinning beta-amyloid aggregation and deposition in order to delay or possibly prevent the AD neurodegeneration. In particular biomarkers expression of AD neuropathology, are actually invoked, thus requiring extensive investigations. In keeping with this observation, many studies have been performed in order to identify and possibly establish early biomarkers of AD pathology.

Sleep disruption is considered a core component of AD, and also a preclinical biomarker, since sleep impairment may emerge before the clinical onset of AD. Moreover, insufficient sleep facilitates the accumulation of β-amyloid, potentially triggering earlier AD neuropathological changes [5]. Therefore, sleep dysfunction has been hypothesized as a risk factor for AD since it may promote and/or accelerate the neurodegenerative AD processes, and thus the cognitive decline [5–8]. In keeping with this supposition, it has been recently hypothesized that poor sleep quality may promote cognitive decline and AD neurodegeneration [9–12]. Accordingly, sleep impairment has been demonstrated as altering the physiological homeostatic brain processes essential to ensure the clearance of toxic substrates which accumulate during wakefulness, such as beta-amyloid [8]. In fact, sleep reduction and dysregulation significantly lessen the demonstrated homeostatic and restorative effects of sleep against neurodegenerative processes, ensured by the glymphatic system [8]. This recently discovered that the macroscopic waste clearance system is active during sleep and alleviates brain from deposition of toxic substrates [8]. Importantly, glymphatic failure may precede β-amyloid deposits, thus representing an early biomarker of AD. However, an animal model study documented that restoring glymphatic inflow and brain interstitial-fluid (ISF) clearance potentially act as therapeutic targets slowing the onset and progression of AD [13]. Hence, restoring sleep in AD may recover the glymphatic system function and thus reduce the progression or possibly stop the ongoing chain of events started with the reduction of β-amyloid cerebrospinal-fluid (CSF) levels and leading to β-amyloid neurotic plaques deposition.

Clinically, AD is characterized by the deterioration of memory, language, and intellect. Although cognitive decline is the main feature of AD, sleep disturbances are a common highly disruptive behavioral symptom associated with AD pathology. Indeed, epidemiological studies have reported that sleep disturbances occur in up to 45% of AD patients [14–16]. The main sleep disorder in AD is obstructive sleep apnea syndrome, although also insomnia is a frequent and disabling disorder affecting AD patients. What is certain is that sleep disorders accelerate AD pathology [17, 18].

The origin of sleep disturbances in AD is thought to be multifactorial. In fact, several hypotheses have been established and tested. In fact, degeneration of suprachiasmatic nucleus, pineal gland, hypothalamus, and brain nuclei containing circadian clock-regulating neurons in basal forebrain and brainstem have been invoked as possible pathological mechanisms causing sleep dysregulation in patients with AD [19]. However, neurodegeneration in these regions does not totally explain sleep impairment occurring in AD patients. In keeping with this need of better identifying the key regions related to sleep dysregulation in AD, investigations have been carried out in order to better explain and possibly treat sleep impairment in AD. Therefore, the possible pathological changes in the hypothalamic regulation of the circadian rhythm have been recently investigated in AD pathology. In fact, hypothalamus, and specifically the lateral hypothalamus containing the orexinergic system, is considered essential in controlling the sleep–wake cycle, since it projects to several crucial brain nodes of the sleep–wake cycle controlling system [20]. Such areas include: locus coeruleus, dorsal raphe, substantia nigra, ventral tegmental area, hypothalamic tuberomammillary nucleus, melanin-concentrating hormone neurons, and basal forebrain.

On these bases, in order to test if the orexinergic system could have a significant impact on sleep in AD neurodegeneration, several reports have investigated the CSF orexin levels in AD patients from the preclinical to the advanced stages of the disease. Moreover, literature proposed the interesting mutual relationship among orexinergic system dysregulation, sleep impairment, and CSF AD biomarkers (tau proteins and beta-amyloid).

2   Orexin and Cognition

The hippocampal formation is principally involved in learning and memory. Alterations in hippocampal structure and function are usually contributors to cognitive dysfunction [21]. Hippocampus receives many inputs from several brain regions; also orexinergic system sends projections to this cognitive-fundamental region. In fact, the orexinergic system has influences on a vast number of homeostatic and physiological behaviors, such as attention, arousal, and cognition [22–24]. Vigilance and daytime activity are necessary components for cognitive performances. It has been also demonstrated that orexin promotes both wakefulness and energy expenditure by interconnecting with the ventrolateral preoptic area, and thus stimulates spontaneous physical and mental activity [25]. Orexin controls hippocampal neurotransmission through direct as well as transsynaptic modulation of various pathways [26]. Therefore, orexin, may play a significant role in hippocampal-dependent cognitive tasks. In particular, orexin-mediated modulation of GABA and glutamate tone in the hippocampus could be a potential contributor to disruption of the sleep–wake cycle as well as cognitive performances [26]. However, not only deficient synaptic activity, but also aberrant networks activities, which can be caused by orexinergic system upregulation [27], may cause cognitive deficits, as already demonstrated in AD animal model studies [28]. In fact, orexin has excitatory properties in both animals and humans by interacting with the mesolimbic pathway and amplifying dopamine release [27]. Therefore, the dysregulation of orexinergic signaling may interfere with cognition, in particular causing hippocampal-related cognitive deficits. This supposition needs to be better addressed also considering that a human studies investigating the impact of orexin receptor antagonists (ORA) on cognition lack. Nevertheless, previous reports in animal models have demonstrated that ORA administration do not alter cognitive tasks [29], but may enhance memory performances [30].

3   Orexin and Alzheimer's Disease

The activity of the orexinergic system can be evaluated by measuring the CSF levels of orexin. Orexin-A (Hypocretin-1) is a neuropeptide produced by the lateral hypothalamic neurons, which regulates the sleep–wake cycle by increasing arousal levels and maintaining wakefulness [20]. Several reports have evaluated CSF orexin levels in AD patients using different techniques, such as radioimmunoassay (RIA) [4], fluorescence immunoassay (FIA) [31], enzyme immunoassay (EIA) [6, 7], and mass spectrometry [32].

Orexin levels in the brain are under a complex regulation. In particular, recent animal studies indicate that the orexinergic system is under the influence of light and present diurnal variation and thus a circadian pattern of release and activity [33]. In humans, it has been also demonstrated that CSF orexin levels vary with season, principally correlating with day length and duration of the light period

[34]. Therefore, the orexinergic system seems to be, like other neurotransmitter systems, subjected to long-term modulation. Moreover, the orexinergic system is affected by the physiological aging, since an overall decrease (averaging 23–25%) in the proportion and density of orexinergic neurons from infancy to older age (0–60 years) has been demonstrated in the human hypothalamus [35]. Finally, the circadian rhythmicity of CSF orexin levels in AD patients and aged controls has been depicted by the well-designed paper from Slats and colleagues examining CSF orexin levels in AD patients and controls at eight individual time points chosen during a 24-h period. Authors demonstrated that in AD pathology the orexinergic system is significantly affected since both the decrease of mean orexin CSF levels and the increase of the orexin circadian rhythm amplitude were observed in the examined AD population compared to the elderly controls. Nevertheless, this study was limited by the lack of polysomnographic recordings, thus not allowing the correlation between orexinergic system dysregulation and the sleep–wake rhythm.

In the last decade, RIA analysis was validated for the quantification of CSF orexin levels in narcoleptic patients [36]. In 2006, for the first time Baumann and coauthors [37] investigated CSF orexin levels in small populations of patients affected by dementia processes documenting normal CSF levels of this biomarker in AD. In 2007, Friedman and colleagues confirmed this finding in a larger group of AD patients. Few years later, Wennstrom and coauthors compared CSF orexin levels among AD patients, Lewy-Body dementia (LBD) patients, and non-demented controls. They detected lower CSF orexin levels in LBD patients compared to both AD patients and controls, whereas CSF orexin level did not differ between AD patients and non-demented controls. However, when dividing AD patients by gender, higher CSF orexin levels were found in females with respect to males. Using FIA analysis, this finding was replicated by Schmidt and colleagues, who analyzed CSF orexin levels in AD patients. Although both groups supposed that female AD patients secrete abnormal orexin levels with a possible higher production rate in respect to males, their discussions did not provide substantial explanations. In fact, the following investigations did not confirm this supposition thus revealing comparable CSF orexin levels between male and female patients [6, 7, 32, 38–40].

Despite no differences found between AD patients and controls [39], a significant increase in CSF orexin concentrations was documented in moderate–severe with respect to mild AD patients [6, 7]. Furthermore, higher CSF orexin concentrations has been reported in mild cognitive impairment (MCI) due to AD patients when compared with both controls [40] and patients affected by other dementing processes [38].

The finding that moderate–severe AD patients as well as MCI due to AD patients present increased CSF orexin levels suggests that the orexinergic neurotransmission system may be dysregulated in the early as well as in the advanced stages of the AD neurodegenerative processes. However, this observation could be somewhat paradoxical and has been drawn from few studies. Hence, further evidence that the orexinergic system impairment persists from the onset throughout the progression of AD is needed. Nevertheless, it is possible to speculate that orexin may play a significant role along the entire progression of AD pathology.

Even though the orexin levels were extensively examined in in vivo CSF samples, only one report interrogated *postmortem* AD brains in order to assess orexinergic neurons and ventricular CSF orexin concentrations. Fronczek et al. documented a 40% decrease of orexin immunoreactive neurons in *postmortem* brain hypothalamic tissues and a modest reduction in orexin-A ventricular CSF levels of AD patients compared to aged controls [41]. Nevertheless, how CSF orexin levels correspond to the number of intact orexinergic neurons in the human brain is difficult to quantify. In rodent models, it was reported that a substantial loss of orexinergic neurons (50–70%) is required before a significant decrease in CSF concentrations of orexin appears [42]. A possible explanation for the finding by Fronczek and colleagues could be achieved from the recent paper by Zhu et al. [43] demonstrating that intermittent short sleep (ISS) produces premature senescence of orexinergic neurons in mice. In fact, chronic ISS, a condition observed in AD patients, causes a significant reduction of orexinergic neurons, which also showed an altered morphology. Moreover, ISS induces the reduction of projections from orexinergic neurons [43], thus possibly causing the increased release of orexin neurotransmitters to ensure the interconnections between orexin and its output terminals. However, this supposition needs to be completely demonstrated experimentally.

On this basis, it could be plausible that the moderate reduction of orexinergic neurons (40%) found in AD patients does not significantly modify CSF orexin levels; then, the increase in CSF orexin levels found in MCI and moderate–severe AD patients may suggest that the dysregulation of the orexin system in AD pathology could be functional and not structural. In fact, it could be hypothesized that the high CSF orexin levels found in patients with AD at the MCI and moderate–severe stages could be the result of increased orexin release, as a compensatory mechanism involving the lateral hypothalamus in the context of the AD neurodegenerative processes [44]. Indeed, the wakefulness-promoting neurons, particularly the basal forebrain cholinergic ones, are principally affected during AD neurodegeneration [45]. This cholinergic neurodegeneration could lead to the upregulation of the other arousal systems, including orexin-producing neurons, not only in the advanced stages but even at early stages, thus contributing to the sleep alteration frequently reported in these patients. However, these suppositions need to be fully demonstrated in experimental model studies.

4 Orexin, Alzheimer's Disease, and the Sleep–Wake Cycle

Once established that CSF orexin levels are normal or slightly increased in both MCI and AD patients, researchers investigated the relationship between the activity of the orexinergic system and the sleep–wake cycle in AD neurodegeneration.

Circadian disruption in AD has been well established. In fact, AD patients show reduced amplitude and period length of circadian rhythm, increased intradaily variability, and a decreased interdaily stability of a rhythm [46]. Pathophysiologic mechanisms underlying dysregulation of the circadian rhythmicity in AD have been identified in suprachiasmatic nucleus (SCN) impairment and loss of pineal gland function [47, 48]. In fact, AD patients show reduced circadian rhythm amplitude of motor activity and delayed circadian phase, which correlate with the number of SCN neurons [49]. However, also orexinergic signaling dysregulation has been invoked in as a possible cause of circadian disruption in AD. Accordingly, a single previous work using actigraphic recordings investigated the sleep–wake cycle and the circadian rest activity of AD patients in relation to CSF orexin concentrations [50]. Since lower CSF orexin concentrations are documented in narcoleptic patients who present diurnal fragmentation with several naps, authors correlated CSF orexin levels with daytime wakefulness in AD patients. Consistently, lower CSF orexin levels were correlated with the higher number and duration of daytime naps in AD patients, thus suggesting that orexin neurotransmission deficiency could be responsible for the daytime napping of AD patients. However, taking into account that CSF levels were in a normal range in all the AD patients evaluated, this supposition remained unconfirmed in the following studies investigating CSF orexin levels exclusively in respect to nighttime sleep.

It is well known that the role of the orexinergic system is not only limited to control the diurnal wake, but also to influence the nocturnal sleep. In fact, orexin seems to primarily reduce REM and slow wave sleep (SWS) and increase wakefulness. Therefore, the orexinergic system shows a wake-on and REM-off pattern of firing, since it physiologically promotes arousal through activation of the wake-active monoaminergic populations and the deactivation of the REM-on cholinergic network [51, 52]. However, the correct orexinergic signaling is considered essential in ensuring the physiological rhythmicity of the entire sleep–wake cycle.

It is well known that Alzheimer's pathology interferes with sleep physiology; in fact, MCI and AD patients suffer from sleep disturbances, such as reduced REM and SWS duration and decreased sleep efficiency, coupled with increased wakefulness after sleep onset (WASO) [14–16, 53]. In detail, increase in fragmented daytime naps, earlier times of sleep onset, and alterations in the timing and frequency of nighttime REM and SWS are usual sleep–wake characteristics of AD patients. However, the most distinct change of sleep architecture in AD neurodegeneration is the reduction of REM sleep, which is featured by longer latency and severe fragmentation [54, 55]. This significant change in REM sleep quality and quantity is already evident in the MCI stage of AD, possibly representing the first sign of sleep impairment in AD pathology [40, 56, 57].

Moreover, also the sundowning, or sundown syndrome, is a disabling disorder occurring in the late phases of the AD process. It is characterized by the appearance of intense behavioural symptoms in the evening (or late afternoon) [58]. It has been related to melatonin secretion impairment, although evidence acout melatonin supplementation treatments for sundowning are scarse [59]. It has been already reported that the impairment of the cholinergic networks in AD neurodegeneration could be responsible for sleep disruption and SWS/REM sleep alteration [60–64]. However, the relationship between sleep impairment and AD neurodegeneration has not yet been fully elucidated. Therefore, taking into account that AD patients show a dramatic impairment of sleep with frequent arousals coupled with the reduction of REM and SWS, recent reports investigated the relationship between CSF orexin levels and sleep macrostructure in AD patients ranging from the mild to the advanced disease stages [6, 7].

In the last decades, actigraphy emerged as a noninvasive tool for determining the sleep-wake circadian rhythm; in fact, it can measure the sleep-wake cycle in individuals going about their usual activities, thus representing an appropriate method of circadian rhythm measurement also for the research questions. However, polysomnography (PSG) remains the gold standard for measuring sleep, since it is a well-validated approach to study and define the sleep architecture.

Two reports investigated and correlated the polysomnographic sleep with CSF orexin concentrations in MCI and AD patients [6, 7, 40]. CSF samples were obtained in the morning, within 1 hour after the morning awaikening in order to better evaluate the CSF nigh-time concentration of orexin. The first investigation correlated the PSGs performed in mild to severe AD patients with CSF orexin concentrations and documented that higher CSF orexin levels correlated with longer sleep latency (SL), higher WASO, and decreased sleep efficiency and SWS. Significantly, the main finding consisted in the correlation between CSF orexin levels and the reduction of REM sleep. In this study, it emerged that higher CSF orexin levels correlated with lower REM sleep and higher sleep latency. Moreover, the additional multivariate regression analysis performed revealed the significant mutual interplay between CSF orexin levels and both SL and REM sleep. Therefore, it appeared evident that the orexinergic system overexpression may result in longer SL and REM sleep impairment in AD patients ranging from mild to severe cognitive decline [6, 7]. The second study demonstrated that orexin system dysregulation is already evident in the MCI stage of AD pathology. In fact, it documented that the orexinergic system overexpression is related to REM sleep impairment and sleep fragmentation in MCI due to AD patients [40]. Notably, by dividing the MCI population into two subgroups on the basis of subjective sleep concerns, authors found that MCI patients with subjective sleep complaints presented higher CSF orexin levels compared to MCI patients without sleep disturbances. Moreover, the further analysis between MCI patients complaining of sleep disturbances and controls affected by similar sleep impairment documented that MCI patients showed higher CSF orexin levels than controls. These findings support the hypothesis that in MCI due to AD patients sleep impairment may be related to the orexinergic system dysregulation, which may cause insomnia, prolonged SL, and nocturnal awakenings. Surprisingly, in the results section authors reported that the highest CSF orexin concentrations were found in two patients who presented a remarkable impairment of nocturnal sleep, with REM sleep suppression. Hence, the already proposed association between orexinergic system dysregulation and REM sleep impairment was early evident in MCI patients [40].

Although the association between orexinergic system overexpression and sleep impairment in the AD pathology have been extensively demonstrated in clinical studies, the mechanisms linking orexin system dysregulation to REM sleep impairment and sleep fragmentation have not yet been investigated in animal model studies. Up to now, it has been largely supposed that the failure of the cholinergic network may represent the main candidate in provoking the derangement of sleep in AD pathology. However, considering the abovementioned studies, the dysregulation of the orexinergic system may also be a factor that induces sleep alteration in the AD pathology. Therefore, the sleep impairment in MCI/AD patients may be caused by the dysregulation of both the cholinergic and the orexinergic systems. In particular, overexpression of the orexinergic neurotransmission system has been suggested owing to the malfunctioning of the damaged cholinergic network, thus resulting in an unbalance between these two systems [65]. Moreover, in vitro intracellular recordings in physiological conditions identified that orexinergic neurons present a depolarized resting membrane potential, with spontaneous firing in the absence of stimuli [52]. Briefly, it could be hypothesized that the absent feedback of the cholinergic network on the orexinergic terminals may produce the spontaneous firing of the orexinergic neurons in a possible AD model. Moreover, in animal models it has been demonstrated that REM sleep deprivation increases CSF orexin levels [66]. These findings suggest that the raised CSF orexin levels found in MCI/AD patients could be also linked to REM sleep impairment, which is related to the cholinergic system failure. Hence, on the basis of the recent evidence linking AD pathology, REM sleep suppression, and orexinergic system dysregulation, it is conceivable to speculate that the upregulation of the orexinergic system present in the AD neurodegeneration could be likely mediated by the lacking deactivation of the wake-on orexinergic neurons due to the derangement of the cholinergic neurotransmission. In particular, this evidence is drawn from studies investigating polysomnographic nocturnal sleep in AD patients and CSF orexin levels. However, based on the observations previously described by [50], further studies evaluating the circadian activity of AD patients (thus investigating both sleep and wake periods) related to CSF orexin levels changes are needed.

5   Orexin and Alzheimer's Disease Biomarkers: Beta-Amyloid

Beta-amyloid pathological aggregation and deposition is the main hallmark of AD pathology. It is widely accepted that beta-amyloid dynamics are altered many years before the onset of clinical symptoms [4]. In fact, the proposed amyloid cascade hypothesis suggests that AD neurodegeneration starts with aggregation of non-soluble monomeric beta-amyloid peptides. In keeping with this biomarker view, it has been demonstrated that low CSF β-amyloid$_{42}$ levels represent a very strong predictor of AD pathology since the preclinical stage [67]. On these bases, in order to target the possible pathological processes promoting AD preclinical neurodegenerative processes, researchers focused their work in understanding the possible mechanisms that early alter beta-amyloid dynamics.

In 2009, a seminal scientific report by Kang and coauthors [68] described that cerebral beta-amyloid dynamics are regulated by orexin, which in turn influences the sleep–wake cycle. By using an animal mouse model, authors documented that intracerebroventricular infusion of orexin, inducing wakefulness in mice, produces the significant increase of β-amyloid concentrations in the brain ISF. To confirm these findings, in a second phase, authors infused for 24 h a dual orexin receptor antagonist (DORA), thus detecting that ISF beta-amyloid levels reduced significantly with the abolishment of beta-amyloid diurnal fluctuations. Moreover, Authors demonstrated that 8 weeks of chronic administration of DORA significantly reduced beta-amyloid plaque formation in several brain areas. Based on these findings, authors proved that perturbations in orexin signaling not only alter the sleep–wake cycle by promoting wakefulness and reducing REM sleep, but also have acute effects on cerebral beta-amyloid dynamics. Hence, it could be hypothesized that the high orexinergic tone may alter the cerebrale beta-amyloid dynamics, since it induces wakefulness, which in turn is related to both the increased the diurnal fluctuation of beta-amyloid ISF levels and the promotion of cerebral beta-amyloid plaque formation [69]. However, Kang and collegues did not measure in their experiments CSF orexin levels after sleep deprivation, thus not confirming this hypothesis.

The impact of sleep deprivation and prolonged wakefulness has been also tested in humans. In fact, it has been documented that sleep deprivation increases CSF β-amyloid$_{42}$ levels, whereas a night of unrestricted sleep leads to decrease of β-amyloid$_{42}$ levels [11]. This finding confirms a previous study documenting that sleep impairment is associated with the diagnosis of preclinical AD [70]. Reading these papers, it is interesting to note that β-amyloid deposition, as assessed by CSF β-amyloid$_{42}$ levels, is described in patients presenting with worse sleep quality and lower sleep efficiency [70]. On the other hand, Ooms and coauthors described a difference of 75.8 pg/mL of β-amyloid$_{42}$ CSF levels between the unrestricted sleep and sleep deprivation groups. Therefore, it appeared evident that sleep impairment and in particular WASO are the main candidates in altering brain β-amyloid dynamics, thus possibly representing risk factors for preclinical AD. However, animal model studies that were subsequently performed tried to determine whether sleep impairment or orexin-mediated nocturnal wakefulness is related to the dysregulation of: (1) β-amyloid metabolism; (2) the increase of ISF β-amyloid levels, and (3) the induction of β-amyloid cerebral deposition. In fact, modulation of sleep, rather than orexin per se, seems to be important in causing AD neuropathological changes. In keeping with this hypothesis, the paper from Roh et al. [71] documented that the stereotaxic injection of a lentivirus for orexin overexpression into the hippocampus of amyloid precursor protein/presenilin 1 (APP-PS1) transgenic mice did no change β-amyloid deposition, also nor changing sleep time. Nevertheless, the injection of the orexin lentivirus vector in the hypothalamus of orexin knockout crossed with APP-PS1 transgenic mice increased the amount of wakefulness as well as increased the amount of β-amyloid deposition.

Considering that sleep deprivation induced β-amyloid pathology also in the absence of orexin, Roh et al. [71] concluded that wakefulness and sleep deprivation concurrently affect β-amyloid clearance and deposition, and that modulation of orexin coupled with its effects on the sleep-wake cycle may change the beta-amyloid dynamics in the brain, rescuing it from beta-amyloid pathology. Therefore, this animal model study totally agrees with clinical evidence of lower CSF β-amyloid$_{42}$ levels in sleep deprived humans. Hence, it appears plausible that the complex interaction between orexin signaling and sleep regulation could alter β-amyloid dynamics. After all, orexin fluctuations are related to the sleep–wake cycle and the diurnal fluctuations of β-amyloid. In agreement with this supposition, Kang and colleagues observed in a small group of healthy volunteers that fluctuations of CSF beta-amyloid levels are present during the day, with reduced levels overnight and increased levels during the wake period with a peak in the evening. Later, Slats et al. [72] by longitudinally collecting CSF throughout a 36-h intrathecal catheter, replicated this study in six AD patients compared to six elderly controls documenting the circadian rhythm of CSF orexin and β-amyloid levels in AD patients. From this experiment, several observations were achieved. Although no differences in CSF orexin levels were observed between AD patients and controls, authors showed that CSF orexin levels were increased during the night and with a higher mean amplitude in AD patients with respect to controls. Significantly, orexin CSF levels changed in relation to β-amyloid levels in both AD patients and controls. Consistently, lower mean CSF beta-amyloid [44] concentrations (consistent with a higher cerebral beta-amyloid burden) were related to both lower orexin levels and higher amplitude of orexin circadian rhythm.

Different from the aforementioned study, the previously documented association between CSF orexin and β-amyloid [44] levels described either in an animal model study or in a small sample of AD patients was not evident in the following reports investigating the orexinergic systems in MCI and AD patients [6, 7, 31, 39, 40, 73]. This lack of correlation has been described as the possible effect of the plateau of low (pathological) CSF β-amyloid [44] levels reached by both MCI and AD patients, which cannot allow correlations with other CSF biomarkers such as orexin [6, 7]. Moreover, in MCI and AD patients, CSF β-amyloid [44] fluctuations disappeared since levels in the CSF were lower due to the β-amyloid deposition in amyloid plaques. Therefore, the significant reduction of CSF β-amyloid [44] levels influences possible interplays between this biomarker and other molecules present in the CSF. In keeping with this observation, in cognitively normal elderly subjects (not showing β-amyloid pathology) it was demonstrated the significant relationship between CSF β-amyloid [44] and orexin levels [74]. However, if considering studies including patients affected by narcolepsy who present the selective damage of the orexinergic system and show both CSF orexin levels dramatically reduced and near to the disease onset the reduction of CSF beta-amyloid [44] levels, probably owing to inflammatory condition and not to beta-amyloid plaque deposition [75], it was evidenti the lack of correlation between CSF orexin and beta-amyloid [44] levels [7]. Therefore, it is possible to speculate that disease-specific alterations (orexinergic system damage in narcolepsy and β-amyloid pathology in

AD) cause the loss of the reciprocal modulation between orexin and β-amyloid$_{42}$ CSF levels. However, it could be very interesting to further investigate the possible in vivo relationship between CSF orexin and β-amyloid levels in larger groups of preclinical AD patients, when it is hypothesized that brain could still be salvageable from AD pathology. In fact, the evaluation of this correlation could be important in preclinical stages of AD in order to better investigate how orexin may modify in vivo β-amyloid dynamics, thus representing a novel therapeutic target.

6 Orexin and Alzheimer's Disease Biomarkers: Tau Proteins

Although the correlation between CSF orexin and beta-amyloid levels appeared controversial in animal model and human studies, in these latter the significant relationship between CSF orexin and tau protein levels has been widely documented either in AD patients or in depressed and cognitively normal healthy subjects [6, 7, 39, 74].

It has been hypothesized that in the AD process the hyperphosphorylation and accumulation of tau proteins appear in a temporal ordering after the accumulation of beta-amyloid plaques [76, 77]. Consistently, tau proteins mark the neuronal injury occurring in AD pathology. In fact, increased CSF tau protein levels correspond to higher NFT pathology [78]. Moreover, higher tau protein levels in the CSF are considered a marker of rapid cognitive decline, since they have been associated with faster and more pronounced neuronal degeneration, significantly supporting the transition from early to more advanced disease stages [79].

Reports from the recent literature suggested that the dysregulation of the orexinergic system, as expressed by the increased CSF orexin levels found in AD patients, was related to a faster and more marked tau-mediated neurodegeneration. This observation was carried out in studies investigating CSF orexin levels in mild and moderate–severe AD patients [6, 7, 39]. In fact, it has been demonstrated that CSF orexin levels directly correlate with CSF tau levels in AD patients, with particular evidence in moderate–severe AD subjects [6, 7]. Explanations have been only suggested. One of them is that the higher neuronal activity mediated by the increased orexinergic function may be responsible for the higher CSF tau protein levels found in AD patients. This supposition is based on the recent report performed in an animal model documenting that neuronal activity could be a regulator of extracellular tau levels [80]. Further alternative explanations are related to the effects of sleep–wake cycle alterations on both orexinergic signaling and tau pathology [74]. In fact, age-related increases in orexin may promote wakefulness and sleep fragmentation, which in turn may promote accumulation of tau proteins [81]. An alternative model of this correlation could be achieved from the results by Davies et al. [82] documenting that application of β-amyloid in cell cultures induced both amyloid plaques formation and tau phosphorylation coupled with the downregulation of orexin receptors thus inducing the possible increase in orexin neurotransmission due to the reduced available receptors. Therefore, AD neuropathology may influence orexinergic function by reducing orexin receptors.

Hence, the mutual relationship between the orexinergic system and tau pathology emerged in human as well as in animal model studies. These findings demonstrated that higher CSF orexin levels corresponded to increased tau pathology. Therefore, new frontiers have been opened for better understanding and possibly counterbalancing the tau-mediated neurodegeneration in AD patients by reducing the overexpression of the orexinergic system. In keeping with this supposition, dual orexin receptor antagonists, recently approved for the treatment of insomnia, should be investigated as a potential preventive or therapeutic measure against AD pathology.

7 Conclusion: Orexin and Alzheimer's Disease Pathogenesis

Researches so far conducted highlighted that the overactivation of the orexinergic system is associated with the dysregulation of the sleep–wake cycle in patients suffering from AD neurodegeneration. Moreover, the orexinergic system dysfunction is also related to β-amyloid and tau protein brain dynamics, either in aging or in AD pathology. However, the associations found among the orexinergic system dysfunction, the biomarkers consistent with AD pathology, and the sleep–wake cycle alteration suggest a mutual relationship of these three factors. The more accredited model, emerged from the recent studies investigating orexin, sleep, and AD, explains these interplays focusing on the effect of the orexinergic system dysfunction on the sleep–wake cycle impairment, which in turn has detrimental effects on β-amyloid and tau proteins deposition.

Therefore, these findings suggest that orexin may be considered as a novel biomarker of sleep impairment in AD pathology, secondarily influencing both beta-amyloid and tau pathologies.

Evidence proposed by animal model and in vivo human studies enforced this hypothesis and inaugurated new potential preventive/therapeutic strategies. Therefore, considering that sleep disruption has been proposed to exacerbate neurodegeneration in AD, results from the recent studies investigating the orexinergic system, sleep disruption, and cognitive decline in MCI and AD patients could propose novel therapeutic approaches to improve sleep by reducing the orexinergic tone. Additionally, taking into account that the dysregulation of the orexinergic system seems to influence AD pathology by acting on the sleep–wake rhythm, it could be hypothesized the use of orexin receptor antagonists as potential preventive, therapeutic, or neuroprotective ways targeting the AD neurodegenerative process in order to improve sleep, slow the cognitive impairment, and thereby hamper the pathological processes at the basis of AD pathology. However, these observations require solid confirmations in AD patients, preferably in the preclinical stage of the disease, in order to test if sleep improvement mediated by orexin receptor antagonisms may stop the ongoing chain of events leading to AD neurodegeneration.

References

1. Alzheimer (1907) Über eine eigenartige Erkan kung der Hirnrinde. Psych Genchtl Med 64:146–148
2. Blessed G, Tomilson B-E, Roth M (1968) The association between quantitative measures of dementia, and of senile change in the cerebral grey matter of elderly subjects. Br J Psychiatry 114:797–811
3. Braak H, Braak E (1991) Neuropathological stageing of Alzheimer-related changes. Acta Neuropathol 82(4):239–259
4. Sperling RA, Jack CR Jr, Black SE, Frosch MP, Greenberg SM, Hyman BT, Scheltens P, Carrillo MC, Thies W, Bednar MM, Black RS, Brashear HR, Grundman M, Siemers ER, Feldman HH, Schindler RJ (2011) Amyloid-related imaging abnormalities in amyloid-modifying therapeutic trials: recommendations from the Alzheimer's Association Research Roundtable Workgroup. Alzheimers Dement 7(4):367–385. doi:10.1016/j.jalz.2011.05.2351 (Review)
5. Mander BA, Winer JR, Jagust WJ, Walker MP (2016) Sleep: a novel mechanistic pathway, biomarker, and treatment target in the pathology of Alzheimer's disease? Trends Neurosci 39 (8):552–566
6. Liguori C, Romigi A, Nuccetelli M, Zannino S, Sancesario G, Martorana A, Albanese M, Mercuri NB, Izzi F, Bernardini S, Nitti A, Sancesario GM, Sica F, Marciani MG, Placidi F (2014) Orexinergic system dysregulation, sleep impairment, and cognitive decline in Alzheimer disease. JAMA Neurol 71(12):1498–1505. doi:10.1001/jamaneurol.2014.2510
7. Liguori C, Placidi F, Albanese M, Nuccetelli M, Izzi F, Marciani MG, Mercuri NB, Bernardini S, Romigi A (2014) CSF beta-amyloid levels are altered in narcolepsy: a link with the inflammatory hypothesis? J Sleep Res 23(4):420–424
8. Xie L, Kang H, Xu Q, Chen MJ, Liao Y, Thiyagarajan M, O'Donnell J, Christensen DJ, Nicholson C, Iliff JJ, Takano T, Deane R, Nedergaard M (2013) Sleep drives metabolite clearance from the adult brain. Science 342(6156):373–377. doi:10.1126/science.1241224
9. Cricco M, Simonsick EM, Foley DJ (2001) The impact of insomnia on cognitive functioning in older adults. J Am Geriatr Soc 49(9):1185–1189
10. Nebes RD, Buysse DJ, Halligan EM, Houck PR, Monk TH (2009) Self-reported sleep quality predicts poor cognitive performance in healthy older adults. J Gerontol B Psychol Sci Soc Sci 64(2):180–187. doi:10.1093/geronb/gbn037 (Epub 9 Feb 2009)
11. Ooms S, Overeem S, Besse K, Rikkert MO, Verbeek M, Claassen JA (2014) Effect of 1 night of total sleep deprivation on cerebrospinal fluid β-amyloid 42 in healthy middle-aged men: a randomized clinical trial. JAMA Neurol 71(8):971–977
12. Peter-Derex L, Magnin M, Bastuji H (2015) Heterogeneity of arousals in human sleep: a stereo-electroencephalographic study. Neuroimage 123:229–244. doi:10.1016/j.neuroimage.2015.07.057 (Epub 26 July 2015)
13. Peng W, Achariyar TM, Li B, Liao Y, Mestre H, Hitomi E, Regan S, Kasper T, Peng S, Ding F, Benveniste H, Nedergaard M, Deane R (2016) Suppression of glymphatic fluid transport in a mouse model of Alzheimer's disease. Neurobiol Dis 93:215–225
14. McCurry SM, Logsdon RG, Vitiello MV, Teri L (2004) Treatment of sleep and nighttime disturbances in Alzheimer's disease: a behavior management approach. Sleep Med 5(4):373–377
15. Moran M, Lynch CA, Walsh C, Coen R, Coakley D, Lawlor BA (2005) Sleep disturbance in mild to moderate Alzheimer's disease. Sleep Med 6(4):347–352 (Epub 31 Mar 2005)

16. Vitiello MV, Prinz PN (1989) Alzheimer's disease. Sleep and sleep/wake patterns. Clin Geriatr Med 5(2):289–299
17. Osorio RS, Gumb T, Pirraglia E, Varga AW, Lu SE, Lim J, Wohlleber ME, Ducca EL, Koushyk V, Glodzik L, Mosconi L, Ayappa I, Rapoport DM, de Leon MJ (2015) Alzheimer's disease neuroimaging initiative. Sleep-disordered breathing advances cognitive decline in the elderly. Neurology 84(19):1964–1971
18. Troussière AC, Charley CM, Salleron J, Richard F, Delbeuck X, Derambure P, Pasquier F, Bombois S (2014) Treatment of sleep apnoea syndrome decreases cognitive decline in patients with Alzheimer's disease. J Neurol Neurosurg Psychiatry 85(12):1405–1408
19. Kondratova AA, Kondratov RV (2012) The circadian clock and pathology of the ageing brain. Nat Rev Neurosci 13(5):325–335. doi:10.1038/nrn3208 (Review)
20. Saper CB, Scammell TE, Lu J (2005) Hypothalamic regulation of sleep and circadian rhythms. Nature 437(7063):1257–1263
21. Rosenzweig ES, Barnes CA (2003) Impact of aging on hippocampal function: plasticity, network dynamics, and cognition. Prog Neurobiol 69(3):143–179
22. de Lecea L, Kilduff TS, Peyron C, Gao X, Foye PE, Danielson PE, Fukuhara C, Battenberg EL, Gautvik VT, Bartlett FS 2nd, Frankel WN, van den Pol AN, Bloom FE, Gautvik KM, Sutcliffe JG (1998) The hypocretins: hypothalamus-specific peptides with neuroexcitatory activity. Proc Natl Acad Sci U S A 95(1):322–327
23. Deadwyler SA, Porrino L, Siegel JM, Hampson RE (2007) Systemic and nasal delivery of orexin-A (Hypocretin-1) reduces the effects of sleep deprivation on cognitive performance in nonhuman primates. J Neurosci 27(52):14239–14247
24. Jaeger LB, Farr SA, Banks WA, Morley JE (2002) Effects of orexin-A on memory processing. Peptides 23(9):1683–1688
25. Mavanji V, Perez-Leighton CE, Kotz CM, Billington CJ, Parthasarathy S, Sinton CM, Teske JA (2015) Promotion of wakefulness and energy expenditure by orexin-A in the ventrolateral preoptic area. Sleep 38(9):1361–1370
26. Stanley EM, Fadel JR (2011) Aging-related alterations in orexin/hypocretin modulation of septo-hippocampal amino acid neurotransmission. Neuroscience 195:70–79
27. Scheurink AJ, Boersma GJ, Nergårdh R, Södersten P (2010) Neurobiology of hyperactivity and reward: agreeable restlessness in anorexia nervosa. Physiol Behav 100(5):490–495
28. Sanchez PE, Zhu L, Verret L, Vossel KA, Orr AG, Cirrito JR, Devidze N, Ho K, Yu GQ, Palop JJ, Mucke L (2012) Levetiracetam suppresses neuronal network dysfunction and reverses synaptic and cognitive deficits in an Alzheimer's disease model. Proc Natl Acad Sci U S A 109(42): E2895–E2903
29. Uslaner JM, Tye SJ, Eddins DM, et al. (2013) Orexin receptor antagonists differ from standard sleep drugs by promoting sleep at doses that do not disrupt cognition. Sci Transl Med 5 (179):179ra44
30. Dietrich H, Jenck F (2010) Intact learning and memory in rats following treatment with the dual orexin receptor antagonist almorexant. Psychopharmacology (Berl) 212(2):145–154
31. Schmidt FM, Kratzsch J, Gertz HJ, Tittmann M, Jahn I, Pietsch UC, Kaisers UX, Thiery J, Hegerl U, Schönknecht P (2013) Cerebrospinal fluid melanin-concentrating hormone (MCH) and hypocretin-1 (HCRT-1, orexin-A) in Alzheimer's disease. PLoS One 8(5):e63136. doi:10. 1371/journal.pone.0063136 (Print 2013)
32. Hirtz C, Vialaret J, Gabelle A, Nowak N, Dauvilliers Y, Lehmann S (2016) From radioimmunoassay to mass spectrometry: a new method to quantify orexin-A (hypocretin-1) in cerebrospinal fluid. Sci Rep 6:25162
33. McGregor R, Wu MF, Barber G, Ramanathan L, Siegel JM (2011) Highly specific role of hypocretin (orexin) neurons: differential activation as a function of diurnal phase, operant reinforcement versus operant avoidance and light level. J Neurosci 31(43):15455–15467
34. Boddum K, Hansen MH, Jennum PJ, Kornum BR (2016) Cerebrospinal fluid hypocretin-1 (orexin-A) level fluctuates with season and correlates with day length. PLoS One 11(3): e0151288

35. Hunt NJ, Rodriguez ML, Waters KA, Machaalani R (2015) Changes in orexin (hypocretin) neuronal expression with normal aging in the human hypothalamus. Neurobiol Aging 36 (1):292–300
36. Ripley B, Overeem S, Fujiki N, Nevsimalova S, Uchino M, Yesavage J, Di Monte D, Dohi K, Melberg A, Lammers GJ, Nishida Y, Roelandse FW, Hungs M, Mignot E, Nishino S (2001) CSF hypocretin/orexin levels in narcolepsy and other neurological conditions. Neurology 57 (12):2253–2258
37. Baumann CR, Hersberger M, Bassetti CL (2006) Hypocretin-1 (orexin A) levels are normal in Huntington's disease. J Neurol 253(9):1232–1233 (Epub 5 Apr 2006, No abstract available)
38. Dauvilliers YA, Lehmann S, Jaussent I, Gabelle A (2014) Hypocretin and brain β-amyloid peptide interactions in cognitive disorders and narcolepsy. Front Aging Neurosci 6:119. doi:10.3389/fnagi.2014.00119 (eCollection 2014)
39. Deuschle M, Schilling C, Leweke FM, Enning F, Pollmächer T, Esselmann H, Wiltfang J, Frölich L, Heuser I (2014) Hypocretin in cerebrospinal fluid is positively correlated with Tau and pTau. Neurosci Lett 561:41–45. doi:10.1016/j.neulet.2013.12.036 (Epub 25 Dec 2013)
40. Liguori C, Nuccetelli M, Izzi F, Sancesario G, Romigi A, Martorana A, Amoros C, Bernardini S, Marciani MG, Mercuri NB, Placidi F (2016) Rapid eye movement sleep disruption and sleep fragmentation are associated with increased orexin-A cerebrospinal-fluid levels in mild cognitive impairment due to Alzheimer's disease. Neurol Aging 40:120–126
41. Fronczek R, van Geest S, Frölich M, Overeem S, Roelandse FW, Lammers GJ, Swaab DF (2012) Hypocretin (orexin) loss in Alzheimer's disease. Neurobiol Aging 33(8):1642–1650. doi:10.1016/j.neurobiolaging.2011.03.014 (Epub 5 May 2011)
42. Gerashchenko D, Murillo-Rodriguez E, Lin L, Xu M, Hallett L, Nishino S, Mignot E, Shiromani PJ (2003) Relationship between CSF hypocretin levels and hypocretin neuronal loss. Exp Neurol 184(2):1010–1016
43. Zhu Y, Fenik P, Zhan G, Somach R, Veasey RX (2016) Intermittent short sleep results in lasting sleep wake disturbances and degeneration of locus coeruleus and orexinergic neurons. Sleep 39(8):1601–1611. pii: sp-00094-16 (Epub ahead of print)
44. Jones BE (2004) Activity, modulation and role of basal forebrain cholinergic neurons innervating the cerebral cortex. Prog Brain Res 145:157–169
45. Francis PT, Palmer AM, Snape M, Wilcock GK (1999) The cholinergic hypothesis of Alzheimer's disease: a review of progress. J Neurol Neurosurg Psychiatry 66(2):137–147
46. Videnovic A, Zee PC (2015) Consequences of circadian disruption on neurologic health. Sleep Med Clin 10(4):469–480
47. Stopa EG, Volicer L, Kuo-Leblanc V, Harper D, Lathi D, Tate B, Satlin A (1999) Pathologic evaluation of the human suprachiasmatic nucleus in severe dementia. J Neuropathol Exp Neurol 58(1):29–39
48. Wu YH, Feenstra MG, Zhou JN, Liu RY, Toranõ JS, Van Kan HJ, Fischer DF, Ravid R, Swaab DF (2003) Molecular changes underlying reduced pineal melatonin levels in Alzheimer disease: alterations in preclinical and clinical stages. J Clin Endocrinol Metab 88(12):5898–5906
49. Wang JL, Lim AS, Chiang WY, et al. (2015) Suprachiasmatic neuron numbers and rest-activity circadian rhythms in older humans. Ann Neurol 78(2):317–322
50. Friedman LF, Zeitzer JM, Lin L, Hoff D, Mignot E, Peskind ER, Yesavage JA (2007) Alzheimer disease, increased wake fragmentation found in those with lower hypocretin-1. Neurology 68 (10):793–794 (No abstract available)
51. Arrigoni E, Mochizuki T, Scammell TE (2010) Activation of the basal forebrain by the orexin/hypocretin neurones. Acta Physiol (Oxf) 198(3):223–235
52. Lee MG, Hassani OK, Jones BE (2005) Discharge of identified orexin/hypocretin neurons across the sleep-waking cycle. J Neurosci 25(28):6716–6720
53. Beaulieu-Bonneau S, Hudon C (2009) Sleep disturbances in older adults with mild cognitive impairment. Int Psychogeriatr 21(4):654–666
54. Bonanni E, Maestri M, Tognoni G et al (2005) Daytime sleepiness in mild and moderate Alzheimer's disease and its relationship with cognitive impairment. J Sleep Res 14(3):311–317

55. Montplaisir J, Petit D, Lorrain D et al (1995) Sleep in Alzheimer's disease: further considerations on the role of brainstem and forebrain cholinergic populations in sleep-wake mechanisms. Sleep 18(3):145–148

56. Maestri M, Carnicelli L, Tognoni G, Di Coscio E, Giorgi FS, Volpi L, Economou NT, Ktonas P, Ferri R, Bonuccelli U, Bonanni E (2015) Non-rapid eye movement sleep instability in mild cognitive impairment: a pilot study. Sleep Med 16(9):1139–1145

57. Naismith SL, Hickie IB, Terpening Z, Rajaratnam SM, Hodges JR, Bolitho S, Rogers NL, Lewis SJ (2014) Circadian misalignment and sleep disruption in mild cognitive impairment. J Alzheimers Dis 38(4):857–866

58. Ferrazzoli D, Sica F, Sancesario G (2013) Sundowning syndrome: a possible marker of frailty in Alzheimer's disease? CNS Neurol Disord Drug Targets 12(4):525–528

59. Lin L, Huang QX, Yang SS, et al. (2013) Melatonin in Alzheimer's disease. Int J Mol Sci 14 (7):14575–14593

60. Grothe M, Zaborszky L, Atienza M et al (2010) Reduction of basal forebrain cholinergic system parallels cognitive impairment in patients at high risk of developing Alzheimer's disease. Cereb Cortex 20(7):1685–1695

61. Kundermann B, Thum A, Rocamora R, Haag A, Krieg JC, Hemmeter U (2011) Comparison of polysomnographic variables and their relationship to cognitive impairment in patients with Alzheimer's disease and frontotemporal dementia. J Psychiatr Res 45(12):1585–1592

62. Mallick BN, Joseph MM (1997) Role of cholinergic inputs to the medial preoptic area in regulation of sleep-wakefulness and body temperature in freely moving rats. Brain Res 750:311–317

63. Perry E, Walker M, Grace J, Perry R (1999) Acetylcholine in mind: a neurotransmitter correlate of consciousness? Trends Neurosci 22:273–280

64. Power AE (2004) Slow-wave sleep, acetylcholine, and memory consolidation. Proc Natl Acad Sci U S A 101:1795–1796

65. Eggermann E, Bayer L, Serafin M et al (2003) The wake-promoting hypocretin-orexin neurons are in an intrinsic state of membrane depolarization. J Neurosci 23(5):1557–1562

66. Pedrazzoli M, D'Almeida V, Martins PJ et al (2004) Increased hypocretin-1 levels in cerebrospinal fluid after REM sleep deprivation. Brain Res 995(1):1–6

67. Blennow K et al. (2016) Cerebrospinal fluid biomarkers in Alzheimer's and Parkinson's diseases-from pathophysiology to clinical practice. Mov Disord 31(6):836–847

68. Kang JE, Lim MM, Bateman RJ, Lee JJ, Smyth LP, Cirrito JR, Fujiki N, Nishino S, Holtzman DM (2009) Amyloid-beta dynamics are regulated by orexin and the sleep-wake cycle. Science 326(5955):1005–1007. doi:10.1126/science.1180962 (Epub 24 Sep 2009)

69. Roh JH, Huang Y, Bero AW et al (2012) Disruption of the sleep-wake cycle and diurnal fluctuation of β-amyloid in mice with Alzheimer's disease pathology. Sci Transl Med 4 (150):150ra122

70. Ju YE, McLeland JS, Toedebusch CD, Xiong C, Fagan AM, Duntley SP, Morris JC, Holtzman DM (2013) Sleep quality and preclinical Alzheimer disease. JAMA Neurol 70(5):587–593

71. Roh JH, Jiang H, Finn MB, Stewart FR, Mahan TE, Cirrito JR, Heda A, Snider BJ, Li M, Yanagisawa M, de Lecea L, Holtzman DM (2014) Potential role of orexin and sleep modulation in the pathogenesis of Alzheimer's disease. J Exp Med 211(13):2487–2496. doi:10.1084/jem.20141788

72. Slats D, Claassen JA, Lammers GJ, Melis RJ, Verbeek MM, Overeem S (2012) Association between hypocretin-1 and amyloid-β42 cerebrospinal fluid levels in Alzheimer's disease and healthy controls. Curr Alzheimer Res 9(10):1119–1125

73. Wennström M, Londos E, Minthon L, Nielsen HM (2012) Altered CSF orexin and α-synuclein levels in dementia patients. J Alzheimers Dis 29(1):125–132. doi:10.3233/JAD-2012-111655

74. Osorio RS, Ducca EL, Wohlleber ME, Tanzi EB, Gumb T, Twumasi A, Tweardy S, Lewis C, Fischer E, Koushyk V, Cuartero-Toledo M, Sheikh MO, Pirraglia E, Zetterberg H, Blennow K, Lu SE, Mosconi L, Glodzik L, Schuetz S, Varga AW, Ayappa I, Rapoport DM, de Leon MJ

(2016) Orexin-A is associated with increases in cerebrospinal fluid phosphorylated-tau in cognitively normal elderly subjects. Sleep 39(6):1253–1260
75. Liguori C, Placidi F, Izzi F, et al. (2016) Beta-amyloid and phosphorylated tau metabolism changes in narcolepsy over time. Sleep Breath 20(1):277–283 discussion 283
76. Hardy J (2009) The amyloid hypothesis for Alzheimer's disease: a critical reappraisal. J Neurochem 110:1129–1134
77. Jack CR Jr, Knopman DS, Jagust WJ et al (2010) Hypothetical model of dynamic biomarkers of the Alzheimer's pathological cascade. Lancet Neurol 9:119–128
78. Blennow K, Hampel H (2003) CSF markers for incipient Alzheimer's disease. Lancet Neurol 2 (10):605–613
79. Kester MI, van der Vlies AE, Blankenstein MA et al (2009) CSF biomarkers predict rate of cognitive decline in Alzheimer disease. Neurology 73(17):1353–1358
80. Yamada K, Holth JK, Liao F et al (2014) Neuronal activity regulates extracellular tau in vivo. J Exp Med 211(3):387–393. doi:10.1084/jem.20131685 (Epub 17 Feb 2014)
81. Di Meco et al (2014) NBA
82. Davies J, Chen J, Pink R, Carter D, Saunders N, Sotiriadis G, Bai B, Pan Y, Howlett D, Payne A, Randeva H, Karteris E (2015) Orexin receptors exert a neuroprotective effect in Alzheimer's disease (AD) via heterodimerization with GPR103. Sci Rep 5:12584